T0140274

Gaming Media and Social Effects

Editor-in-Chief

Henry Been-Lirn Duh, Hobart, Australia

Series editor

Anton Nijholt, Enschede, The Netherlands

For further volumes:
http://www.springer.com/series/11864

Anton Nijholt
Editor

Playful User Interfaces

Interfaces that Invite Social and Physical Interaction

Editor
Anton Nijholt
Computer Science
University of Twente
Enschede
The Netherlands

ISSN 2197-9685 ISSN 2197-9693 (electronic)
ISBN 978-981-10-1192-4 ISBN 978-981-4560-96-2 (eBook)
DOI 10.1007/978-981-4560-96-2
Springer Singapore Heidelberg New York Dordrecht London

Printed on acid-free paper

Springer is part of Springer Science+Business Media (www.springer.com)

Preface

This book is about user interfaces to applications that can be considered as "playful." The interfaces to such applications should be "playful" as well. The application should be fun, and interacting with such an application should, of course, be fun as well. Maybe more. Why not expect that the interface is persuasive, engaging, challenging, and aims at helping to provide the user with fun, trying to keep the user motivated, not frustrated or bored, or, in terms of "flow theory," in a state where there is a balance between skills and challenges? Obviously, we can introduce playful interfaces to boring tasks and tasks that require efficiency in the first place. Also such tasks can profit from interfaces that introduce playful elements, for example, performance statistics and competition elements, or personalized and motivating conversational agents. But of course, we can expect that most useful applications of "playful interfaces" appear where users have to interact with computers, sensor-equipped environments, social robots, wearables, and mobile devices that are embedded in smart environments that support our general daily-life activities and that are not directly aimed at efficiency. Gamification of society aims at introducing playful elements in our digitally supported daily activities, whether it is about home activities, work activities, public space activities, or recreational activities. Playful interfaces, that is, interfaces that allow playful interactions with such activities, are then required.

Playful interfaces are designed to invite playful, social, and physical interaction. Users should feel challenged and persuaded to engage in the interaction with the particular application and the interaction should be fun. This does not necessarily mean that the application has been designed for providing fun only. Nothing wrong with that, but playful interfaces can also be interfaces to educational material introducing physics, mathematics, and informatics to a student. Or they can be interfaces to simulation environments that are meant to train professionals in decision-making situations or performing tasks in riskful situations. In addition to training and educational applications there can be aims such as playfully supporting rehabilitation activities or activities aimed at improving physical and mental health. Artists interested in digital art and entertainments have introduced—and will continue to do so—art installations with sensors and actuators that invoke playful user participation to experience their art.

Advances in interaction technology have allowed us to talk about ubiquitous and pervasive computing. That is, sensors and actuators embedded in environments, (mobile) objects, and wearables, have made it possible to extend the view of human–computer interaction where the user is attached to mouse, keyboard, and monitor with a graphical user interface to a reactive and proactive environment that surrounds a user and where the computing power is not necessarily addressed in an explicit way by the inhabitants of the environments. Such smart environments allow the sensing of their inhabitants, including the interpretation of their verbal and nonverbal behavior, their bodily behavior, and their physical activities in this sensor-equipped environment. And, of course, they allow the sensing of how inhabitants of these environments interact with each other. Examples of sensors are cameras, microphones, position and proximity sensors, acceleration meters, augmented reality glasses, augmented, and immersive virtual reality headsets, and physiological body sensors, including brain–computer interfaces for monitoring and stimulating brain activity. Smartness embedded in the environment makes it possible to offer playful interaction possibilities to inhabitants of these environments.

These developments allow users to interact with objects and devices that are part of their natural physical environment. Information presentation, information exchange, and information manipulation can be done in a context where the environment knows about the user and its preferences, and its moods and emotions. Digital multimedia can be employed to augment physical reality and what we see, hear, feel, and smell can be manipulated by artificially evoked events. These events can take place in physical, augmented, and virtual reality environments where users can interact with tangible or virtual objects, including social robots in a home environment, embodied agents in conversational environments, or avatars and semi-autonomous actors in video game environments. Clearly, these developments allow a transition from video game environments to game and entertainment applications that are part of a digitally augmented physical world. That is, videogames enter the real world.

In the chapters of this book we discuss playful interfaces. We discuss new interaction technologies and applications that require these new and playful interaction technologies. We survey the present state-of-the-art research and future developments in the area of playful user interfaces. Many chapters in this book discuss designs and applications of playful interfaces that will only become available in commercial applications 5 years or later from now. In this book, we see the introduction of many prototypes of potentially interesting human–computer interfaces and their connection with their applications. Persuasive, social, and tangible interfaces are among the topics discussed in the chapters of this book.

In the first chapter ("Playful Interfaces: Introduction and History") of this book, there is a short introduction to the history and the state-of-the-art research in playful interfaces. Introduction and survey are short. After that there are five parts with chapters that introduce state-of-the-art-research on Playful Interfaces.

These five parts are (1) Public and Mobile Entertainment, (2) Indoor and Outdoor Playgrounds, (3) Games for Change, Personalization, and Teaching, (4) Health and Sports, and (5) Learning by Creating. The chapters in these parts provide a state-of-the-art survey of the current research on playful interfaces and provide a look into the future of playful interfaces.

Anton Nijholt

Contents

Playful Interfaces: Introduction and History

Anton Nijholt

Abstract In this short survey, we have some historical notes about human–computer interface development with an emphasis on interface technology that has allowed us to design playful interactions with applications. The applications do not necessarily have to be entertainment applications. We can have playful interfaces to applications that have educational goals or that aim at behavior change, whether it is about change of attitude or opinion, social behavior change, or physical behavior. For the developer and the designer of these applications and their interfaces, there is no need any more to assume that, in addition to focusing on the application, the user has to pay attention to manipulating a mouse, using the keyboard and monitoring the screen. Smart sensors and actuators embedded in a user's physical environment, objects, wearables, and mobile devices can monitor a user, detect preferences and emotions and can re-actively and proactively adapt the environment and the behavior of its actuators to demands of the user or changing conditions. In this way, interface technology can be employed in such a way that the emphasis is not on offering means to get tasks done in the most efficient way, but on presenting playful interaction opportunities to applications that provide fun, excitement, challenges, and entertainment. Clearly, many applications that have more serious goals than "just" providing fun can profit from this interface technology as well. In this introductory chapter, we shortly survey the chapters in this book that show the many applications of these playful interfaces.

Keywords Playful interfaces · Human–computer interaction · Pervasive computing · Ubiquitous computing · Entertainment computing · Games · User experience · Tangible interfaces · Mobile devices · Persuasion · Behavior change · Exertion interfaces · Whole body interaction · Gesture interfaces

A. Nijholt (✉)
University of Twente, Enschede, Netherlands
e-mail: a.nijholt@utwente.nl

A. Nijholt (ed.), *Playful User Interfaces*, Gaming Media and Social Effects,
DOI: 10.1007/978-981-4560-96-2_1, © Springer Science+Business Media Singapore 2014

1 Introduction

This introductory chapter to "Playful User Interfaces: Interfaces that Invite Social and Physical Interaction" is meant to introduce and discuss user interfaces to applications that have been designed to invite users to engage into playful interactions. Obviously, the applications should allow playful interaction. Moreover, the interfaces we want to look at should also allow playful social and physical interaction. The interfaces are "playful," that is, users feel challenged or are otherwise persuaded to engage in social and physical interaction because they expect it to be fun. However, both from the point of view of the users and that of the designers, there can be more than fun that has inspired the design of the application and characteristics of the user interface or that are meant to motivate the user. Users do not necessarily be aware of that. A video game can be fun to play, but maybe it has also been designed to teach mathematics or history, have the user learn about art, or the game was aimed at enhancing cognitive or social capabilities, or at changing an unhealthy life style. Recreational activities can now be digitally supported and enhanced. Solving puzzles, reading books, playing chess, maintaining collections, providing information to social media and consuming information from social media, picture and video processing and collecting and retrieving sports events and results are some examples that come into mind.

Whether it is just about providing, supporting, and enhancing fun activities or whether there are additional educational or change of behavior, attitude, and opinion motives involved, designers can now also use physical, sensor-equipped environments, to design such games and entertainment applications where the user is not condemned to sit on a chair, using keyboard and manipulating mouse or joystick and following actions on a monitor. That is, games, entertainment, and educational applications can be designed where the user, or maybe several users, can be physically engaged in an application, and where, when there are more users, whether they are co-located or distributed, users can compete and collaborate or inform others about their whereabouts and activities. Competition and collaboration can take place in home and office environments, "arcade-like" public spaces, or public spaces in general, for example in the case of urban games. Sensors and actuators in wearables and mobile computing devices will add to the possibility to design a playful interface to the physical world and its inhabitants. These added possibilities to have playful interfaces will extend application areas and approaches to application areas, such as passive and active recreation, education, behavior change, training, and sports.

2 Exploring Playful Applications: Early History

The assumption that only in recent years or in the last decade ideas about playful applications of computers and computer supported environments emerged is very wrong. Already in the early years of computer science (1950 and 1960s), applications were predicted, and sometimes even designed and implemented that focused on non-scientific, non-administrative and non-industrial use of computers. Alan Turing, Norbert Wiener and later many Artificial Intelligence (AI) researchers considered such applications. However, at that time the focus was mainly on the application, not on how users, that is, the general audience, could interface in a convenient or attractive way interface with the application. Understandable of course, the users were computer scientists and intellectual challenges such as can we make the computer play chess were more important than having a "user-friendly" interface to a chess playing program. And, of course, the general public did not have access to computers. Computers became available for scientific, administrative, and industrial (process control) applications, computer time was expensive and only professionals were able to feed the computers with programs that were executed in "batch processing," without interactivity between computer and professional user. That is, hand over the program and see how it has been processed by the computer the next day. Most probably there was an error message. Having a computer more efficiently running a program was worth the extra human effort. Soon there were attempts to provide users with a language that could be interpreted by the computer and that helped them to control how their programs had to be executed without human intervention.

New applications and more and other groups of users required more direct access to available commercial computer power. It required also more interactivity to control processing of collections of interacting programs and to provide user data. Interactivity in the late 1960s and early 1970s meant having access to a Teletypewriter (TTY) that allowed interactively changing commands in your program, resubmit your program, and evaluate results (and error messages) in real-time. Communicating with computers in real-time and from a distance, rather than offering a pack of punch cards to a receptionist of a computer center, became common practice. Having a "dialogue" with the computer about tasks that had to be processed became a point of view when using computers. Two additional points of view, not really in the main stream of computer science and its applications, came from Artificial Intelligence (AI) research and from artists that explored computer applications from an artistic viewpoint. These views are explained below.

- AI Research: AI researchers explored whether and how computers could perform tasks that required intelligence, that is, when performed by a human being. Early AI research in natural language processing looked at machine translation systems, question-answering systems and database retrieval interfaces. Performance, not efficiency was the issue. And although useful applications could be foreseen, the applications did not necessarily address societal, business or industrial problems. But of course, the political situation in the 1950 and 1960s did steer some of the interests. Eliza, a conversational program developed by Joseph Weizenbaum in the 1960s, allowed users to chat, using natural language, about any topic (Weizenbaum 1966). Although performed in a rather primitive way, this research can be considered as a first attempt to understand the user and to offer feedback based on that understanding. Moreover, the application did not in any way ask for efficiency in the interaction. Users took their time (more than the system) to think about the questions posed by the system and to formulate their answers. In the same period there were other attempts to design natural language interfaces to applications that were meant to amuse the user or to provide information about a user's sports and entertainment interests rather than about his or her computer-supported professional needs for handling information.

- Artistic Research: Artistic applications, starting with drawings of pin-ups (ASCII art) using pen-and-ink plotters and matrix printers, were added to the domain of applications. Other input and output modalities than text were investigated in the interaction between humans and computers. Camera's that provided information about the user's presence, movements and activities, and allowing the computer to manipulate this information before giving feedback, were certainly among the main tools used by many interactive art artists. That is, the user or the audience played an active role in the creation of interactive drawings, paintings or music. Less-known than these applications are the artistic efforts of composers, musicians, brain researchers and computer scientists to use brain activity as input to artistic computer applications. Although in the early years computer science did not yet offer advanced (digital) signal processing, machine learning methods, or even the possibility to store data for future analysis, there nevertheless was much artistic activity to use brain signals in order to create and modify visual, auditory and audiovisual landscapes.

AI research, interest of artists, and interest of computer scientists that came with ideas to use the computer for recreational purposes and to support their own daily activities (including their recreational activities) with this new technology helped to draw attention from the general audience (starting with amateur engineers and computer hobbyists) to the use of computers for tasks that were in the interest of a particular user in his or her home and interest environment rather than in his or her task-oriented office or industrial environment. However, many investigations and developments in computer science research labs and institutes remained unknown for the general audience until their results became part of a wide-scale employment in the context of the advent of the personal computer. Long before the introduction of the personal computer we see research institutions experimenting with graphical user interfaces (GUIs), with devices (indeed, the mouse) to interact with such interfaces, and with devices that allow users to use input devices for the computer to compose drawings and sketches, that is, presenting the computer with non-textual and non-command-like information that has to be processed and transformed. Workstations with GUIs appeared in the early 1970s at Xerox's Palo Alto Research Center, commercial workstations with GUIs followed and Apple introduced the GUI in the personal computer in the 1980s. In the same period, that is, before the introduction of the personal computers, we see the introduction of virtual reality environments and devices (head-mounted displays) that provide access to these environments, including the possibility that the environment adapts to the user's view.

3 Arcade Systems, Home Consoles, and Personal Computers

When the first personal computers were introduced in the 1970s by computer hobbyists, it was often the case that the abilities of these hobby and "garage" computers were shown with simple games or other properties that showed how simple software could perform on this simple hardware. But already in these and later years we see that small commercial companies developed playful applications. An interesting view on the development of the early personal computers can be found in (Markoff 2005). Companies developed software and hardware, for hobby and personal computers that was meant to attract users, other than hobbyists and (very) early "professional" personal computer users, to buy and use software and special-purpose hardware that allowed them to play games. An independent development was the advent of text-based adventure games, often made in the spare time of computer science researchers and distributed through the ARPAnet (early 1980s). Multiuser games (for example, MUD: Multi-User Dungeon), first available on local computer networks of universities and research institutes, became also accessible for external users through the ARPAnet.

First home console and entertainment systems (Atari, Nintendo) appeared at the end of the 1970s and early 1980s (Wolf 2008). At the same time small companies took the initiative to develop playful applications, applications that allowed users to consider his or her "hobby computer" as a device that was there to have fun with. Examples could also be drawn from arcade video and electro-mechanical games. The interfaces to arcade games such as Pac-Man and Space Invaders were extremely playful, persuasive, sometimes humoristic, providing sounds, animations, and force feedback and doing this in such a way that not only the gamer, but also his or her friends and possibly other audience could become engaged in this social activity (Smith 2006). Human–computer interaction researchers took notice of this development (Malone 1982). Simple keyboard and mouse controlled graphical user interfaces appeared. But other devices, allowing speech or pen input were developed as well.

Interestingly, during the 1980s we see the development of software and hardware for game computers that allow the design of games and input modalities that make use information obtained from measuring physical movements or changes in physiological information from the user. Arcade games moved to the personal computer, even when the graphics, the sounds, and the animations were not or hardly comparable with what could be experienced in arcade environments. In the 1980s and early 1990s of the previous century we can see applications that were designed from the point of view from bodily interaction (gestures, movements) and a point of view that involved physiological information to control an application or, but certainly less obvious at that time, adapt an application to a particular user. This burst of creativity and interest in bodily interaction did not remain. Many of the ideas disappeared until they reappeared some decades later in the twenty first century when cheap sensor technology to measure physical and physiological user information became available.

4 From ARPAnet to the Worldwide Web

Already in the 1960s it became possible to offer programs to a mainframe computer for execution or communicate with a distant computer using telephone lines. ARPAnet made it possible to make the transition from distributed input devices connected to mainframe computers to the possibility to access a network of worldwide connected computers. Messages between computer users could be exchanged and documents and programs could be transferred from one user to the other. Internet, as it existed since its early exploitation in the late sixties and early seventies, remained the domain of scientists at research institutes and universities for some decades. Internet facilities such as file transfer, electronic mail and, later, news and discussion groups only slowly entered the world of personal computer users during the 1990s of the previous century.

Standards to format documents for standardized exchange, editing and retrieval using distributed databases and computers connected together through the Internet were also first developed in a scientific environment and for scientific purposes. Tim Berners-Lee at the CERN laboratory in Geneva developed the technologies that made World Wide Web possible between 1989 and 1991 (Berners-Lee and Fischetti 1999). This technology was made publicly available some years later and made attractive for a broader audience with graphical browsers. They allowed ubiquitous use and commercialization through an increase in start-up companies in the late nineties and early 2000s. Web research and new web technologies that included the use of audio, pictures, video and animations made it possible to have entertaining and playful web applications. Users extended their presence on the Internet from a linear address to personal webpages and by becoming present in social media displaying personal information, preferences, opinions, and daily activities.

5 Ambient, Ubiquitous, and Pervasive

During the early years of computing, in parallel with the more mainstream developments that focused on improving efficiency of hardware, software, and interface technology in general, there were experiments in research laboratories that aimed at introducing special purpose hardware, software and interaction technologies. We already mentioned AI applications, mainly software-oriented (with the exception of special symbol processing machines) and game hardware, software and interaction devices, allowing players to have more natural interaction, based on the game-activity provided by the application, than made possible by keyboard, mouse, windows and menus. Distributed collaboration issues had already gotten early attention (Hiltz and Turoff 1978), just as virtual and augmented reality, and haptic applications with new interaction possibilities (data gloves, headsets, haptic devices). A well know example from the early haptics history is the Tactile Vision Substitution System (TVSS) (Bach-y-Rita et al. 1969). Images of a television camera were converted in vibrations with different frequencies of 400 pens that were put in the back of a chair. A person, for example a blind person, could then experience (or "see") the image while sitting in this chair.

In the early 1990s, Mark Weiser introduced his vision of ubiquitous computing (Weiser 1991). Weiser based his views on three forms of ubiquitous devices that became available in research laboratories: tabs (wearable centimetre sized devices), pads (hand-held decimetre-sized devices), and boards (metre sized interactive display devices). In the years that followed interconnectivity and the use of Internet became more visible. This led to similar concepts, sometimes emphasizing the role of the environment (ambient intelligence), the use of small sensors (pervasive computing) or the interconnectivity of devices (Internet of Things). Presently it is difficult to distinguish these "different" views.

Although there was quite some of interest in the ubiquitous computing view and similar views but with different names, most research efforts related to Human–Computer Interaction, went to Internet, the World Wide Web, Multimedia, Computer-Supported Collaborative Work, and Information Retrieval. There were certainly great, useful, and successful attempts to lay the foundations of the field by developing methods and methodology for interaction design, for requirements engineering, for usability research, user experience design (Hassenzahl and Tractinsky 2006), and persuasive technologies (Fogg 2003). The foundations were also laid for interaction research based on virtual and augmented reality and, starting with speech, natural language, and pointing gestures, multimodal interaction research. Again, as always, once there is a clearly visible new development, it is always possible to trace it back to some ideas that were introduced some decades before. Successful development of new interaction technology very much depends on the possibility to have it integrated with existing technology and to being able to develop an infrastructure that helps to make this technology attractive and affordable. The latter obviously depends on mass production or massive use of a new technology.

6 Tangibles, Smart Materials, and Wearables

In Weiser's view the tabs, pads and boards were assumed to be wirelessly connected; devices such as tabs (and pads) can move around and proximity can be detected. But there is still lot of attention for large, medium, and small-sized displays on these devices to present information. A more rigorous break with the tradition of graphical user interfaces appeared in the work of Hiroshi Ishii in the

MIT Media Lab (Ishii and Ulmer 1997). The emphasis in this work is on physical objects that have sensors and actuators and that invite physical interaction with digital content represented by the object. This view does not exclude interconnectivity between objects as we discussed in the previous section. Neither does it exclude the ambient intelligence view where it may be the case that although the user focuses on the interaction with a physical object, ambient media are there at the periphery of human perception to shift a user's attention. But certainly, in this view the focus is on objects in the physical world that can be grasped and spatially manipulated. These Tangible User Interfaces (TUIs) can be seen as a way to implement Weiser's view of computers that disappear in the environment by coupling digital information and information processing capability to everyday physical objects. This view was illustrated with a physical implementation of a GUI that included the possibility to move physical objects (phicons) on a desk surface to control the computation.

Commercial interactive surfaces (tabletops, multitouch tables) became available in later years and found their use in collaborative work and entertainment applications. Tangible tabletops allow the movement and manipulation of tangible objects on their surface and therefore also the manipulation of digital content as it is projected on the surface. But many other tangible user interfaces appeared. A tangible tabletop is about objects that can be moved and manipulated on a fixed surface with a graphical and touch interface and a perceptual coupling between these physical objects and the dynamic representation of content on the surface. But, to mention another extreme, tangible user interfaces can also be about interconnected physical objects with sensors and actuators that can be thrown from one player to another player, keeping track of speed, position, and individual or team player activity. Players can be informed of the play or interaction knowledge collected, integrated and interpreted by the tangibles. Players can change their behavior based on such information, the play, as it is implemented in the tangibles and the environment where the play takes place, can adapt its parameters to the players and the progress of the play. Again, we see a close, synchronous and real-time coupling of real-world activity involving physical objects and a digital model of a play and players' activities. Educational and entertainment applications appeared and domestic applications have been investigated. In the next section, rather than exploiting a user's or player's activity from the point of view of interacting with tangibles, we will look at measuring human activity, behavior and bodily expressions with multiple sensors embedded in the environment, including sensors embedded in physical objects, to better understand the actions and intentions of a user (the human computing view).

In a next edition of the view on tangible user interfaces it was observed, for example in (Ishii et al. 2012), that the tangibles, that is, the objects that invite physical interaction and their physical manipulation represents manipulation of digital content, despite actuators that provide sound and light effects or information on an embedded display, do not really change their (natural) physical appearance. Is it possible to have tangibles that dynamically change their appearance and behavior in sync with changes in digital content? We can, for example, think of objects that have motors and gears and investigate them in order to make a transition from, as mentioned in (Ishii et al. 2012), the transition from static/passive to kinetic/active tangibles. This view assumes a bidirectional coupling between dynamically controllable deformable and reconfigurable physical objects or physical material and an underlying computational model. In particular nanoscience research on material property changes has made it possible to introduce smart material interfaces that change their appearance because of changes in underlying digital content based on changes induced by interacting users (Vyas et al. 2012).

Other views on tangible user interfaces take into account "wearables." That is, devices that are integrated in our clothes, or, dependent on the definition of wearables, devices that we wear on our body, and in our pockets (Mann 2013). These devices know about our activities, and they can also inform others about our activities. A similar observation can be made about devices that measure physiological information, including information about brain activity. Such kind of information provides knowledge about the emotional and cognitive state of a user and how he or she wants to provide input to the system. That is, if there is involuntary input, based on monitoring a user's mental state or a user's reaction on externally evoked feedback, or voluntary provided input, such as motor imaginary input.

7 The Human Computing View

Weiser's view did not include, at least not explicitly, the measurement and interpretation of human behavior and human activity. Neither does the work of Ishii. Obviously, humans are part of the physical worlds that are accommodated with embedded sensors, actuators and intelligence. There are traditional displays, but also tangibles and smart material interfaces as explained in the previous section. In these digitally supported physical worlds, new interaction modalities or new integrations of interaction modalities have to be investigated. This can be done from the point of view of the characteristics of a particular device or tangible that allows other than remote control input devices such as mouse and keyboard, but it can also be done from the point of view of being able to sense human activity, human behavior, human (body) movements, and to sense (neuro-) physiological information when performing tasks or otherwise being active in such an environment.

Although it is not impossible to detect some aspects of a user's mental state from his or her mouse and keyboard use, in particular when the mouse has some physiological sensors, more information related to natural human activity, behavior, and movements need to be extracted and interpreted in order to provide satisfactory reactive and pro-active support by an environment. For specific applications, including games that require bodily activity, other interaction devices are of course available. Haptic devices, devices that capture movements, eye trackers and other, now sometimes considered to be exotic interaction devices such as thread mills to experience virtual reality, were already introduced decades ago, but usually in a context of a human-device interaction (one human, one device, one particular application). These devices capture one particular natural human physical activity and transform it into the control of an application. Cameras to capture human behavior did not yet connect to computers to analyze this behavior. Applications based on measurement and analysis of human vocal sounds (speech processing) got more attention.

In contrast, intelligence embedded in environments and in physical and virtual objects is meant to allow interaction with users in pro-active and reactive ways and therefore requires more knowledge about their users and their activities. With the exception of the just mentioned input devices, in the past, knowledge about the user had to be collected from keystrokes and mouse movements and the tasks and contents that were accessed. Current sensor technology and the embedding of intelligence in environments, physical objects and clothes and devices on our body allows other and more comprehensive ways of knowing about the user, including his or her preferences, abilities, and emotions. There are many ways to have sensors track human behavior and have this information integrated in order to allow such information to be used in a playful way. Gestures, body poses, body movements, and moving around in an environment or in front of an application can be thought of as explicit commands, or as ways to provide information (produced voluntarily or involuntarily) to the environment and its objects, just as we do in interaction with our human partners. Clearly, microphones and cameras are among the sensors that are embedded in environments and objects and that can measure such behavior. Eye movements and facial expressions provide information about interest or boredom or about focus of attention. And, obviously, when interacting with a social robot or virtual (embodied) agent, our verbal and nonverbal behavior should have meaning to them in order to make interaction more natural. In addition there are applications where an environment or its objects is required to know about and understand the interaction between its human inhabitants. Human computing (Pantic et al. 2008) and social signal processing (Vinciarelli et al. 2009) are research areas that have emerged to serve such applications. Computer-supported play, games and sports in the physical world with two or more players can be designed in which such information is exploited, whether it is for making interactions more natural or for making interactions more challenging, and whether it is for competition or for cooperation (Nijholt et al. 2012).

Physiological sensors, including sensors that measure brain activity, can complement the information generated from other sensors, or, depending on the application, be used separately to feed an application with information about the physical or mental state of a user. It can be used to inform the user about this physiological state, asking or persuading him or her to change current activities or long-term behavior, for example for health or fitness reasons. Based on physiological information from the user an application can also adapt to the user, asking for more or less effort, asking for other input modalities, or providing different feedback. In particular games that require physical effort can profit from such information, but also videogames can use it to adapt the level of the game to measured frustration, interest or boredom. There are also playful applications where the user is asked to manipulate aspects of his (neuro-) physiological state. This is in particular true for brain-computer interfacing, where human—computer interaction researchers are now experimenting with interfaces that expect, maybe in addition with other modalities, brain activity input that is evoked by external stimuli or by voluntary mental activity that is transformed to a command to a computer or other device in the environment (Nijholt et al. 2008).

8 Design Your Own Playful Interfaces for Your Entertainment

Logo (Papert 1980) was a child-friendly programming language that was based on Piaget's constructivist educational philosophy. It allowed children to construct their knowledge through experience. "Turtle graphics," that is, simple animations could be programmed by children. There were also possibilities to "program" physical objects. Logo programming environments for teaching purposes were developed, including programming the control of sensors, motors, and lights in physical objects ("Programmable Bricks," later called LEGO Mindstorms). Teaching and learning was also the objective of the Alice environment developed by Randy Pausch and colleagues. "Drag and drop" enabled students to create programs and get familiar with programming constructs (Cooper et al. 2000). Programming environments for children and students have been further developed into environments that allow designing, in a playful way, interactive stories, animations, music and art applications. Environments can provide examples that can be "remixed" to introduce other characters, animations and storylines. An example of such a visual programming environment is Scratch (http://scratch.mit.edu/).

We already mentioned the programming of physical objects. Nowadays, commercially available micro-controller boards such as Arduino allow the reading of sensors, the control of motors and the behavior of actuators. Microcontrollers, sensors (location, proximity, and movement) and actuators (changes of appearance, location, or movement) are becoming affordable and can be used to design playful tangibles, including the control of natural objects in an educational or home environment. Simple tools such as Makey MaKey make it possible to construct tangible interfaces. Hence, in addition to creating possibilities for constructivist learning for educational purposes, interactive entertainment can be constructed using commercial off the shelf technology (cheap sensors, Kinect, Arduino, Makey Makey, etc.). And, creating entertainment and playful interfaces, especially when done with others, can be as much fun or even more than playing a commercial videogame.

9 More About This Book

This first chapter with background information about playful interfaces is followed by five sections. The first section is devoted to Public and Mobile Entertainment. The chapters in this section provide a view on playful interaction in various situations using different technologies. The chapters discuss interaction with large displays in public environments, using playful whole body and location-based interaction detected with cameras (Chap. 2) and mobile phones (Chap. 3), and interaction with small displays on mobile devices (mobile phones, smart phones, tablets) that allow the user to play ubiquitous games, wherever the user is (Chap. 4).

- In Chap. 2, "Public Systems Supporting Non-instrumented Body-based Interaction" by Dimitris Grammenos and colleagues, three technologies for body tracking are demonstrated in three public systems for culture and marketing. These camera-based technologies are location-tracking, body-shape tracking and skeleton tracking. The applications use wall-projected 2D and 3D game and virtual worlds and all three allow multiple users. They concern information presentation in an exhibition room, an "advergame," and a public system to explore timelines using hand and leg gestures. Design considerations and user evaluations are discussed.
- In Chap. 3, with the splendidly fitting title "Playing with the Environment" by Pedro Centieiro and colleagues a persuasive location-based multiplayer game is introduced that aims at inducing or increasing a pro-environmental attitude. Players use mobile phones to interact with a large public display. The application requires players to physically walk around and collect (virtual) litter on their phones and drop it in correct virtual recycle bins on the public display. Environmental information is displayed to players and their audience. In addition to raising environmental awareness and aiming at a pro-environmental attitude change, social and collaborative activities are stimulated in an entertaining and awarding way. The authors discuss the design methodology and present their user studies, including observations on the persuasive abilities of their system.
- In Chap. 4 on "Designing Mobile and Ubiquitous Games and Playful Interactions" Paul Colton discusses a development not really foreseen by Weiser and others: the transition from portable phones to feature phones and to smart phones, where the latter have operating systems that allow the integration of computing capabilities, connectivity and multimedia options and many on-board sensors that can collect information about location, position and movements. Primitive versions of traditional console games were recreated on early mobile phones. However, presently game and entertainment applications can be developed that are built on knowledge of the environment, including maps, positions of other players or users, real-time recordings (pictures, audio, video) of the environment, and knowledge about nearby objects. And, of course, there is the possibility to communicate with others in a multiplayer setting. Behavioral and physiological information are other knowledge sources that can be exploited in games and entertainment applications. Colton surveys characteristics of

mobile games, in particular the on-board sensors that allow different kinds of interaction and therefore also different kinds of mobile and ubiquitous game play. These developments are illustrated with examples.

The second section of this book is devoted to interfaces that are not only playful but also have educational purposes. Development of social, cognitive, and physical skills is a goal that is addressed. Persuading users to perform physical activity by doing some exercises can be a main aim of design, but it can also be a side effect of the playful applications discussed in the chapters of the first section of this book. In this second section we focus on playful interfaces to applications that are aimed at providing children (but adults are invited to join) with opportunities to engage in physical and social play in interactive indoor and outdoor environments. The chapters in this section discuss interactive playgrounds that provide fun and that invite play employing social and physical interaction. Design of playgrounds where sensors and actuators are embedded in the environment is discussed in Chap. 5; design of playgrounds where sensors and actuators are embedded in player devices is the topic of Chap. 6; in Chap. 7 a player device is introduced that has its own play intelligence, but performs in an environment that can monitor and change its behavior in the interaction with players.

- In Chap. 5, "Interactive Playgrounds for Children," Ronald Poppe and colleagues discuss design considerations of interactive room-sized playgrounds with sensors and actuators. They focus on playgrounds where technology supports open-ended play. That is, play without pre-defined rules and goals and where children can have ad-hoc competition or cooperation. Children can introduce their own rules or borrow and adapt rules from games they already know. Design challenges are discussed from the points of view of context-awareness, personalization and adaptiveness. The role of various types of sensors and actuators is discussed, with an emphasis on cameras that determine position and movements and floor or wall feedback using projections. The chapter concludes with observations on future interactive playgrounds.
- Chapter 6, "Designing Interactive Outdoor Games for Children" by Iris Soute and Panos Markopoulos focuses on the design process for outdoor games. As in the previous chapter, players are assumed to be collocated, but rather than assuming sensors and actuators embedded in the environment, children have mobile player devices (physical objects) with several modes of interaction and the possibility of communication between devices. These games that distinguish themselves from games that rely on screen interaction are called Head Up games. The authors discuss the role of brainstorming sessions to generate ideas and how and when to involve children in the design process. Various methods for early user requirements gathering are discussed, including the positive and negative experiences the authors had with these methods. Playtesting of prototypes with children can help to introduce rules in the game that they understand or they think that are fair. Playtesting with adults, in addition to testing with children, can also lead to insights in usability problems and to useful

feedback to designers. The chapter concludes with a list of recommendations for designing Head Up games.

- Chapter 7, "Smart Ball and a New Dynamic Form of Entertainment" by Sachiko Kodama and colleagues introduces a tangible object, a smart ball, that has embedded sensors and actuators, and that is wirelessly connected to a more powerful computing device (a personal computer) in the environment. Sensors can be embedded in toys, or more generally, devices that can move around or be moved around in a physical environment. Among them are play, entertainment, and sports devices and equipment that are used in physical play. Wireless connection to a computer makes it possible to process and integrate sensor data coming from these devices and augment it with other context information to adapt the behavior of the object or to adapt the environment to the behavior of the object. In this chapter on smart balls, the authors discuss various implementations of smart balls and games that rely on specific properties of these balls. Embedded sensors detect the "state" of the ball (not moving, being grasped, thrown or rolled), LEDs in the ball can be actuated, and sensor information can be processed by a wirelessly connected computer that decides how to add sounds and graphical effects to the ball's behavior, for example, when and where it bounces on the field. Cameras are used to track the position of the ball on the playfield or, using a high-speed camera, the speed of the ball. Experiences obtained at exhibitions with various implementations involving one or more players are discussed.

The third section of his book is devoted to games that aim at a change of opinion, attitude or behavior (Chap. 8), playful interfaces that help in collaborative decision making (Chap. 9), and playful interfaces that help teachers of autistic children (Chap. 10). All the multiuser applications in the chapters of this section run on a multiuser touch table.

- In Chap. 8, "Games for Change: Looking at models of Persuasion through the Lens of Design" by Alissa Antle and colleagues the authors start off by reminding us that there is little evidence that Games for Change are effective. These digital games aim at changing players' opinions, attitudes, or behavior. In this chapter, the focus is on games that address the issue of sustainability. The authors discuss models of persuasion. The underlying idea of the Information Deficit model for example is that when learning about facts and consequences people will change their opinion, attitude, or behavior related to an issue such as climate change. In the Procedural Rhetoric model, when implemented in a game, the players experience the consequences of their assumptions and actions during game play and, again, it is assumed that this will lead to an awareness of the problem and the necessity of a behavior change. In addition to such existing models of persuasion the authors introduce a new model called Emergent Dialogue that puts emphasis on enabling participation in discussions about information, decisions and personal values. In an analysis of several Games of Change design markers are identified that can provide evidence of the persuasive model(s) that have been used in a game. A tabletop game on sustainable land use is introduced that incorporates the author's Emergent dialogue model. Guidelines based on the design markers that support behavior change through Emergent Dialogue are provided.

- Chapter 9, "Individual and Collaborative Personalization in a Science Museum" by Betsy van Dijk and co-authors, investigates how a multitouch table that provides playful access to information about a museum's exhibition can be used to enhance the experience of a museum visit. The table can of course be considered as a tangible interface. Children have touch interaction with the table, but they certainly can continue verbal and nonverbal interaction, discussing and negotiating with the other players, while interacting with the table. Clearly, this is different from what we saw in several previous chapters where users could freely move around in an environment with sensors and computing power to give meaning to their positions and movements, or where users interact with their mobile player devices. In this application, based on the information presented to them, a small group of children can discuss and integrate their interests in a collaborative interaction with the table. They are then provided with their "collaboratively personalized" route through the museum. The authors report results of experiments that aimed at measuring aspects such as enjoyment and collaboration during the multitouch interaction with the table and the effect on their visiting experience when following their suggested route and answering questions about objects (the "quest").
- In Chap. 10 "No Problem! A Collaborative Interface for Teaching Conversational Skills to Children with High-Functioning Autisms Spectrum Disorder" Massimo Zancanaro and colleagues introduce a multiuser interface to teach children with autism spectrum disorder social conversation and social interaction skills. They built their work on techniques of cognitive behavioral therapy. These techniques include role-playing to learn about various social situations and observational learning, where the latter is implemented in such a way that children can observe themselves in videos. Several social settings are provided by the system; two children, assisted by a facilitator can choose settings and their conversations can be recorded. Authoring tools to design settings and stories to introduce them were developed for the facilitator. Example conversations can be provided and can be compared with the conversation the children choose to have in a particular setting. In experiments the multitouch table implementation was compared with a multimice implementation on a desktop computer. From the experiments it could be concluded that the No Problem! system was usable, enjoyable, and the therapeutic goals could be achieved.

The fourth section of this book is devoted to health and sports applications. It should be noted that also in many of the previous chapters playful interfaces were designed in such a way that they required physical activity of their users. Apart from developing interesting games and entertainment that is "just" fun and provides enjoyment, many authors, including authors of chapters in the previous sections, also motivate their research from a point of view of developing cognitive, social or physical skills, and, when physical activity is involved, make references to encouraging a healthy life style and attacking sedentary behavior of children who are playing traditional video games. In this section, we have two chapters that explicitly address these issues. That is, we have a chapter on designing interfaces that invite social and physical interaction, with an emphasis on exertion games, that is, on games that require intensive physical efforts and interfaces that help users to be successful with their efforts (Chap. 11), and on designing interfaces that know how and when to interrupt user activity in order to persuade the user to engage in some physical activity (Chap. 12).

- Chapter 11 on "Designing for Social and Physical Interaction in Exertion Games" by Florian (Floyd) Mueller and colleagues a decade of research on exertion games is summarized with the aim to provide future developers with a set of design themes and recommendations. Exertion games require intense physical activity of the user, but this activity can be performed in a playful environment. In this chapter, a representative case study is presented (*Table Tennis for Three*) that allows investigations in social and physical behavior of players, where players can be in physically distant locations. Video recordings and questionnaires were used to analyze behavior and to gather user provided input to questionnaires. From this qualitative analysis, some salient themes emerged that facilitate social and physical exertion play, such as the availability of shared virtual objects play, being able to anticipate a player's next action, supporting players in expressing themselves using their bodies, have the opportunity to "bend the rules" of a game, and, utilizing the uncertainty that arises in physical exertion play.
- Chapter 12 is about "Designing Games to Discourage Sedentary Behavior" by Regan Mandryk and colleagues as authors. Games, as mentioned in the title, are called "energames." The authors define energames as "*… games that reduce sedentary time by requiring frequent bursts of light physical activity throughout the day.*" The authors start with making a useful distinction between being physically active and anti-sedentary behavior. Persons can be physically active and nevertheless spend most of the day sitting. The negative effects of a sedentary lifestyle can apply to physically active persons. The authors discuss and compare existing guidelines for physical activity and anti-sedentary behavior. The latter aim at introducing frequent, low-intensity physical activity into daily routine, rather than demanding intense physical effort. Barriers to physical activity and nonsedentary lifestyles are discussed. Guiding principles for exertion games (exergames) design are extended to energames design and additional principles for energames are introduced. Casualty, motivation and persuasion are some of the issues that are addressed in these principles. Examples of energames and a comparison with exergames with a focus on casualty and accumulated activity are also discussed.

In the final and fifth section of this book, we find two chapters that are about creating games and tangible interfaces to games by children or teenagers using specialized tools, game design platforms (for example, Scratch), low-cost tangible interface construction kits (for example, Makey Makey) and multitouch tables. Low-cost tools such as Arduino and GoGo Board, sensors and actuators also appear in the final chapter where students are provided with such tools to build physical and virtual models for science learning.

- Chapter 13, "Playing in the Arcade: Designing Tangible Interfaces with MaKey Makey for Scratch Games" by Eunkyoung Lee and co-authors is about how they guided children (10–12 years) in setting up a game arcade with games and tangible (touch-sensitive) interfaces that were constructed using the Makey Makey construction kit, Play-Doh, or made from whatever materials that were available. They also learned the basics of creating circuits. The interfaces that were built connected to remixed on-line available games from the Scratch game design platform. The authors describe the two workshops they organized, one focused on game and controller design, the second added the playing in the arcade experience. All activity in the workshops was recorded (observation notes, photographs, and video recordings) and analyzed. In remixing the Scratch games, the children added functionality and multimedia effects and spent time on game mechanics and aesthetic features. Tangible game controllers for these remixed games were designed and gender specific characteristics of these designs were noted. Insights on creating learning opportunities (design, programming, control) for children are reported.
- Game and interface design and implementation are also the topics of Chap. 14, "Playful Creativity: Playing to Create Games on Surfaces", by Alejandro Catalá and colleagues. In this chapter tabletop systems are explored on dimensions such as fostering creativity, development of computational thinking, and game and interface design. The focus is on teenage students who have to collaborate in creating games and the assumption is again that learning to create games is more effective from the point of view of design, computational thinking, and, more generally, creativity, than "just" playing a game. The authors discuss the various tools that are available to create games and interfaces, but they conclude that existing tools support single-user interaction, rather than supporting a group process that is aimed at fostering creativity and learning. A tabletop interface and software platform is introduced that supports non-programmers in designing game environments. Results of experiments with teenage students are reported.
- The final chapter (Chap. 15), "Bifocal Modeling: Promoting Authentic Scientific Enquiry through Exploring and Comparing Real and Ideal Systems Linked in Real Time" is by Paulo Blikstein. The chapter aims at improving STEM (Science, Technology, Engineering, and Mathematics) education. This is done by providing students with tools to connect real world physical models with computer simulated systems in real time. This is called bifocal modeling. Real-time sensing and computational modeling are brought into the classroom and are connected in real time. The exploration of this synergy is the main aim of this chapter. Tool kits such as Arduino and GoGo board are provided to students to build the sensor-equipped physical models. Computational models of

certain phenomena such as bacterial growth or heat transfer are built using game and other modeling platforms. The chapter provides a taxonomy for modes to merge sensors, actuators and models for science learning. Examples and case studies of bifocal modeling are presented. Among them are studies concerned with biology (bacterial growth), physics (Newton's laws), and chemistry (gas laws study). Experiments involving many students are reported and analyzed. The real world may be too messy; the virtual world may be too perfect. How to provide students with software and hardware tools to playfully explore incongruities and contradictions is one of the aims of this chapter.

10 Predictions and Conclusions

The chapters in this book do not only provide the current state of art in design, technology and use of playful interfaces, they also provide a view of the future of playful interfaces. Obviously, new technological developments will happen and new playful interfaces will appear. Any attempt to be complete at one particular moment will fail. Some of the developments reported in the chapters of this book could not or hardly have been predicted ten years ago, even when the basic technology was already available. Many ideas that were already available in the 1980s were not followed up until thirty years later when basic analogue and digital technology could be integrated in products that became interesting for mass production. That has happened before. In 1928, in his essay "The Conquest of Ubiquity," Paul Valéry wrote (Valéry 1928):

> Just as water, gas, electricity are brought into our houses from far off to satisfy our needs in response to a minimal effort, so we shall be supplied with visual or auditory images, which will appear and disappear at a simple movement of the hand, hardly more than a sign.

and,

> Just as we are accustomed, if not enslaved, to the various form of energy that pour into our homes, we shall find it perfectly natural to receive the ultrarapid variations or oscillations that our sense organs gather in and integrate to form all we know. I do not know whether a philosopher has ever dreamed of a company engaged in the home delivery of Sensory Reality.

Valéry's enthusiasm was caused by inventions that made it possible to reproduce art such as photography, motion pictures and phonograph recordings, and the possibility to manipulate pictures and recordings. Obviously, this was written long before families possessed a photo camera, let alone many photo cameras. Valéry did not predict and could not foresee a world with wireless security cameras or Wi-Fi digital cameras for private use, or smartphone cameras that can send pictures and recordings "with a simple movement of the hand" to wherever the user wants. And at that time certainly no one would predict that separate nineteenth century inventions such as photography, telephone, phonographic recordings, and motion pictures in the future could be integrated in one device.

Many know also the first sentences with which Mark Weiser started his famous *Scientific American* article (Weiser 1991) in which he introduced the notion of "ubiquitous" computing:

> The most profound technologies are those that disappear. They weave themselves into the fabric of everyday life until they are distinguishable from it.

Weiser developed his view of "ubiquitous computing" by extrapolating from the computing devices (tabs, pads and boards) that were researched in his computer science laboratory at the Xerox Palo Alto Research center. He envisioned rooms with hundreds of computers, but most or all of them "invisible to common awareness," that is, computers embedded in the everyday world. In later years slightly updated views were denoted by terms such as ambient intelligence and pervasive computing. In (Nijholt et al. 2004a; Nijholt 2004b), we discussed some problems when having to interact with computers that have disappeared in the environment. How do we recognize how to interact (Gibson 1977)? The impact of smartphones as computing devices was not foreseen by the computing research community. Due to developments in technology research into social media, social robots, and affective computing has become much more important than 20 years ago could have been foreseen.

There now is a foreseeable impact of wearables in general, including devices embedded in clothes, body, and brains. Detecting and interpreting human physical and mental behavior with the aim to pro-actively support humans in their daily and professional activities (Pantic et al. 2008; Vinciarelli et al. 2009) has made human–computer interaction an interesting research area for behavioral scientists.

Many of these developments in research and technology underlie the design and implementation of the playful interfaces that are discussed in this book. Future playful interface will also profit from the possibility of having brain-computer interfaces (Nijholt et al. 2008; Gürkök and Nijholt 2012), due to the cooperation of neuroscientists with HCI researchers. Developments in nanoscience and the development of smart materials will lead to increased interest in smart material interfaces (Vyas et al. 2012) and the cooperation between HCI researchers and nano-scientists. Playful interfaces that also make use of smart materials and that can reactively and proactively interact with us knowing about our physical and cognitive activity through wearables and sensors in the environment are something to look forward to. Playful interfaces will enter our homes and weave themselves into the fabric of everyday life.

References

Bach-y-Rita P, Collins CC, Saunders FA, White B, Scadden L (1969) Visual substitution by tactile image projection. Nature 221:963–964

Berners-Lee T, Fischetti M (1999) Weaving the web: the original design and ultimate destiny of the world wide web. Harper, San Francisco

Cooper S, Dann W, Pausch R (2000) Alice: a 3-D tool for introductory programming concepts. J Comput Sci Coll 15(5):107–116
Fogg BJ (2003) Persuasive technology: using computers to change what we think and do. Morgan Kaufmann Publishers, San Francisco
Gibson JJ (1977) The theory of affordances. In: Shaw R, Bransford J (eds) Perceiving, acting and knowing: towards an ecological pssychology. Wiley USA
Gürkök H, Nijholt A (2012) Brain-computer interfaces for multimodal interaction: a survey and principles. Int J Hum Comput Interac 28(5):292–307
Hassenzahl M, Tractinsky N (2006) User experience—a research agenda. Behav Inf Technol 25(2):91–97
Hiltz SR, Turoff M (1978) The network nation. Human communication via computer. Addison-Wesley Publishing Company Inc., Reading, Massachusetts
Ishii H, Ullmer B (1997) Tangible bits: towards seamless interfaces between people, bits and atoms. In: CHI '97 Proceedings of the ACM SIGCHI conference on human factors in computing systems, ACM New York, USA, pp 234–241
Ishii H, Lakatos D, Bonanni L, Labrune J-B (2012). Radical atoms: beyond tangible bits, toward transformable materials. Interactions 19(1):38–51
Malone TW (1982) Heuristics for designing enjoyable user interfaces: lessons from computer games. In: CHI'82 Proceedings of the ACM SIGCHI conference on human factors in computing systems, ACM New York, USA, pp 63–68
Mann S (2013) Wearable computing. In: Soegaard M, Dam RF (eds) The encyclopedia of human-computer interaction, 2nd edn. The Interaction Design Foundation, Aarhus. http://www.interaction-design.org/encyclopedia/wearable_computing.html
Markoff J (2005) What the dormouse said: how the sixties counterculture shaped the personal computer industry. Penguin Books, New York
Nijholt A, Rist T, Tuinenbreijer K (2004a) Lost in ambient intelligence. In: CHI '04 Proceedings of the ACM SIGCHI conference on human factors in computing systems, ACM New York, USA, pp 1725–1726
Nijholt A (2004b) Where computers disappear, virtual humans appear. Comput Graph 28(4):465–476
Nijholt A, Tan D, Pfurtscheller G, Brunner C, Millan J, Del R, Allison B, Graimann B, Popescu F, Blankertz B, Müller K-R (2008) Brain-computer interfacing for intelligent systems. IEEE Intell Syst 23(3):72–79
Nijholt A, Arkin R, Brault S, Kulpa R, Multon F, Bideau B, Traum D, Hung H, Santos E Jr, Li D, Yu F, Zhou L, Zhang D (2012) Computational deception and noncooperation. Trends & Controversies IEEE Intell Syst 27(6):60–75
Pantic M, Nijholt A, Pentland A, Huang T (2008) Human-centred intelligent human-computer interaction (HCI^2): how far are we from attaining it? Int J Auton Adapt Commun Syst 1(2):168–187
Papert S (1980) Mindstorms: Children, computers, and powerful ideas. Basic Books, New York
Smith J (2006) Digital dance hall: the fan culture of dance simulation arcade games. Chapter 10 In: O'Hara K, Brown B (eds), Consuming music together: social and collaborative aspects of music Springer, Berlin, pp 193–209
Valéry P (1928). The Conquest of Ubiquity. First published as "La conquête de l'ubiquité." in De La Musique avant toute chose. Editions du Tambourinaire. English version in Aesthetics. Translated by Ralph Manheim. Pantheon Books, Bollingen Series, New York, 1964
Vinciarelli A, Pantic M, Bourlard H (2009) Social signal processing: survey of an emerging domain. Image Vis Comput J 27(12):1743–1759
Vyas D, Poelman W, Nijholt A, de Bruijn (2012) A. smart material interfaces: a new form of physical interaction. CHI '12 Proceedings of the ACM SIGCHI conference on human factors in computing systems, ACM New York, USA, pp 1721–1726
Weiser M (1991) The computer for the 21st century. Sci Am Spec Issue Commun Comput Netw 265:94–104

Weizenbaum J (1966) ELIZA—a computer program for the study of natural language communication between man and machine. Commun ACM 9(1):36–45
Wolf MJP (ed) (2008) The video game explosion: a history from pong to playstation and beyond. Greenwood Press, London

Part I
Public and Mobile Entertainment

Public Systems Supporting Noninstrumented Body-Based Interaction

Dimitris Grammenos, Giannis Drossis and Xenophon Zabulis

Abstract Body-based interaction constitutes a very intuitive way for humans to communicate with their environment but also among themselves. Nowadays, various technological solutions allow for fast and robust, noninstrumented body tracking at various levels of granularity and sophistication. This chapter studies three distinct cases showcasing different representative approaches of employing body-based interaction for the creation of public systems, in two application domains: culture and marketing. The first case is a room-sized exhibit at an archeological museum, where multiple visitors concurrently interact with a large wall projection through their position in space, as well as through the path they follow. The second example is an "advergame" used as a means of enhancing the outdoor advertising campaign of a food company. In this case, players interact with the wall-projected game world through a virtual, two-dimensional shadow of their body. Finally, the third case presents a public system for exploring timelines in both two and three dimensions that supports detailed body tracking in combination with single-hand, two-hands, and leg gestures. Design considerations are provided for each case, including related benefits and shortcomings. Additionally, findings stemming from user-based evaluations and field observations on the actual use of these systems are presented, along with pointers to potential improvements and upcoming challenges.

D. Grammenos (✉) · G. Drossis · X. Zabulis
Foundation for Research and Technology-Hellas (FORTH), Institute of Computer Science, Heraklion, Greece
e-mail: gramenos@ics.forth.gr

G. Drossis
e-mail: drossis@ics.forth.gr

X. Zabulis
e-mail: zabulis@ics.forth.gr

A. Nijholt (ed.), *Playful User Interfaces*, Gaming Media and Social Effects,
DOI: 10.1007/978-981-4560-96-2_2,

Keywords Body-based interaction · Body tracking · Gesture-based interaction · Public information systems · Large displays · Cultural information systems · Advergames

1 Introduction

Human beings employ their bodies, implicitly or explicitly, to interact with their environment but also among themselves. At present, there are various technological solutions available that allow for fast and robust, noninstrumented body tracking at various levels of granularity and sophistication. The term "noninstrumented" denotes that users do not have to carry any object pinpointing their location or the position and pose of their head, arms, or legs. Thus, more natural interactions can be supported also through very simple usage "logistics," as there are no physical items that users have to carry or fasten on themselves—a fact which in the case of public systems can be crucial for their success.

Up to now, three of the most popular approaches for taking advantage of noninstrumented body-tracking technology when developing systems for public use include:

1. *location tracking*. Only the position of the user's body in Cartesian space is taken into account, much like the location of the mouse cursor on a computer screen (e.g., Zabulis et al. 2010).
2. *body-shape tracking*. A two-dimensional (*typically, body-part "agnostic"*) image of the user's body is projected on an interface or a virtual world and used to interact with it through collisions (i.e., overlaps) with virtual entities (e.g., buttons, characters, moving objects) (e.g., Grammenos et al. 2012).
3. *skeleton tracking*. The location and orientation of the user's individual body parts, as well as predefined gestures, are identified and mapped into a virtual world (e.g., Drossis et al. 2013).

Each of the aforementioned approaches comes with its own advantages and drawbacks, thus rendering itself appropriate for a specific range of applications and contexts of use. This chapter provides three case studies—one for each distinct approach—offering related design considerations along with findings stemming from user-based evaluations and field observations on their actual use in practice.

2 Background and Related Work

The notion of *body-based interaction* (though at a more rudimentary level) has been around since Myron Krueger's artificial reality work experiments in the 1980s (Krueger et al. 1985). In the beginning of our century, it has been popularized by Sony's EyeToy and more recently revolutionized by Microsoft's Kinect. One of the earliest application examples of noninstrumented body tracking is the

KidsRoom (Bobick et al. 1999), an interactive playspace simulating a children's bedroom where young children are guided through an adventure story. More recently, Laakso and Laakso (2006) developed a multiplayer game system using one top-view camera where player motion is mapped to digital character 2D motion. Another very popular contemporary example are interactive floors—physical sensor-based, like Magic Carpet (Paradiso 1997) or vision-based, e.g., iGameFloor (Grønbæk et al. 2007)—which are mainly being used for playing games. In the domain of museum applications, Kortbek and Grønbæk (2008) explored three different ways for supporting location-based interaction: (a) a coarse-grained passive infrared sensor; (b) pressure sensors embedded in the floor and a small staircase; and (c) camera tracking. In "Immersive Cinema" (Sparacino 2004), one ceiling-mounted camera is used to track a user's position on a floor segmented in five areas. A different, but quite interesting approach was followed by Robertson et al. (2006) in Bystander. They employed a ceiling-mounted IR camera to track users' position and motion, which are subsequently combined into flocking behavior used to drive the browsing of collections of photographs and texts.

The pose of a user's body can also be tracked and used for selective interaction with the environment (Jaimes and Sebe 2007). For instance, Papadopoulos et al. (2012) use defined body poses recognition in order to allow navigation in 3D environments. Additionally, gestural interaction is widely studied in the literature, but also one of the most popular approaches due to its intuitiveness, as gesture constitute a typical way of interaction among humans and their environment (Nickel and Stiefelhagen 2003; Sangsuriyachot et al. 2011; Yoo et al. 2010). Furthermore, hand gestures can be used to augment systems and allow supplementary interactions (Hilliges et al. 2009) when combined with other means of interaction, such as single- or multitouch. Apart from single-hand gestures both users' hands may be used in combination to enrich the set of available gestures (Fikkert et al. 2010). Finally, the use of feet gesturing (Sangsuriyachot et al. 2011; Valkov et al. 2010) and foot tapping (Crossan et al. 2010; Ronkainen et al. 2007) are occasionally conceded as supplementary helpful interaction tools.

3 Noninstrumented Body Tracking in Practice: The Case Studies

This section presents three cases of different representative approaches to body-based interaction for the creation of public systems, in two application domains: culture and marketing. The first case, Macrographia (Zabulis et al. 2010), is a room-sized exhibit at an archeological museum, where multiple visitors concurrently interact with a large wall projection through their position in space, as well as through the path they follow. The second example, Paximadaki (Grammenos et al. 2012), is an "advergame" used as a means of enhancing the outdoor advertising campaign of a food company. In this case, players interact with the wall-projected game world through a virtual, two-dimensional, shadow of their body. Finally, the third case, TimeViewer (Drossis et al. 2013), presents a public system for exploring timelines in both two and three dimensions that supports detailed body tracking in combination with single-hand, two-hands, and leg gestures.

3.1 Case A. Macrographia: Multiuser Location-Based Interaction with a Room-Sized Exhibit

In 2010, the Institute of Computer Science of the Foundation for Research and Technology-Hellas (ICS-FORTH) and the Archaeological Museum of Thessaloniki (AMTh) collaborated toward the creation of a special exhibition of prototypical interactive systems with subjects drawn from ancient Macedonia, named "Macedonia from fragments to pixels".[1] The exhibition (Zabulis et al. 2011) comprises seven interactive systems based on the research outcomes of ICS-FORTH's Ambient Intelligence Programme. The digital content of the systems includes objects from the Museum's permanent collection and from ancient Macedonia in general. The largest exhibit is *Macrographia* (Zabulis et al. 2010; Zabulis et al. 2012) a system that presents very large images, which visitors can explore by walking around in a room (Fig. 1). The images are projected on a screen and are analyzed part-by-part depending on the location of each visitor in the room. Macrographia presents the "Wall-painting of the Royal Hunt" from the tomb of Philip II at Vergina, ancient Aigai, the largest ancient Greek painting that has been found to date, its length exceeding 5.5 m. Widely admired as a rare masterpiece of ancient Greek art, the painting shows ten hunters chasing five different animals in a complex landscape.

The exhibit includes a camera network that observes multiple humans in front of a very large display. This network enables the observation of visitors from multiple views (see Fig. 2). The acquired views are used to volumetrically reconstruct and track the humans robustly and in real time using the method of (Argyros and Lourakis 2004), even in crowded scenes and challenging human configurations. The system includes one computer that acquires the corresponding images, processes them, and extracts a spatial representation of the persons in the room. Given the frequent and accurate monitoring of humans in space and time, a dynamic and personalized textual/graphical annotation of the display can be achieved based on the location and the walk-through trajectory of each visitor.

The digital representation of the wall-painting is conceptually separated in five zones perpendicular to the display, based on a semantic interpretation of the themes that appear in it. The room is also conceptually separated in four rows

[1] http://www.makedonopixels.org

Fig. 1 Macrographia: a room-sized interactive exhibit at the Archaeological Museum of Thessaloniki with more than 150,000 visitors up to now

parallel to the display, which correspond to different distances of observation. To prevent from continuous alternation in the case of visitors lingering across the boundary of a cell of the grid, the cell size is assumed magnified by 10 % comprising a grid of partially overlapping slots.

At the room entrance, signs guide English-speaking visitors to enter the room by moving rightwards and Greek-speaking visitors to enter the room by moving leftwards. The corresponding textual annotation for each distance is presented at the bottom of the screen (see Fig. 3). Within the context of a zone, the presented content is varied graphically and conceptually according to the distance of observation. When a visitor enters a zone, the presented content matches the viewing distance. The visitor has the capability to explore the corresponding theme by stepping back to get a more abstract view or step closer and focus on the details of the exhibit. When idle and upon visitor entrance, the system presents the wall-painting in its current state. As one or more visitors approach the display, graphical outlines are superimposed to the corresponding region(s) of the display reviving the deteriorated forms. In the next row, the system presents a fully restored version of the painting. In the closest row, the restored version is grayed out, and a specific detail is highlighted, using a combination of color and animation. When multiple visitors stand in the same zone, the person closest to the display determines the content of the presented textual annotation. When this person leaves, the next in line (if any) becomes the closest one to the display. By tracking visitors and assigning a unique identifier to each one, the system also retains attributes for each

Fig. 2 Person localization and tracking. (*Top*) original images from two out of eight views, (*below*) person segmentation and 3D reconstruction. Person tracking results are rendered as circles, superimposed on the ground plane of the 3D reconstruction. Circle colors correspond to track ids. Tracking retains the correct id for all persons although it is often that visitors may come in contact or occlude each other to the cameras

Fig. 3 Sample of the projected image as presented when three distinct persons are located in the room at different positions. The leftmost and rightmost parts are presented in their original state, as no one is standing in front of them. The second from the left part presents a sketch and a descriptive title, the next one a fully reconstructed view and a detailed description and the next part highlights a detail of the image, also offering related information

of them. Using an attribute for the language, which is set upon visitor entrance, the textual components of the presented content are provided in the language selected by each visitor. Furthermore, using collected data regarding the user's trajectory in space, as well as the time spent on each slot, the system has the ability to present additional content (this feature is currently used experimentally at an installation of the exhibit at the premises of ICS-FORTH).

As a means of providing real-time visual feedback, except the highly visible changes that happen to the image, a more subtle—but continuous—cue is also provided. Underneath the image, to the left of the information box, there is a small triangle that directly maps to a user's distance from the wall (see bottom part of Fig. 3); i.e., when a user moves toward the wall, the triangle moves downwards and vice versa. The area upon which this arrow moves is separated in three parts, which correspond to the available three levels of information. The currently selected level is drawn in green color, while the rest in red.

During the development of the system, ethnographic field methods (Blomberg et al. 2003) were employed, using a combination of the "observer participant" and "participant observer" approach. During a 6-months period, more than 200 persons have participated. Using the final version of the system, summative evaluation sessions took place both at the premises of ICS-FORTH, as well as at the Archaeological Museum of Thessaloniki (Zabulis et al. 2012). For the purposes of evaluating the experience of the users with the system, a 13-item attitude Likert scale questionnaire was created which was based on Brook's System Usability Scale (SUS) questionnaire (Brooke 1996). Evaluation at ICS-FORTH took place in a laboratory room that was set-up in a way to resemble as close as possible the actual museum exhibit where this application is currently being housed. Twenty-two volunteers participated (13 male, 9 female). The average age of the participants was 31.7 years old, the youngest being 18 and the oldest 41 years old. No specific instructions were given to the participants as to how the Macrographia system actually worked or displayed the information in order to examine if they were able to understand how the system actually worked and how to retrieve the information that corresponded to each section. At the Archaeological Museum of Thessaloniki 22 questionnaires were filled out by 15 visitors and 7 guards. Of those, 11 were males and 11 were females. Their average age was 34.6 years, the youngest being 21 and the oldest 56 years old.

The overall usability of the system was rated high in both studies: 82.8 % in-house and 80.8 % at the museum. The questionnaire results were also supported by data collected by analyzing the recorded sessions or interviewing the participants for the in-house evaluation, and by hand-written comments of the museum visitors. Overall, the comments made by the all participants were highly positive. Most of them were impressed with the system's ability to track accurately their position in the room and display the information in the language that was chosen. Even though very little instruction was given to them before they entered the room, they all managed to understand that their movement was tracked and that the information changed according to their distance from the screen. As a result, all users were able to read all the information that was presented in each section of the Macrographia. They also offered some suggestions on how to improve it. For

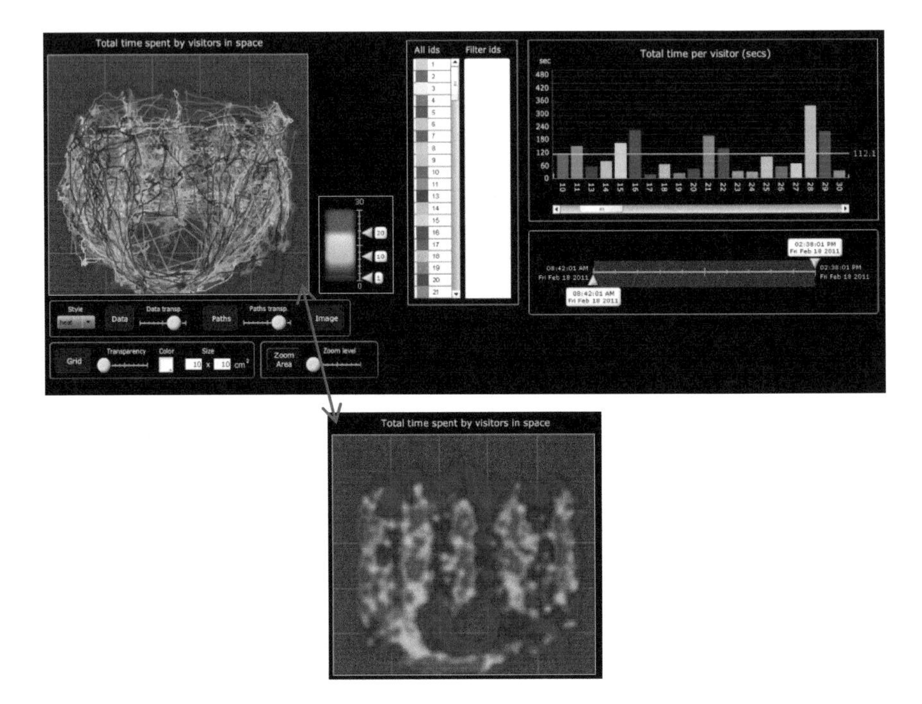

Fig. 4 Interactive visualization component for analyzing visitor data *top* showing visitor paths; *bottom* heat map (partial view of the interface)

example, a few users suggested that it would be better if the text font size changes from larger to smaller as the user approaches the display.

An interactive visualization component which presents information about the exhibit's visitors was also developed (Stamatakis et al. 2011). The main part of the visualization component's user interface comprises a top-down view of the room and its interactive slots. On this view, user paths are presented (using a different color for each distinct user—Fig. 4, top). Additionally, accumulated user time spent on a specific area is illustrated in the form of a heat map (Fig. 4, bottom). Quantitative information about each point of the heat map can be seen by positioning the mouse pointer over it. Additionally, several view parameters can be modified, for example, changing the transparency level of the paths or the heat map, hiding the background image, overlaying a grid, and changing the coloring scheme of the heat map. Furthermore, zoom is supported in any part of the view at different levels. The interface also presents the total time spent by each visitor in the room, the average time for all visitors and provides the means to filter out data by selecting a subset of IDs. Finally, there is an interactive control for setting a time period of interest that can be used for filtering data according to the time of their creation. The visualization component can work with both real time and stored data. Up to now, the analysis of the collected data from the museum installation has not revealed any prevalent patterns regarding the trajectory of visitors in the room. By examining the generated heat map views, it becomes evident that although the interactive slots are not marked on the floor, there is a significant concentration of visitors' "footprints" near their centers. Also, as expected, the furthest slots, where the presented information is minimal (just a title), have a much smaller aggregate visit time than the rest.

3.2 Case B. Paximadaki: An Advergame Public Installation Controlled Through Virtual Body Shadows

Paximadaki (Grammenos et al. 2012) is an advergame installation targeted to promoting, in exhibition spaces and key points of sale, the products of a food company producing various types of traditional Cretan rusks. The game, entitled "Paximadaki" (*small rusk* in Greek) is a Kinect-based PC exergame (Mueller et al. 2003) projected against a large surface, involving physical activity as a means of interaction. The main reasons for selecting Kinect were that it allows for noninstrumented game control through natural movements, which also afford higher levels of engagement and social behavior (Lindley et al. 2008), performs well under various environmental conditions, and comes at a very low cost. In this respect, it was decided to just use the depth camera's image in order to render a virtual shadow of the players, instead of tracking body skeletons. The rationale is two-fold. On the one hand, it is easier for people, especially "non game-players", to identify with their shadow rather than with an avatar, thus achieving a higher level of control and immersion (Sweetser and Wyeth 2005). On the other hand, this approach allows for maximum flexibility regarding the number, posture, and size of players, as well instantly joining and leaving the game, thus maximizing the opportunities for social interaction (Sweetser and Wyeth 2005). The downside is that people with larger body sizes have an advantage, and there also exists the possibility of accidental "intrusions" in the play area.

The gameplay[2] is simple, straightforward, and has very clear goals (Sweetser and Wyeth 2005). Players perceive their bodies as shadows projected on a brick wall (Figs. 5 and 6). On the startup screen there are buttons for setting the game's difficulty and starting a one- or two-player game. Players can select these buttons by overlapping any part of their shadow on them for a specific period of time (i.e., dwell), which is indicated by a countdown that appears right upon the button. Depending on the players' number, there may be one or two baskets at the two bottom sides of the wall (Fig. 6). A "rainfall" of rusks starts. Players must use their shadows to put the rusks into their basket. Rusks that fall on the floor are broken into pieces. The game ends when a certain number of rusks have fallen. In order to deliberately create a "memorable moment" (Jenkins 2002), when the end of the game approaches, a huge amount of rusks suddenly start to fall. To notify

[2] Videos of indicative play sessions can be found at: http://www.youtube.com/user/icsforthami.

Fig. 5 Paximadaki: an advergame installation promoting the brand and products of a food company, installed at five different public events and played by more than 1,500 people with ages ranging from 2 to 76 years old

players about this event, a bleating goat appears, providing a humorous note. Additionally, the music shifts to a faster tune. During gameplay, at moments that are likely to provoke interesting players' poses the game automatically takes photos of them. Additionally, when the game ends, a photo countdown ("smile moment") appears allowing players to pose. The photos are presented on screen when the game finishes and can be sent to the players via e-mail. Each game session has a unique serial number that appears on the bottom right corner of the screen. When the game finishes the players can write down (on a tablet or a paper pad) their e-mail address on a list next to the game's serial number in order to receive the photos.

Up to now, the game (with some variations) has been installed with remarkable success in five different public events in key locations in Athens, Greece, where it has been played by more than 1,500 people of ages ranging from 2 to 76 years old. When a complete version of the game was available, more than 50 h of playtesting along with the employment of observational usability assessment methods took place in order to fine-tune the gameplay and also debug the game, in a realistic installation at the premises of ICS-FORTH, with more than 30 players of both

Fig. 6 Screenshot of indicative gameplay showing a player's virtual shadow against the wall

genders, with ages ranging from 4 to 52 years old. Detailed usability evaluation results can be found in (Grammenos et al. 2012).

Subsequently, the game was installed in an exhibition of traditional food products, in the central metro station of Athens. In the course of three days, the game was played by 173 players (127 female, 46 male). Their age ranged from 3 to 76 years. All players stated that they liked the game and that they enjoyed playing it. More than 100 people used words like "fantastic," "great," "very good," or another synonym. As a means of collecting qualitative data about the game and its impact, a 20-item online questionnaire was created. 25 people answered it (18 female, 7 male) with ages ranging from 18 to 60. Among the things that respondents mentioned were that they liked: the responsiveness of the game; the (easy) way it was controlled; its high quality; its originality; the fun they got out of it; the fact that their whole family could play it; the music.

The game was then installed at the Zappeion Exhibition Hall, in Athens. In three days, the game was played by 337 players (195 female, 142 male) from 2 to 75 years (played against his 70-year-old mate). In 20 cases, the game was concurrently played by three or more players. The maximum number of concurrent players was 6. When more than two players played the game, their positions in space were usually dictated by their age and personal relationships. For example, friends or children of the same age would stand next to each other (see Fig. 5), while parents would mostly stand behind their children trying to discretely help them. When there were up to four players, they would usually form 2 teams. When there were more, people would not pay much attention to the score or in forming teams, focusing mainly on having fun through the physical aspects of play.

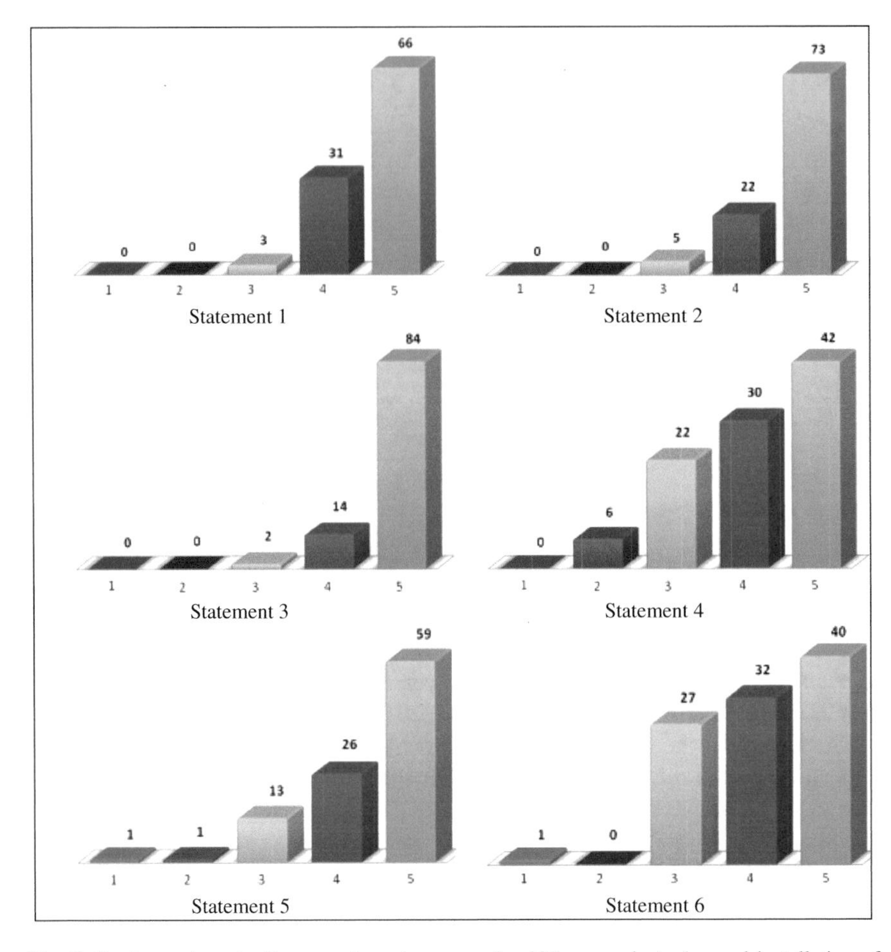

Fig. 7 On-the-spot evaluation questionnaire scores by 100 respondents (second installation of the game—Grammenos et al. 2012)

In 55 cases children, played along with their (grand)parents—sometimes both of them were adults. This time, a shorter questionnaire was employed and a total of 100 people responded (61 female and 39 male). The questionnaire included a few demographic-related questions and six statements.

1. I liked it.
2. It was easy to learn how to play.
3. It was fun.
4. It helped me familiarize with the brand *<brand name>* and its products.
5. It contributed toward creating a positive image about *<brand name>* and its products.
6. It positively affected me toward purchasing products of *<brand name>* .

Questionnaire scores are illustrated in Fig. 7. Interestingly, there were 28 respondents who scored 5/5 (i.e., best score) all statements. In short, players believed that the game was indisputably very fun and very easy to learn how to play. The related statements in all questionnaires were the ones with the highest average ratings and the least standard deviations. Additionally, as was deducted from both the analysis of the questionnaires (Grammenos et al. 2012) and on the spot questioning of the players, the game offered a highly entertaining experience while, at the same time, positively contributing to the marketing of the company and its products, reinforcing previous research findings.

The third installation of the game was in Athens Plaython, a city play event. The most interesting fact in this case, was that out of the 150 game sessions that took place in the course of 2 days 127 were played by more than two players, with an average of four players and a maximum of 12 players! In the latter case, players created a compact mass with several protruding hands and were mostly interested in just hitting the falling objects (e.g., much like playing volleyball on the beach), irrespectively of whether these would land in a basket or not. It should be noted that, due to appropriate design considerations, even in such extreme cases neither game performance nor play time were negatively affected. No formal evaluation was conducted in this installation.

Subsequently, there have been two more installations of the game at food exhibitions, while since July 2010, a new version of the game, targeted to familiarize visitors of the island of Crete with local products, is permanently installed at the arrivals hall of Heraklion international airport "Nikos Kazantzakis" as part of *Creative Crete,*[3] a project realized by the Institute of Computer Science of FORTH under the initiative of the Region of Crete.

3.3 Case C. TimeViewer: A System for Exploring Timelines in Two and Three Dimensions Through Hand and Body Gestures

TimeViewer (Drossis et al. 2013) is a system allowing the modeling, storing, visualization, and multimodal interaction with timelines (Fig. 8). For example, up to now it has been used to present historical information (e.g., the history of computers, the history of a city), the life trajectory of famous personalities, or even time threads comprising important events, work, and achievements of various entities. A principal characteristic of TimeViewer is the provision of an immersive display of temporal information supplementary to a view that favors the provision of an overview. In this direction, TimeViewer supports two distinct co-existing modes (i.e., views) for visualizing information, respectively (Fig. 9): the 'Tunnel' and 'Classic2D'. Users can freely and seamlessly switch between the two modes at

[3] https://www.facebook.com/creativecrete

Fig. 8 TimeViewer: A system allowing the modeling, storing, visualization, and noninstrumented body-based kinesthetic interaction with timelines. Up to now, it has been evaluated in-house with 16 representative users

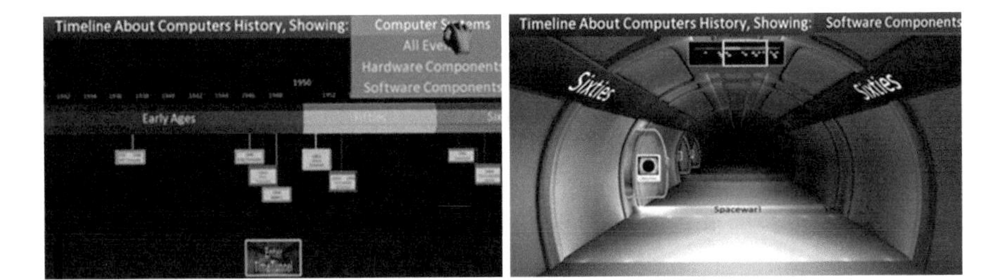

Fig. 9 The Classic2D (*left*) and Tunnel (*right*) visualization modes

any time. Classic2D mode is used to provide an overview of the available information in a manner familiar to the user, easy to understand, and convenient for navigation (Fig. 9, left). The Tunnel mode (Fig. 9, right) is targeted to supporting the task of sequential, exhaustive, exploration of the presented information, allowing the user to physically experience the time dimension as well as the unambiguous display of each event. The time dimension is visualized as a "tunnel" along which all information is integrated. The tunnel walls serve a dual role: they considerably constrain navigation in 3D space, also offering rich orientation cues, and can host contextual information, such as event-related items. The user's perception of the time currently inspected is enhanced by the presence of a minimap in the form of a railway wagon, which is placed at the tunnel's roof, covering the whole timeline extent. The minimap rendered is the Classic2D representation along with a frame that provides live feedback regarding the equivalent position in the other view of TimeViewer, while also acting as a gateway to Classic2D view.

TimeViewer supports a variety of multimodal interaction techniques, even in combination, so as to offer natural interaction in a wide range of hardware set-ups. Apart from common desktop-based interaction techniques (i.e., point and click or multitouch gestures), TimeViewer supports noninstrumented kinesthetic interaction using Microsoft's Kinect, targeted to navigation and manipulation of 3D virtual environments. Mimicking a person's actions in the real world, users can employ their feet to move and their hands to reach out to virtual elements. More specifically.

- *User Position.* In Classic2D mode the user can stand off-center, toward either side of the display, in order to scroll the projected view toward the respective direction. Furthermore, the system interprets the distance of the user from the display as a zoom mechanism. In the Tunnel mode, the system maps the actual position to the place from which the user looks in the tunnel, creating the illusion of "being there".
- *Virtual Hand.* A cursor, in the form of a three-dimensional virtual hand, appears when the user extends his or her hand toward the display, following and mapping the user's real hand. Items' selection is accomplished by keeping the hand cursor over them for a short period of time.
- *Hand Gestures.* In several cases, when the usage of the virtual hand is not optimal or appropriate, TimeViewer additionally supports single- and dual-hand gestures. The simplest single-hand gesture supported is the movement of a hand for scrolling lists of multimedia elements. Another example involves raising a hand while turning the torso in the same direction, to rotate the projected view around its vertical axis. Dual-hand gestures include pushing and pulling information objects in order to select/deselect them (i.e., bring them closer or send them away), moving between events in Tunnel mode, or examining detailed information about an event in Classic2D mode.
- *Leg Gestures.* TimeViewer adopts the notion of stepping: when a user steps (with one leg, holding the other still) right, left, forward, and backward, navigation is achieved in the corresponding direction. In Classic2D mode, stepping forward or backward results in increasing or decreasing the scale of the timeline respectively, while stepping left and right results in exploring the timeline's display at the respective direction. In Tunnel mode, stepping forward or backward allows navigation along the tunnel, while stepping at the side results in displaying the tunnel as if the user is looking from the equivalent side.

To assess TimeViewer in terms of usability and user experience but also to compare the various supported interaction methods, an evaluation study (Drossis et al. 2013) was conducted with 16 participants (7 female and 9 male) the age of whom varied from 20 to 40 years old. Participants were instructed to express their thoughts aloud (Think Aloud) throughout the completion of each task. In addition, at the end of the evaluation they were requested to fill out a Likert type of questionnaire.

The enthusiasm of the users was apparent during their interaction with TimeViewer when using body tracking, as only one user stated his preference to interact with the system using more traditional modes. Users were captivated by the remote handling of the system achieved without the use of any wearable component.

Hand-related interaction was regarded as very natural, as indicated by the comments during the evaluation and the completed questionnaires. All users found the gestures representative of their functions and the only concern of a small percentage of them (4/16) was that they might result in fatigue after prolonged interaction. Gestures were also considered easy to learn, as all users were generally able to accomplish them on their own after being shown by the evaluator and largely preferred them to the use of the virtual hand cursor, the manipulation of which also came easily to the users. A common remark involved the cursor's (in)stability, which created some accuracy problems with small-sized items. Furthermore, some of the users stated that they would prefer shorter dwell times.

Leg gesturing was almost unanimously accepted. Only one user had trouble with it, as Kinect failed to successfully recognize the exact placement of his legs due to the material of his trousers. Especially nonexpert users were more in favor of leg gestures as they felt more comfortable with handling the system more naturally and less unambiguously. Participants did not face any problem understanding the related conceptual model and found it efficient and fascinating, a fact which is clearly indicated in the corresponding questionnaire answers.

In conclusion, although kinesthetic interaction was considered as fractionally less responsive and more tiring to use, both hand and leg gesturing were deemed as preferable to touch interaction. Despite the higher user error rates in comparison to interaction through touch, users were almost unanimously (15 out of 16) in favor of kinesthetic interaction. This is mainly attributed to the fact that the fun factor was considered as more important than higher accuracy.

4 Discussion

If we had to use just a couple of words to describe each of the aforementioned approaches to body tracking, based on the users' reaction to the respective application then these words would be:

- Macrographia (*location tracking*): Surprising and Impressive
- Paximadaki (*body-shape tracking*): Fun and Intuitive
- TimeViewer (*skeleton tracking*): Versatile and Powerful.

Macrographia made a memorable impression to its users. The moment people walked in the room (especially when there was no one already in) and saw the "wall" responding to their movement, their facial and verbal expressions articulated the feeling of positive surprise and excitement. This phase was followed by a "curiosity" period during which they tried to discover and explore all the available pieces and modes of information. Then again, Paximadaki had a more "cozy" effect. The fact that people saw their own shadow on the virtual wall, as well as their face in the photographs at the end of the game, created an affinity toward the game, which was subsequently directed toward the people running the installation (see Grammenos

Table 1 Advantages and limitations of the 3 presented approaches to employing body tracking

Type of tracking	Advantages	Limitations
Location	• Robust • Multiuser (number depending on available space) • Can work well with very large displays • Can work in large spaces • Requires limited user mobility	• Small "interaction vocabulary" • "Midas touch" • Can support coarse-grained interaction • Limited number of users • Complex to implement • Requires (considerable) space • Limited potential applications • Requires ability to move in space • Indirect user control • Higher cost
Body-shape	• Very robust • Very intuitive • User perceives own image • Multiuser (virtually unlimited number of concurrent users) • Very flexible, works for highly diverse user characteristics • Very quick response • Easy to implement • Low cost	• Small "interaction vocabulary" • "Midas touch" • Limited potential applications (mostly games) • Requires ability for (gross) body movement • Prone to interferences (from objects or other people in range)
Skeleton	• Rich interaction vocabulary • Multiuser (limited number of concurrent users) • Highly versatile, essentially unlimited application types supported • Can provide very natural interaction • Can support tasks requiring high accuracy • Low cost	• Lless robust • "Midas touch" (in some cases) • More 'rigid' (more strict user motion), especially when gestures are used • May not work (properly) for some body types, clothes, accessories • Requires ability for fine body movement and control • More complex to implement • Harder to learn and use • (Can be) less intuitive, may require training

et al. 2012). Furthermore, the selected interaction approach was very intuitive indeed. In all installations 30–40 % of the players had never played any type of video game before and had no problem mastering game control after the first few seconds. Probably the best example illustrating this fact was 2-year-old Maria who, after playing the game and being all excited about it, would wander around explaining to anyone standing nearby how to play: "Big basket, rusks put in—many many rusks, yeaah!" Finally, TimeViewer provides much richer and fine-grained interaction supporting a large "interaction vocabulary" which allows designers to apply the most appropriate interaction technique to each task, and, often, users to select among redundant techniques according to their preferences. Additionally, maybe also due to the novelty of kinesthetic interaction, users showed significant preference and liking toward it, despite its potential fatigue and accuracy issues.

Table 1 provides an overview of the individual advantages and limitations of each approach.

In principle, all three approaches can accommodate multiple users, but at different scale. Skeleton tracking, currently works best for a very small number of users. Body shape tracking has no technical limitations (we have seen up to 12 players at Paximadaki) but in practice, there is a certain point where a crowded shadow image "overflows" the game terrain, rendering interaction meaningless. Location tracking can scale up, but at the cost of scaling up the interaction space, as well as of the technical equipment used.

The first two approaches have a very limited "interaction vocabulary", i.e., distinct "commands" that users can issue through them. Skeleton tracking, especially when hand and leg gestures are used, can have a very rich and context-related "interaction vocabulary" but at the cost of ease of use, intuitiveness, and requirement for user training.

All three approaches have considerable user mobility requirements with skeleton tracking being the most demanding one, both in terms of gross motor skills (i.e., movement of the arms, legs, feet, or entire body) and endurance (can become quite tiring after a while).

A common problem of all approaches is what is usually referred to as "Midas touch," or, in other words, identifying when users want to perform an action to an interactive object that is currently under their "reach" (e.g., a slot that corresponds to their location in Macrographia, a button that overlaps with their virtual shadow in Paximadaki, or an object located underneath the hand cursor in TimeViewer). In all cases, a common solution is "dwell time," where the respective action is triggered after a predefined time period. Another possible solution that can work in some cases is to require a specific combination of overlaps in time and space which would be hard to happen by accident. For example, in order to start the game, a player may have to sequentially "touch" 3 different objects or even the same object 3 times, but every time it is touched it moves to a different location. Such approaches are appropriate for actions that require confirmation and do not have to be frequently performed. Additionally, if a richer interaction vocabulary can be supported (like in the case of TimeViewer), more elaborate solutions can be provided (e.g., activation gestures).

5 Conclusions

All collected pieces of evidence stemming from the evaluation and observational data of the three case studies converge to the fact that people of all ages and backgrounds indisputably like and enjoy body-based interaction with public systems. On the one hand, it feels more natural and intuitive, while one the other hand it "frees" them from the highly static and rather mundane interaction with desktop and mobile computing platforms. Furthermore, in general, body motion is conceived as more "fun," irrespectively of the actual application's content. For instance, unlike Paximadaki, both the Macrographia and the TimeViewer systems provide "serious" (educational) information, but most users commented that they were fun to use. Furthermore, in several cases, a considerable number of concurrent users can be supported, thus effectively allowing for group interactions (coordinated or not), which are highly valued (especially among families and friends) and are currently underserved (if at all) by most public systems. In more practical terms, another important aspect of noninstrument body-based interaction with public systems is that users do not come in physical contact with any part of them, thus avoiding dirt, wear and (intended or not) damage of the equipment, contributing to its longevity but also solving several public hygiene issues.

On the downside, body-based interaction also comes with some inevitable weaknesses. First of all, as requirements for body motion and control increase, so do the potential accessibility problems for people with mobility limitations or diversifications, due to circumstantial events (e.g., carrying something), temporary (e.g., a broken leg) or permanent (e.g., paraplegia) disabilities, age (e.g., young children), etc. Then, as the domain is still emerging, there are no widely used and established conventions—a common "interaction vocabulary"—that users are familiar with. As a result, quite often users may not be aware of what they are able to do with a system and some training may be required (even for apparently "intuitive" interactions), and, in some cases, interaction approaches of different systems can be completely inconsistent. To this end, the level of instruction that should be provided in public systems, as well as, how this should be achieved, is a key open issue. Finally, due to the large diversity of human bodies, the items that may be residing on them (e.g., clothes, accessories, devices), as well as of specific environmental conditions, there will always be cases were a particular body-tracking approach may not (properly) work.

A very interesting aspect of noninstrumented body-based interaction is that it inherently supports various forms of formal and informal social interaction. As it takes place in considerable space, a number of actors can be easily involved, in various roles. For example, in the case of location-based interaction (Macrographia) typical "museum-related" social behavior (e.g., politeness—avoid to stand in front of another person) was observed, but also playful activities (e.g., a couple hugging trying to share the same slot), including ad hoc group forming (e.g., a number of strangers standing next to each other in order to reveal all the layers of information of a specific level). In the case of Paximadaki, except the types of social interaction already described in the related section (e.g., cooperation, competition, unstructured group play), another noteworthy behavior was the occasional instantaneous "intrusion" of spectators in the play area, for example to tease a friend by blocking a basket, or extending a helping arm to aid the weaker player. Finally, as the third system presented was targeted for use by a single user, the respective social behavior was limited to people switching roles by going closer or retreating from the screen. Still, the described approach can be (and is currently) extended to support both synergetic and parallel interaction by at least two users.

Acknowledgments This work has been supported by the FORTH-ICS RTD Programme "Ambient Intelligence and Smart Environments".

References

Argyros AA, Lourakis MIA (2004) Real time tracking of multiple skin-colored objects with a possibly moving camera. In: Proceedings of the European conference on computer vision (ECCV'04), vol 3. Springer, Prague, Czech Republic, pp 368-379, 11–14 May 2004

Blomberg J, Giacomi J, Mosher A, Swenton-Wall P (2003) Ethnographic field methods and their relation to design. In: Participatory design: principles and practices. Lawrence Erlbaum Associates, pp 123–155

Bobick AF, Intille S S, Davis JW, Baird F, Pinhanez CS, Campbell LW, Ivanov YA, SchütteA, Wilson A (1999) The KidsRoom: a perceptually-based interactive and immersive story environment. Presence: teleoper. Virtual Environ 8(4):369–393

Brooke J (1996) SUS: a quick and dirty usability scale. Taylor and Francis, London, pp 189–194

Crossan A, Brewster S, Ng A (2010) Foot tapping for mobile interaction. In: Proceedings of BCS '10, British Computer Society Swinton, pp 418–422

Drossis G, Grammenos D, Adami I, Stephanidis C (2013) 3D visualization and multimodal interaction with temporal information using timelines. In: proceedings of INTERACT 2013, lecture notes in computer science, vol 8119. Springer, Heidelberg, pp 214–231

Fikkert W, van der Vet P, van der Veer G, Nijholt A (2010) Gestures for large display control. In: Proceedings of GW'09, LNCS 5934. Springer, Heidelberg, pp 245–256

Grammenos D, Margetis G, Koutlemanis P, Zabulis X (2012) 53.090 Virtual rusks = 510 real smiles using a fun exergame installation for advertising traditional food products. In: Nijholt A, Romão T, Reidsma D (eds) Advances in computer entertainment, LNCS 7624. Springer, Heidelberg, pp 214–229

Grønbæk K, Iversen OS, Kortbek KJ, Nielsen KR, Aagaard L (2007) iGameFloor: a platform for co-located collaborative games. Proceedings of ACE '07, vol 203. ACM, New York, pp 64–71

Hilliges O, Izadi S, Wilson A, Hodges S, Mendoza AG, Butz A (2009) Interactions in the air: adding further depth to interactive tabletop. In: Proceedings of UIST '09. ACM, New York, pp 139–148

Jaimes A, Sebe N (2007) Multimodal human–computer interaction: a survey. Journal computer vision and image understanding archive, vol 108(Issue 1–2). ACM, New York, pp 116–134

Jenkins H (2002) Game design as narrative architecture. In: Harrington P and Frup-Waldrop N (eds) First person. MIT Press, Cambridge, pp 118–130

Kortbek KJ, Grønbæk K (2008) Interactive spatial multimedia for communication of art in the physical museum space. In: Proceeding of MM '08, pp 609–618

Krueger MW, Gionfriddo T, Hinrichsen K (1985) VIDEOPLACE—an artificial reality. In: Proceedings of CHI'85, San Francisco, pp 35–40

Laakso S, Laakso M (2006) Design of a body-driven multiplayer game system. Comp Entertain 4(4):7

Lindley SE, Le Couteur J, Berthouze NL (2008) Stirring up experience through movement in game play: effects on engagement and social behaviour. In: Proceeding of CHI '08, pp 511–514

Mueller F, Agamanolis S, Picard R (2003) Exertion interfaces: sports over a distance for social bonding. In: Proceedings of CHI '03, pp 561–568

Nickel K, Stiefelhagen R (2003) Pointing gesture recognition based on 3D-tracking of face, hands and head orientation. In: Proceeding of ICMI '03. ACM, New York, pp 140–146

Papadopoulos C, Sugarman D, Kaufmant A (2012) NuNav3D: a touch-less, body-driven interface for 3D navigation. In: Proceeding of IEEE VR 2012. IEEE, pp 67–68

Paradiso J, Abler C, Hsiao K, Reynolds M (1997) The magic carpet: physical sensing for immersive environments. In: CHI '97 extended abstracts, pp 277–278

Robertson T, Mansfield T, Loke L (2006) Designing an immersive environment for public use. Proceedings of PDC '06. ACM, New York, pp 31–40

Ronkainen S, Häkkilä J, Kalev S, Colley A, Linjama J (2007) Tap input as an embedded interaction method for mobile devices. In: Proceeding of TEI'07. ACM, New York, pp 263–270

Sangsuriyachot N, Mi H, Sugimoto M (2011) Novel interaction techniques by combining hand and foot gestures on tabletop environments. In: Proceeding of ITS '11. ACM, New York, pp 268–269

Sparacino F (2004) Scenographies of the past and museums of the future: from the wunderkammer to body-driven interactive narrative spaces. In: Proceeding of MM '04, pp 72–79

Stamatakis D, Grammenos D, Magoutis K (2011) Real-time analysis of localization data streams for ambient intelligence environments. In: The proceedings of AmI 11: international joint conference on ambient intelligence, Amsterdam, Springer, Berlin, Heidelberg, pp 92–97, 16–18 Nov 2011

Sweetser P, Wyeth P (2005) GameFlow: a model for evaluating player enjoyment in games. Comp Entertain 3(3):1–24

Valkov D, Steinicke F, Bruder B, Hinrichs K (2010) Traveling in 3D virtual environments with foot gestures and a multitouch enabled world in miniature. In: Proceeding of VRIC 2012. IEEE, pp 171–180

Yoo B, Han J-J,Choi C, Yi K, Suh S, Partk D, Kim C (2010) 3D user interface combining gaze and hand gestures for large-scale display. In: Proceeding of CHI '10. ACM, New York, pp 3709–3714

Zabulis X, Grammenos D, Argyros A, Sifakis M, Stephanidis C (2011) Macedonia: From Fragments to Pixels. ERCIM News Special Theme: ICT Cult Heritage 86:25–26

Zabulis X, Grammenos D, Sarmis T, Tzevanidis K, Argyros AA (2010) Exploration of large-scale museum artifacts through noninstrumented, location-based, multi-user interaction. In: Proceedings of VAST'2010, Palais du Louvre, Paris, France, pp 155-162, 21–24 Sept 2010

Zabulis X, Grammenos D, Sarmis T, Tzevanidis K, Padeleris P, Koutlemanis P, Argyros AA (2012) Multicamera human detection and tracking supporting natural interaction with large scale displays. Mach Vis Appl J 24(2):319–336. Feb 2013

Playing with the Environment

Pedro Centieiro, Teresa Romão and A. Eduardo Dias

Abstract Games as entertainment mechanisms can be exploited to promote education, social relationships and behaviour changes. In this chapter, we discuss how mobile multiplayer games that explore interaction with public displays stimulate engagement, persuasion and social interaction. We developed Gaea, a persuasive location-based multiplayer mobile game, which prompts people to recycle virtual objects within a geographical area. The goals of this prototype were to instruct, inform and persuade users to recycle their wastes. Four key requirements guided Gaea's design and development: making use of a natural environment; reaching a large number of people; promoting entertainment; and fostering social engagement. With these ideas in mind, Gaea was designed to take advantage of the intrinsic characteristics of mobile devices (allowing players to freely move around while communicating and receiving information) and public displays (allowing audience awareness and players' social engagement). Players use a smartphone to locate and collect the virtual litter in their surroundings, which should then be dropped into the correct virtual recycle bin, available for selection when approaching the public display. Gaea raises users' awareness to the impact of their actions on our planet's natural resources. It also promotes users' physical activity, social interaction and environmental behaviour changes. User studies were performed, revealing encouraging results that are also described in this chapter.

P. Centieiro (✉) · T. Romão · A. E. Dias
CITI Centro de Faculdade de Ciências e Tecnologia, Universidade Nova de Lisboa, 2829-516 Caparica, Portugal
e-mail: pcentieiro@gmail.com

T. Romão
e-mail: tir@fct.unl.pt

A. E. Dias
bViva International, B. V, Romanovhof 9, 3329 BD Dordrecht, The Netherlands
e-mail: aed.fct@gmail.com; edias@bviva.com

A. Nijholt (ed.), *Playful User Interfaces*, Gaming Media and Social Effects,
DOI: 10.1007/978-981-4560-96-2_3,

Keywords Mobile interaction · Public displays · Persuasive technology · Contextual awareness · Multiplayer mobile games · Recycling · Educational entertainment

1 Introduction

For years, human beings have been consuming Earth resources without even thinking about the consequences of their behaviours. In order to preserve our planet's resources for a long period of time, we need to develop a natural equilibrium with the surrounding environment. This balance can be achieved in several ways, such as by recycling the waste we produce or by saving (and also recycling) daily consumptions (like water, energy, gas or diesel) and consequently reducing pollution. However, not all citizens are proactive in terms of recycling.

In this chapter, we show how new technologies can be used to change people's behaviours towards the environment. In this context, we approached three different areas: persuasive technology, mobile computing and interaction with public displays.

The intention of persuasive technology is to change people's attitudes or behaviours, or both, through persuasion and social influence (without using coercion or deception) in an interactive way (Fogg 2003). It focuses on the study of Human Computer Interaction, investigating how people are motivated or persuaded when interacting with computer devices. Take, for instance, the Nike+ Running application that motivates joggers' to do better, by presenting them facts about previous performances and giving them words of encouragement.

Mobile phones are the most personal technology in the world, since they are always by our side when we need them, they advise us and entertain us anytime and anywhere we need them (Fogg 2008). Thus, through mobile systems, it is possible to persuade users at the right time and at the right place (Fogg 2003). And since virtually everyone owns a mobile phone, it is also possible to provide citizens with games that engage them in social activities, alert them to environmental problems and stimulate them to adopt more appropriate habits.

Finally, in order to really make a difference, we need to get the attention of a large group of people. One way to achieve this is by using public displays, which can contribute significantly to the dissemination of the message being conveyed, due to their high visibility. Thus, by applying persuasive technology concepts to mobile games, and by exploring innovative forms of interaction between mobile devices and public displays, we can stimulate citizens to become more aware of our planet's environmental problems, and we can also study how this new kind of persuasion influences people's conscience and everyday life. To this end, we created a mobile multiplayer game called Gaea that helps to motivate citizens to become more responsible towards the environment, engaging them in environmental preservation activities. Gaea promotes education, social interaction and behaviour changes, by inviting players to recycle virtual objects spread throughout a geographical area and providing them with information about recycling. Players use a mobile phone to locate and collect the virtual objects (litter), which should then be placed into the correct virtual recycle bin, available on the mobile screen when approaching the public display. Players who recycle more objects and learn with the tips given by Gaea, will end up winning the game.

The work presented in this chapter was developed in the scope of the DEAP project (Developing Environmental Awareness with Persuasive systems), which was funded by Fundação para a Ciência e Tecnologia (FCT/MEC).

This chapter is organised as follows: Sect. 2 presents motivation and some related work; Sect. 3 describes Gaea, its design process and features and the evaluation procedures and finally, Sect. 4 presents our conclusions.

2 Having Fun While Helping the Environment

There is always a crucial challenge when creating educational or persuasive applications: users will not use them unless they are fun and users can have a good time using them. Thus, the prototype presented in this chapter is based on three main goals: promote pro-environmental behaviours, foster social interactions among users and between them and members of the audience and entertain players.

2.1 Environmental Behaviours

The Living Planet Report 2008 (2008) shows that we are demanding too much from our planet's living resources, and Humanity's Ecological Footprint exceeds the planet's regenerative capacity in about 30 %. If we do not balance our consumption with the Earth's capacity we risk doing irreversible damages. Actions to prevent that catastrophe must be taken on a global level, making the majority of the human beings change their behaviours in order to save, reuse and recycle the Earth's resources.

However, changing behaviours is not an easy task. It follows a group of principles, concepts and design decisions that need to be applied to different types of audiences. Steg and Vleg (2009) stated that environmental quality strongly depends on human behaviour patterns and that individuals can contribute significantly to achieve long-term environmental sustainability by adopting pro-environmental behaviour patterns. They proposed a general framework for understanding and promoting pro-environmental behaviour, comprising: (1) identification of the behaviour to be changed, (2) examination of the main factors underlying that behaviour, (3) design and implementation of interventions to change behaviour to reduce environmental impact and (4) evaluation of those interventions' effects.

Some examples of applications that promote pro-environmental behaviours are UbiGreen Transportation Display Boston (2009), PerCues (Reitberger 2007) and SmartBins (Lobo et al. 2009). UbiGreen Transportation Display is a mobile application prototype that semi-automatically senses and reveals information about transportation behaviour, aiming at engaging users in the goal of increasing green transportation and reducing pollution. PerCues is a persuasive system intended to reduce pollution through the use of public transportation. If people choose to use the bus instead of the car, the system informs users, through their mobile phones, about the consequences of their choices. Educational games like SmartBins try to educate users to correctly recycle waste at a predefined place and with all the objects to be recycled at hand.

When we designed Gaea, we believed that the idea of physically walking around a spatial (green) area, along with other users, while picking up several virtual objects (litter) could immerse participants in an educational and enjoyable activity. If we could take advantage of the social interactions between users playing the game, this activity could be both fun and entertaining. As we will discuss in the next sections, we were not wrong, as the results showed that users had a great time playing Gaea, while becoming physically active, assimilating the environmental information provided by the application (which induces pro-environmental behaviours) and learning how to do a correct waste recycling.

2.2 Social Interactions

It is possible to distinguish two types of interaction with public displays: *overt* and *covert* interaction (Kaviani et al. 2009). *Overt* interaction is a consequence of using devices like a mouse or a keyboard to directly interact with a display. In this case users' actions can be watched by the audience, which often causes social embarrassment, therefore discouraging user interaction with the display. *Covert* interaction refers to the use of mobile phones or implicit physical movement to interact with public displays. *Covert* interaction provides users with more privacy and confidence during the interaction process. Although it does little to inform about the interaction with the display, it still allows the audience to benefit from the information provided by the large display. Gaea explores *covert* interaction, in order to restrict direct interaction to players but, at the same time, allow the audience to follow the game, receive environmental information and be entertained.

Gaea's design also followed the guidelines proposed by Rogers and Brignull (Rogers and Brignull 2002) in order to allow the audience to easily follow the game, and to know what they were supposed to do if they wanted to take part in the activity.

2.3 Entertainment

Games may influence players to take action through gameplay, as described by Ian Bogost in his theory on how videogames make arguments and influence players (Bogost 2007). Games simulate experiences, deliver messages and may become rhetorical tools for persuading players. People are more willing to perform activities when they are fun and entertaining. Through the use of scores and statistics, it is possible to entertain and promote competition between users, in order to motivate them to reach their goals (Fogg 2003). Whatever may be the competition that people enter, they enter to win. They want to prove that they are the best amongst their friends or strangers, even if they are not rewarded with material prizes. Thus, computer technology can motivate users to adopt a target attitude or behaviour by leveraging the human being's natural drive to compete. Take for instance trivia games, which became very popular in the last years in the video game market with games like Buzz selling over 6 million copies (Eurogamer 2008). This type of game promotes social interaction and learning at the same time. We thought that Gaea could also benefit from having a quiz, after the play area became clean, and the evaluation results confirmed that assumption.

3 Gaea

Gaea aims to raise users' awareness of the consequences of their actions on the natural resources of our planet. It was designed as a multiplayer mobile game intended to instruct, inform and persuade players to perform a correct recycling of their wastes. Users are asked to recycle virtual objects (litter) spread over a geographical area, by using a mobile device to locate and collect those objects, which should then be placed into the correct virtual recycle bins, which become available on the mobile screen when near the public display.

In order to better explain how a Gaea game session unfolds, imagine a garden, a natural park or any other open space with virtual waste scattered all over it. These virtual objects do not really exist, but we can locate them and see them on a map (Fig. 1) when using a mobile phone running the Gaea application.

The virtual objects (litter) are addressed to geographical coordinates (generated by a server), and in order to pick them up the user has to go to their location. Once near one of these objects, the user must grab it by touching the object on the map displayed on the mobile device. Then the user must drop it into the correct virtual recycle bin, which becomes available for selection on the mobile device when the user approaches the public display (which is also associated to a geographic coordinate) and faces it (Fig. 9). When the user selects the right recycle bin, he is rewarded with points and receives recycling tips and facts. However, if he misplaces an object, he is informed by the application so as not to make the same mistake again, and consequently he does not receive any points or tips. During the game, the public display exhibits all the information relevant to the players and

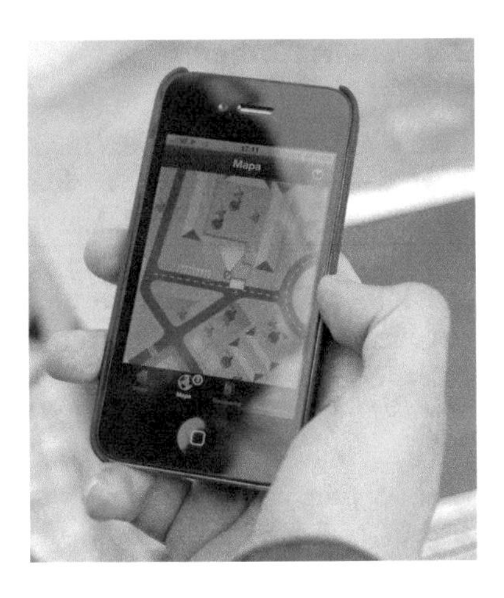

Fig. 1 Gaea showing the map of the game area

audience, such as game instructions, a map with the virtual objects and players' locations (so the audience knows what is going on), feedback given to the users about their actions (if they placed the object into the correct recycle bin) and scores, as explained later in Sect. 3.3. Also, in order to urge the users to recycle and to inform them about the status of our planet resources, every time users drop an object into the correct recycle bin, they receive on their mobile device, specific information about that object in the context of our planet resources. If users mark this information as read, they are rewarded with extra points. After the play area is clean, a quiz appears on the public display with questions about the information that was previously presented to the users when they recycled. The users are then prompted to use their mobile devices to answer these questions, by selecting one of three possible answers. The audience can also follow the quiz.

During a Gaea game session, multiple users compete and each user struggles to recycle the largest number of objects and to achieve the best possible score. A collaborative version of Gaea was also created, allowing teams of users to be in charge of picking up one different type of waste (paper, plastic/metal or glass). In this version, Gaea becomes a collaborative/competitive social activity that persuades a group of people to do a single action together. Obviously they are not directly helping the environment, because they are collecting virtual objects instead of real objects, but they are learning better ways of recycling and they are being informed about the consequences and gains of their actions through this activity. Gaea also stimulates social and collaborative activities and may influence people to have the same kind of positive behaviour in their daily lives.

3.1 Using Persuasive Technology as a Design Factor

Since one of the main goals of Gaea is to promote pro-environmental behaviours, we decided to use persuasive technology concepts to motivate and influence people to become more environmentally responsible. Thus, during the design process that devised Gaea's features, many of the decisions taken, were based on persuasive technology concepts, which ended in establishing the fundamental elements of the system.

We started by analysing persuasive technologies tools (Fogg 2003). From these concepts, four emerged as key factors to shape Gaea's gameplay: suggestion (to intervene at the right time and at the right place), conditioning (to give positive reinforcement to shape complex behaviours), tunnelling (to ease the process to perform tasks) and self-monitoring (to allow people to monitor data about themselves).

Taking into account the previous concepts, we defined four distinctive sections in the mobile application: Home (which allowed the users to register on the system, as well as to read the activity's instructions), Map (shows the location of the virtual objects), Gaeapedia (provides information regarding environmental aspects) and Performance (that records and presents the user's performance).

In a subtle way, we also want Gaea to act as a persuasive media (Fogg 2003), creating situations that reward and motivate people to adopt a target behaviour, allowing rehearsal and showing the consequences of users actions.

Finally, the social cues taken into account when computers act as persuasive social actors (Fogg 2003) include the physical cues (use of visually attractive elements) and the language cues (the way applications communicate with the users).

In the next sections, we present more details on the design decisions that helped to shape Gaea.

3.2 Design Methodology

In order to achieve our goals, an iterative design process was followed. We performed a user and task analysis process, and after it we developed a paper prototype that went through two iteration cycles. Later, according to the feedback we received from the evaluation of the paper prototype, a computational prototype was implemented. Finally, we analysed the data obtained during the evaluation of the computational prototype in a live environment on two different events, which are described on Sect. 3.4.

Gaea is primary addressed to teenagers and young adults, and secondarily to children. Obviously, older people can also participate, especially parents with their children (which helps promoting relationships between them). The application can also be used to support school activities, which promote knowledge towards recycling and competition between students.

The paper prototype was an interactive paper mock-up used by participants to execute tasks, by using their fingers to point to the several hand-drawn buttons while they were frequently encouraged to think aloud. It was evaluated by several users over two iteration cycles and revealed several problems, some of them

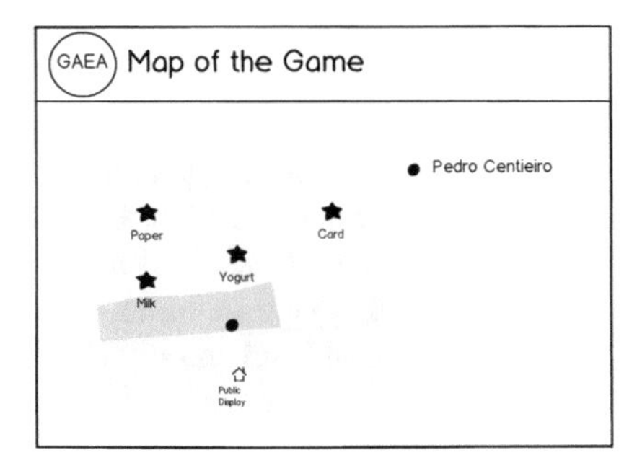

Fig. 2 Map section on the public display

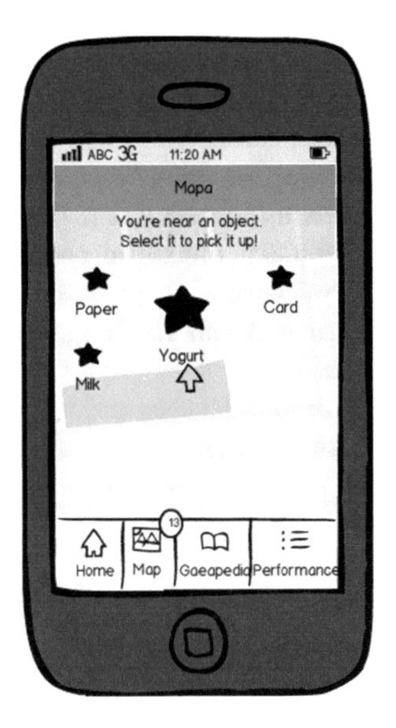

Fig. 3 Map section on the mobile application

required major overhauls of the interface, including the simplification of the scoring and the registration systems.

In order to have a glimpse of what the paper prototype looked like, Figs. 2 and 3 show the final design of the Map sections, changed according to the user feedback.

3.3 Computational Prototype

According to the results obtained during the evaluation of the paper prototype, a computational prototype was developed, and the following features were set:

- *Three Phases*. Gaea activities comprise three phases: registration, recycling and quiz (each one of them with their own interfaces), which happen automatically by this order. Users must start by registering on the system, entering their personal data (first and last name) or entering their Facebook account data (e-mail and password). The recycling phase is the main and longer phase. During this phase, users can catch virtual objects from the play area and bring them to the recycle bins accessible when they are near the public display. Whenever they do that, they receive information regarding the recycled object. The virtual objects were chosen accordingly to their frequency of use at people's houses (based on the recycling rules of the Sociedade Ponto Verde website[1]). Finally, during the quiz phase users answer several questions based on the information provided by the system when they have placed the objects in the recycle bins. Having individual interfaces for each phase, allowed users to be focused on their current goal in each of these phases.

- *Scoring System*. There are three ways of earning points: recycling an object (100 points), marking the information provided by the Gaeapedia as read (50 points) and correctly answering a question during the quiz (50 points). The primary objective is to recycle the objects on the play area, therefore a higher number of points is awarded to this action. The other two actions are also rewarded with points in order to persuade users to read and memorise the information provided by Gaeapedia.
- *Two Play Modes*. Gaea is a multiplayer game that can be played individually or in team mode, where the score of a team is the sum of the scores of all its players. In team mode, there are three teams of different colours (blue, yellow and green) each one in charged of picking up the objects matching the recycle bin of the corresponding colour.
- *Object's Locations*. The virtual objects' locations can be generated in two ways: in a grid where each cell contains an object (at most), or by setting a different number of possible locations for objects. The first approach is more appropriate when the play area is a wide space with no buildings nor obstacles. In this case, to automatically generate the grid and the virtual objects' positions, the application only needs to know the area's northwest and southeast geographic coordinates and the number of objects to be included in the activity. The second approach is best suited when there are buildings and obstacles that do not allow objects to be placed in certain positions. It demands the insertion of all the possible locations where objects can be placed on the play area. Then, the system randomly chooses the different locations, from all the possible options, where the objects are going to appear.
- *Map*. To create a more immersive experience, it was decided to use a custom map designed specifically for this application (not use Google Maps or any other web mapping service application) to show the objects and users' locations. On the mobile device, the map only shows the objects' location. On the public display, objects and users' location are dynamically shown in real time to

[1] http://www.pontoverde.pt

entertain and engage the audience. Also, when the compass is being used, the map rotates accordingly to the direction the user is pointing to.

- *Facebook*. To try to achieve a massive number of people, an external application was created on Facebook (allowing users to share messages about their experiences on their Facebook Walls from the mobile application), as well as a Facebook page for the DEAP project, which contained information about Gaea and other prototypes. Shared messages were created to be tailored to a specific person, instead of creating messages that sounded generic. This was made to create a persuasive experience in order to achieve mass interpersonal persuasion. For example, the user's score, genre and the city where the activity took place are taken into account to personalise the message to be shared on the user's Facebook Wall.

The development of Gaea can be divided in its two components: the public display (server application) and the mobile device (client application). The following two sub-sections detail the user interfaces of both Gaea's components, which provides an overview of the whole system.

3.3.1 Public Display

Regarding the design of the public display application, an overhaul of its initial design was needed in order to encompass the dynamic characteristics of the public display, which would help to create a "honey-pot" effect. These characteristics consist on a rotating cube where each face corresponds to a specific feature of the application: Home, Map, Instructions, State of the Game, Gaeapedia and Gaeaquiz. As the game unfolds, the cube rotates in order to present the appropriate face (screen) to the audience and players. The activity starts on the Home screen and as soon as all the players set themselves as ready (on the client application), the cube rotates to the Map screen. During Gaea recycling phase, the public display mostly alternates between three screens: Map, Instructions and State of the Game. When someone drops an object into a recycle bin, the public display shows the Gaeapedia screen, and then switches to the State of the Game screen. If it is the last object on the play area, the public display will then switch to the Gaeaquiz, and after all the questions have been presented, the final ranking is shown and the activity ends. Figure 4 presents the public display screen flow.

This way we created a consistent interface with a well-defined information flow, where players and audience could easily follow the game and predict (if they were watching the activity for a while) what was going to be shown. Ultimately, it led users and audience to have a better experience by learning how the activity unfolds.

A more detailed description of some of the system's features is presented next:

- *Map*. The purpose of the Map is to give users and the audience an overview of what is happening. Users can use the Map as a strategic element to know where the other players are, in order to choose which objects they can pick up more

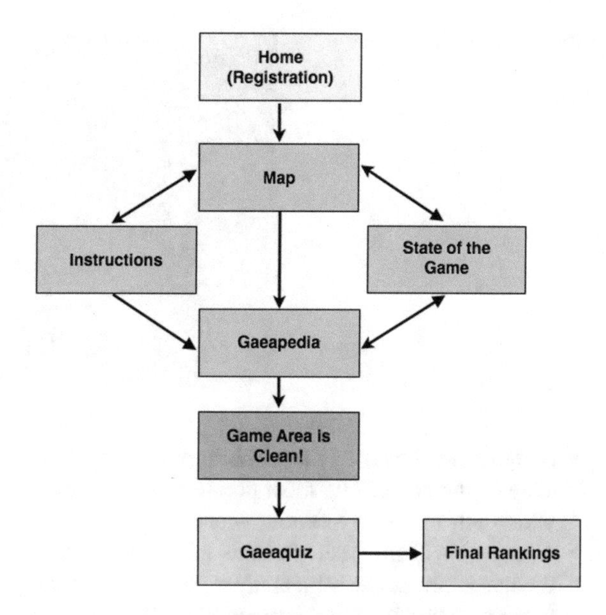

Fig. 4 Screen flow of the public ambient display

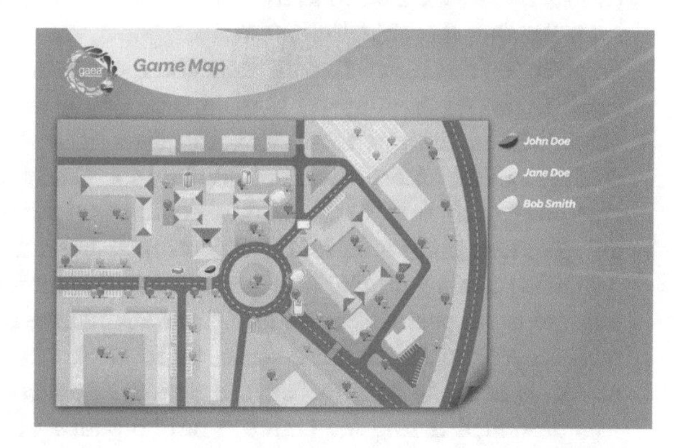

Fig. 5 Map screen

quickly. As for the audience, the Map tries to produce a "honey-pot" effect, since the players' movements can be seen on the public display in real time, which arouses curiosity, interest and makes people socialise, by exchanging opinions and comments about what is happening. The Map screen is depicted in Fig. 5.

- *Instructions*. As mentioned in Sect. 2, it is crucial to give instructions to the audience on how to interact on a covert interaction (Kaviani et al. 2009). So, this feature is mainly aimed for the audience, because users are supposed to know the instructions already (they read them on the mobile application before the activity started). Therefore, by reading the instructions, the audience will know how the game unfolds and what the objectives are. Figure 6 shows the Instructions screen.

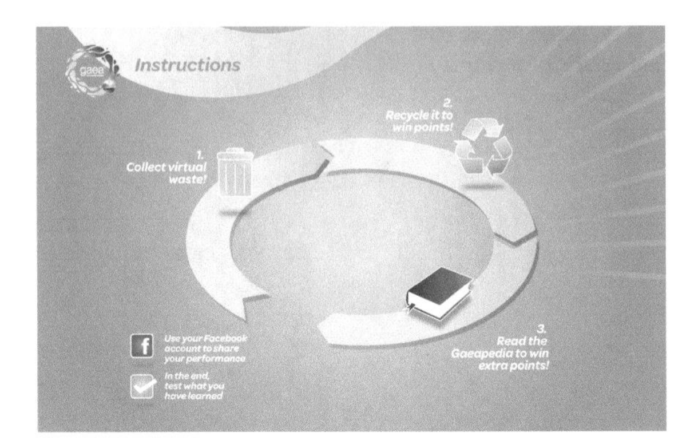

Fig. 6 Instructions screen

- *State of the Game*. As the name implies, this screen is meant to show the current state of the game. The most popular information it displays is the classification, which allows users to know where they stand in the ranking.
- *Gaeapedia*. This screen presents three important pieces of information. First, the feedback about the object that the user just dropped into the recycling bin (which allows the user to know if he dropped the object into the correct recycle bin). Second, the object's type (paper, glass, plastic/metal) in order to provide the user and the audience with facts about environmental resources. And third, a small tip about the players' performance on the activity.
- *Gaeaquiz*. Is the definitive way to check if the users really read and paid attention to the information provided by the Gaeapedia throughout the activity. By telling users from early on, that a quiz about the information on Gaeapedia is going to take place after all the objects have been dropped into the recycle bins, it makes them aware that they need to pay attention to that information. This is how we try to persuade users to be aware about the environmental topics related with recycling. This feature works as follows: after all players have set themselves as ready to start the quiz, a question appears (about an information that was presented during the activity) on the public display, and users must select the right answer on their mobile device. If a user gets the answer right, he is awarded with points, otherwise he does not get any. Figure 7 depicts the Gaeaquiz on the public display.

3.3.2 Mobile Application

The client application was the component that required more time and work due to the technology involved in the Map section (GPS, compass and camera). Its main features include:

- *Map*. While on the Map section, the user can be guided in two ways, with or without compass, to support navigation. This feature provides the users with a better perception of the whole game environment, showing them where they are

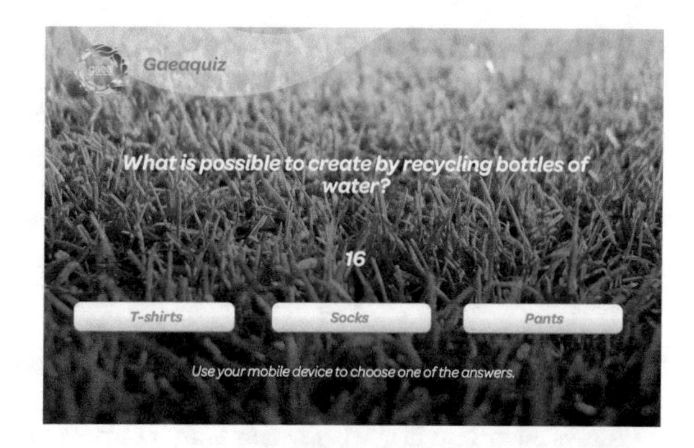

Fig. 7 Gaeaquiz screen presenting a question

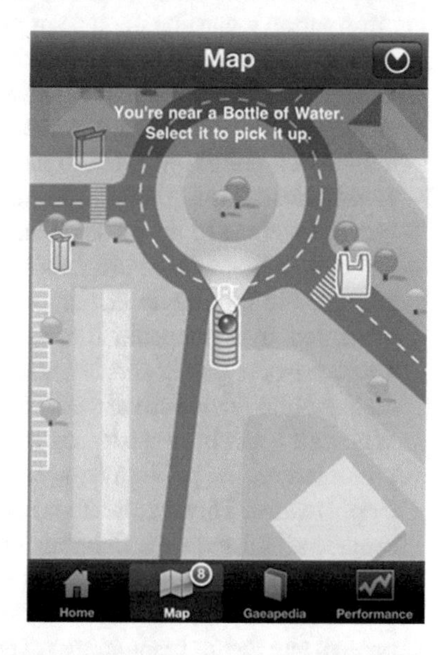

Fig. 8 Map view (user is near an object)

Fig. 9 User recycling an object, using the augmented recycle bins on the mobile phone

and where the virtual objects are. To collect an object, the user needs to get near the object and select it from the Map. Then, when he arrives near the public display and wants to drop the object into a recycle bin, he just needs to rotate the mobile device, which will activate the camera. Based on the concepts of augmented reality, the user can point it to the public display and the recycle bins will appear on the mobile phone screen. The user then selects one of the recycle bins and the public display changes to the Gaeapedia screen, while the mobile device shows a loading view informing the user to check the public display. Figures 8 and 9 present some of these actions.

- *Gaeapedia*. In the mobile application, the Gaeapedia records all the information provided to the user when dropping objects into the recycle bins. This information is presented in a list, where each entry contains an object icon, environmental information regarding the type of object (paper, glass or plastic/metal) and a mark button. This mark button allows users to mark the information as read. It is only enabled after a few seconds (the estimated time needed to read the information). When one entry is marked as read, an image of a medal appears, indicating the number of points awarded. This acknowledges the user that by marking the environmental information as read he is being awarded with points, motivating him to continue to do so. The Gaeapedia screen is presented in Fig. 10.
- *Gaeaquiz*. This feature presents users with the possible answers to the questions proposed in the public display during the final quiz, regarding the information presented by Gaeapedia throughout the activity. Each time a new question appears in the public display, users can select one of three possible options on their mobile devices to answer the question (also shown on the public display). When all users have answered the question (or the time expires) the results are presented on the public display. The Gaeaquiz screen can be seen in Fig. 11.
- *Performance*. This feature does not involve any user interaction, but it allows the users to monitor their actions and their scores, eliminating the cognitive effort that they would have if they would need to remember their performance. In the Performance section, the following information is presented: number of recycled objects and the corresponding points awarded, number of information items read on Gaeapedia and the corresponding points awarded and the overall score.

3.3.3 Architecture

Gaea was developed to be compatible with iOS 4.2 (or higher) running on most iOS devices, while the public display is compatible with Mac OS X 10.6 or higher. Gaea is based upon a client–server architecture. This architecture fits in the two types of settings where the activity can take place: in an area where there is a wide wireless network (in order for the clients to connect to the server) and in an area with no wireless network. In the first case, all the traffic between the clients and the server is handled through the local network (LAN), while in the second one the traffic is channelled through the Internet (WAN).

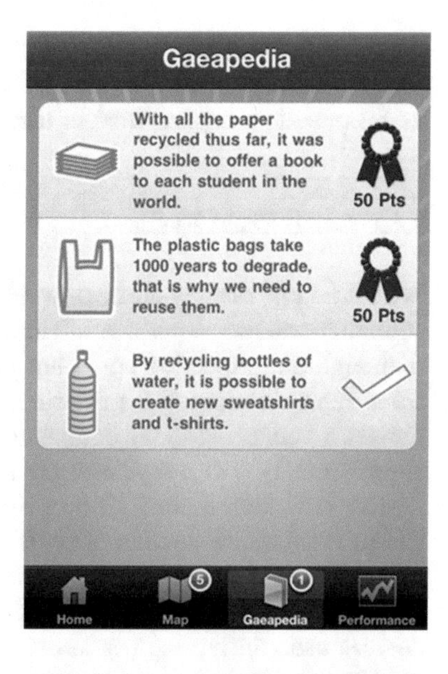

Fig. 10 Gaeapedia on the mobile application

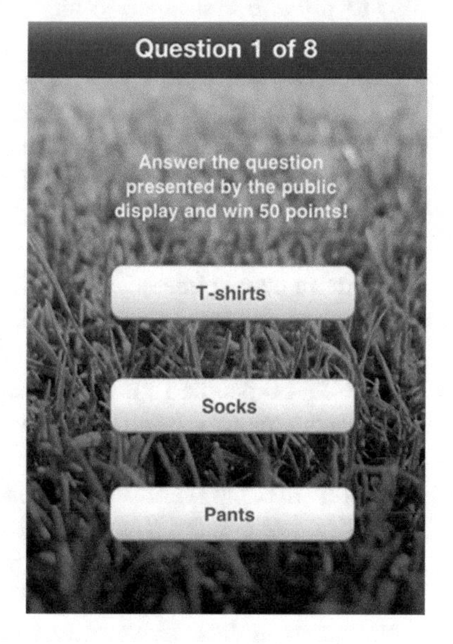

Fig. 11 Gaeaquiz screen presenting the possible answers on the mobile device

3.4 Evaluation and Analysis of Results

After the development of Gaea, two evaluation procedures were conducted to assess how people would use Gaea, and how they would be affected by it. The first time we evaluated Gaea, we ran a small-scale event where we focused on evaluating Gaea's concept, usability and persuasive abilities, in order to understand if users learnt with the game, and if they became more motivated to recycle. A second user study was conducted later, during a large event, to assess Gaea's usability and gameplay. Both of these studies took place in our University campus.

3.4.1 Small-Scale Event

We set a large display during a weekday afternoon (3–6 pm) at the entrance of the computer science department. The projection screen was 4 m wide and 3 m tall (although the projection could not fill the entire screen), and the text was legible from approximately 5–8 m during daylight, and 12–15 m during evening. The projector connected to the computer running the server application was positioned approximately 6 m away from the projection screen.

The tests took around 10 min each, and were made in groups of three or four users (based on the number of devices available) and all the tests (except one) were played in the individual mode, where each user played by himself. A team member stood next to the computer to install the mobile application on the several mobile devices used during the test sessions, to give instructions to the participants, and also to provide assistance to any problems that users might face. Four research team members observed the way that the tests unfolded, and how users and audience reacted to what was happening. Figure 12 illustrates the system's setup.

Fifteen users aged 23–44 (average of 32.2) participated in the evaluation event (12 male and 3 female). All of them were familiarised with new technologies, using computer, mobile phone and the Internet on a daily basis. Eleven of them rarely use a game console, two used it on a weekly basis, one on a daily basis and one did not use it.

As more people interacted with Gaea, more other people came along around the display, and the rate of participation increased over time as the evening went on, until there were not any users and audience left. In the end, the audience included around 25–30 people (not simultaneously). Finally, at the end of each game session, each user was asked to answer a questionnaire to evaluate Gaea's usability and persuasive ability.

Results and Discussion. The questionnaire presented to the users was based on the USE questionnaire (Lund 2001). Users were asked to rate their agreement with the statements presented using a five-point Likert scale ranging from strongly disagree (1) to strongly agree (5). The questionnaire included questions regarding users' personal data, Gaea general aspects, persuasive concepts, usability, as well as users' opinions towards recycling and emotional involvement.

When we asked users to rate their overall experience while playing with Gaea, feedback was very positive, demonstrating that users liked to use Gaea, that it was easy to use and to learn. The general feedback was that Gaea did not seem complex. Users stated that the application required few steps to perform each task, and that each one of these tasks were easily executed, leaving no doubts about the modus-operandi. This was intended to evaluate the tunnelling, suggestion and conditioning persuasive concepts and the results show that the objectives were

Fig. 12 System setup

quite achieved. The next two statements presented to users, reflected that the instructions given by the application were informative and useful (another important aspect that was supposed to be studied). Finally, the response time to the different user's actions were evaluated, which also received a very good score. Figure 13 shows the statements previously mentioned, and the frequency of scores for each statement.

Next, users were asked how easy it was to use Gaea specific features. This time the lowest score in the Likert scale meant it was difficult to use (1), while the highest score meant it was easy to use (5). As the graph on Fig. 14 shows, the user feedback was very positive. However, it is necessary to take into account the least, yet, quite positive scores related with: marking the information on Gaeapedia as read and recycling an object.

It was not clear to the users that there was a new notification available on the Gaeapedia section, Thus, we decided to insert a new pop-up message that prompts the users to check the Gaeapedia section, after they recycle an object for the first time. Also, we changed the text message informing users how they need to position the mobile phone to access the recycle bins when near the public display, in order to be more clear.

To evaluate the users' opinions towards recycling, the questionnaire also included some questions on how people felt about the topics promoted by Gaea, and whether the experience with this prototype was meaningful and motivated them to change their attitudes and behaviours towards recycling. Based on the results, it was possible to conclude that Gaea managed to convey the intended message. Everyone expressed that the topics addressed by Gaea were important, and they all think that Gaea increases the awareness for the need to recycle. Almost all users stated they learned with the information provided by Gaea (87 %). Most of the users (83 %) were already used to recycle before using Gaea and stated they felt more motivated to continue recycling after using Gaea. One of

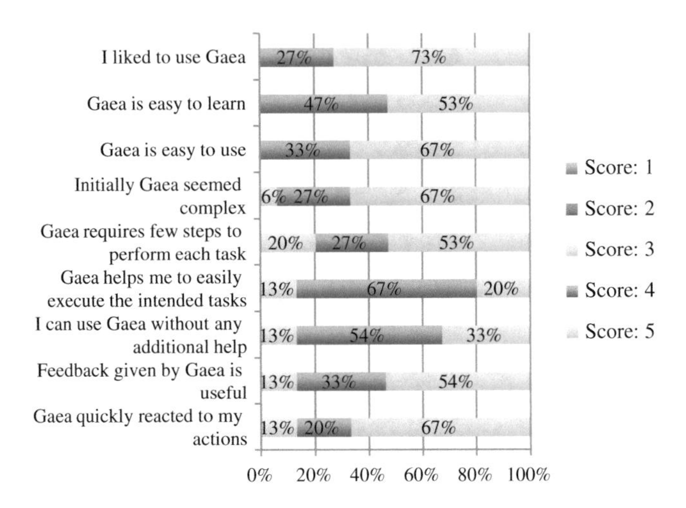

Fig. 13 Evaluation of several general aspects of Gaea during the small-scale event

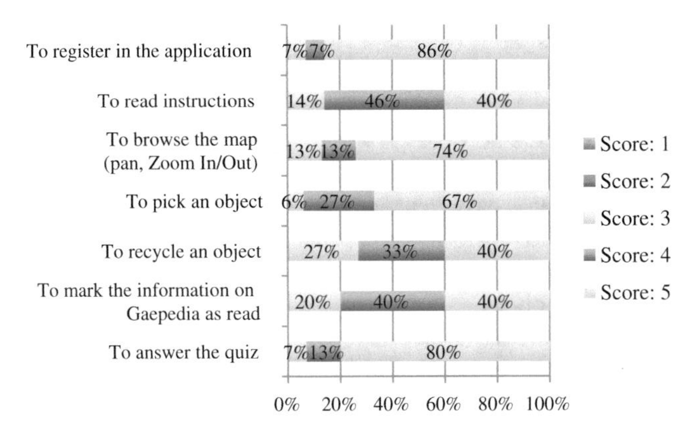

Fig. 14 Average rate of ease of use about specific features

the users that did not recycle before experiencing Gaea, said that he will now start to recycle, and another said that maybe he will start recycling.

The last part of the questionnaire was based on the Microsoft "Product Reaction Cards" Benedek and Miner (2013) . The purpose of using this method was to collect feedback on desirability and to measure the users' emotional involvement during the test. Thus, the users were asked to choose the words that best describe their experience, from a selected set of words. Figure 15 shows the obtained results.

The majority of users felt that Gaea was pleasant and fun, which is very positive, since Gaea was designed exactly to encourage the audience participation. More than half the users felt that Gaea was simple, attractive and motivating, other

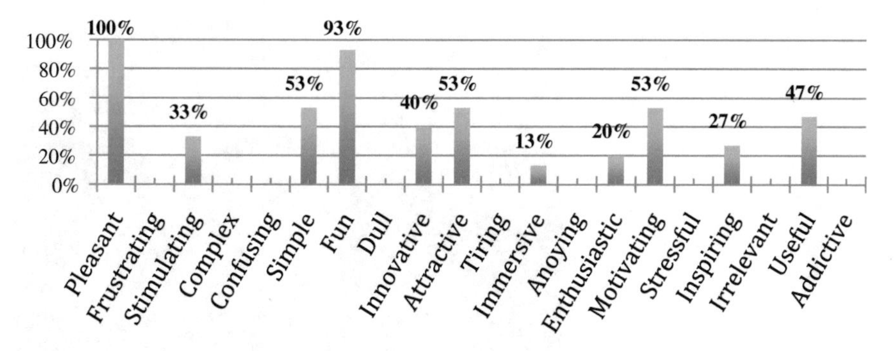

Fig. 15 Users' emotional involvement with Gaea during the small-scale event

key aspects about Gaea design. Almost half of the users said that Gaea was useful and innovative.

Preliminary Impact Test. In a first attempt to study the impact of Gaea on the users' environmental attitudes, one week after the user tests, we asked the participants to answer another questionnaire. It was intended to evaluate their behaviours and attitudes, some time after experiencing Gaea, to acknowledge if they were still stimulated to make recycling a part of their everyday life, and if they were still aware of the environmental issues presented by Gaea.

Once again we asked users to rate two statements using a five-point Likert scale ranging from strongly disagree (1) to strongly agree (5). In the first statement, 86 % of the users rated the statement "Gaea increased my awareness about the need to recycle" with a score of 4 or above. On the second statement, 27 % of the users rated the statement "my motivation to recycle after using Gaea has increased" with a score of 4 or above. Although positive, these indicative results need to be confirmed by more extensive users studies.

The results of these first user studies were encouraging. After we have made some improvements to the prototype according to the users' feedback, further tests were performed.

3.4.2 Large-Scale Event

During the second user studies, we wanted to focus on the primary target users and further evaluate Gaea's usability and gameplay. Thus, Gaea was deployed during a large event for teenagers, an open day at our University campus attended by 7,000 high-school students. We setup Gaea in a very similar way to the previous event, but this time we used an LCD screen instead of a projection (due to the lightning and weather conditions). Figure 16 provides a glimpse of the setup.

Like in the previous tests, a team member also stood next to the computer to provide assistance for any problems that users might face. Since it was expected a higher number of participants, five research team members accompanied the users

Fig. 16 Setup used on the large-scale event

while they were moving around to collect the virtual waste, observing and taking notes regarding what they were doing and how they reacted.

Unlike the small-scale event, where each person used one iPhone, this time 2 to 3 persons shared an iPhone, making it a team game, where one person handled the iPhone and the others exchanged comments and opinions. This happened because usually friends wanted to play together and there were not enough available devices for all of them. Although this was not planned, the ending result was quite positive, because there was a lot of social interaction, especially during the quiz phase, between the friends sharing the iPhone and the other friends playing against them.

At the end of each game test session, participants were asked to answer a short questionnaire to evaluate Gaea's usability and persuasive ability. Thirty-seven users aged 16–28 (average of 17.2) voluntarily participated in the evaluation procedures (21 male and 16 female) and answered the questionnaire. Almost all of them (92 %) were familiarised with new technologies. More than half the users (52 %) used Facebook on a daily basis, 23 % used it weekly, 3 % used it on a monthly basis and another 3 % rarely used it. The remaining 19 % did not use Facebook at all. Since most people passing by the setup were groups of students, we often had a lot of people for the same game session (either as users or in the audience). In total, there were around 150 people in the audience during this test sessions (not simultaneously).

Results and Discussion. This time we decided to use a smaller questionnaire mainly because teenagers would not like to fill-in long questionnaires. It addressed personal data, general aspects of the interface, the topics promoted by Gaea and the users' experience.

Users started by rating four statements, using a five-point Likert-type scale, which ranged from strongly disagree (1) to strongly agree (5). Once again, users' feedback was very positive (better than in the previous tests), demonstrating that users liked to use Gaea, that it was easy to use and to learn. Like in the previous tests, users did not considered Gaea to be complex. Figure 17 shows these results.

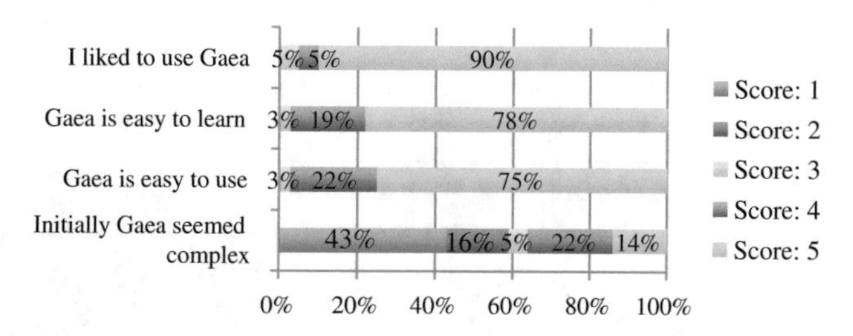

Fig. 17 Evaluation of general aspects of the system during the large-scale event

Next, we asked questions about recycling, and how the experience with Gaea-motivated users to change their behaviour towards it. Almost all users (95 %) stated that Gaea increased their awareness regarding recycling and almost everyone learned with the information provided (97 %). More than half of the users (54 %) stated they use to recycle, 40 % said they recycled once in a while and 6 % said that they never recycle. When asked whether they felt more motivated to start or continue recycling, 58 % of the participants answered positively, 37 % said probably and 5 % answered negatively. Based on these results and the ones from the previous user tests, we can conclude that Gaea has the potential to promote pro-environmental behaviours in an effective way.

Again, based on the Microsoft "Product Reaction Cards", we asked participants to best describe their experience while playing with Gaea. The results obtained (Fig. 18) were similar to the ones we got on the first event.

The majority of users felt that Gaea was fun, pleasant and useful. Almost half of the users had an enthusiastic experience, others felt that it was innovative and attractive and finally more than one third felt it was a stimulating and motivating experience.

Several participants (56 %) used the Facebook Connect to login to Gaea and one of them even said he wanted to participate in the activity, because he wanted to share his accomplishments on Facebook. This demonstrates that the power to share personal achievements through social networks may, in some cases, be enough to persuade people to participate in an activity or perform a specific action. The remaining 44 % participants did not use Facebook, due to several reasons: they did not have a Facebook account, confidentially reasons or slow authentication. The participants' posts on Facebook, during the week after this event, resulted in 1510 views on our project's page.

Regarding the observations that we took during the event, participants were very enthusiastic during the activity, and they were very competitive, running through the game area to collect the virtual garbage, laughing and teasing each other. This was a clear sign that they were having fun and committed to recycle the objects, in order to achieve the best possible score. Also, the quiz phase made it possible to instruct and inform the players and the audience about the consequences and gains of recycling. Moreover, we observed that players and the audience following the game interacted and socialised constantly. We also noted, a

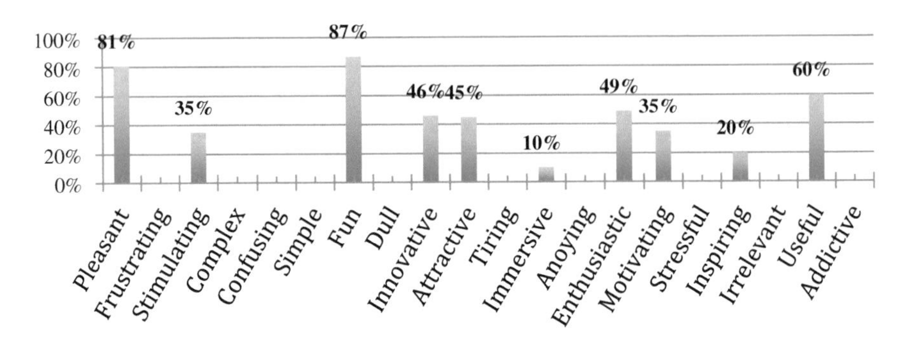

Fig. 18 Users' emotional involvement with Gaea during the large-scale event

high enjoyment between groups of friends, when someone would correctly answer a question and the others would not. Gaea also seems very appropriate as a school activity, allowing a large number of children to play, interact, exercise and learn.

4 Conclusions

This chapter describes the use of persuasion technology through mobile phones and public displays. It presented a prototype, called Gaea, which introduces an innovative way of playing a multiplayer game by exploring the interaction between mobile devices and public displays. Through the detection of users' locations and actions on a specific geographical area, Gaea seeks to exploit a natural environment, to promote users' physical activity and social interaction. Gaea applies the concepts of persuasion technology with the intent of affecting human attitudes and behaviours towards a better environmental consciousness.

Gaea is a persuasive location-based multiplayer mobile game, which prompts people to recycle virtual objects on a specific geographic area in a fun and appealing way. During the evaluation events, almost everyone felt enthusiastic and had a good experience playing with the game. Gaea also foments social interaction between users and members of the audience, and allows the dissemination of information to a large number of people, through both the public display and the Facebook.

The evaluation results were very positive and promising, regarding both the gameplay and Gaea's persuasive ability. By structuring Gaea in three different phases (registration, recycle and quiz), each one with specific goals and user interfaces, members of the audience and users followed the game in a clear way. The distribution of the information presented on the different devices (mobile and public display) helped to keep the players and the audience informed and engaged. For example, the instructions were presented both on the mobile device and the public display, which played a crucial role to guide users and to inform new members of the audience about what was happening.

Although it is difficult to evaluate behaviours since they are complex and non-linear, we believe, from the feedback we had, that Gaea can help to shape people's attitudes and behaviours towards a better environmental conscience.

Acknowledgments This work was funded by Fundação para a Ciência e Tecnologia (FCT/MEC), Portugal, in the scope of project DEAP (PTDC/AAC-AMB/104834/2008) and by CITI/DI/FCT/UNL (PEst-OE/EEI/UI0527/2011). The authors thank Bárbara Teixeira for her contribution to the graphic design.

References

Benedek J, Miner T (2013) Product reaction cards. Available via http://www.microsoft.com/usability/UEPostings/ProductReactionCards.doc. Accessed 28 May 2013

Bogost I (2007) Persuasive games: the expressive power of videogames. MIT Press, Cambridge

Eurogamer (2008) PS3 has outsold Xbox 360 in Europe. Available via http:// www.eurogamer.net/articles/ps3-has-outsold-xbox-360-in-europe. Accessed 27 Aug 2013

Fogg BJ (2008) Mass Interpersonal persuasion: an early view of a new phenomenon. In: Proceedings of 3rd international conference on Persuasive Technology (Oulu, Finland, June 4–6, 2008). Persuasive '08. Springer, Berlin, pp 23–34

Fogg BJ (2003) Persuasive technology: using computers to change what we think and do. Morgan Kaufmann, San Francisco

Froehlich J, Dillahunt T, Klasnja P, Mankoff J, Consolvo S, Harrison B, Landay J (2009) UbiGreen: investigating a mobile tool for tracking and supporting green transportation habits. In: Proceedings of 27th international conference on human factors in computing systems (Boston, USA, 4–9 April 2009). CHI '09. ACM Press, New York, pp 1043–1052

Kaviani N, Finke M, Lea R (2009) Encouraging crowd interaction with large displays using handheld devices. In: Proceedings of crowd computer interaction workshop at CHI 2009, Boston, USA, 4–9 April 2009

Lobo P, Romão T, Dias A, Danado J (2009) Smart bins: an educational game to encourage recycling activities. In: Proceedings of IADIS international conference interfaces and human computer interaction (Carvoeiro, Portugal, 20–22 June 2009). IHCI '09. IADIS Press, Portugal, pp 155–162

Lund A (2001) Measuring usability with the USE questionnaire. Usability interface, 8, 2 (Oct. 2001). Available via http://www.stcsig.org/usability/newsletter/0110_measuring_with_use.html

Reitberger W, Ploderer B, Obermair C, Tscheligi M (2007) The PerCues framework and its application for sustainable mobility. In: Proceedings of 2nd international conference on persuasive technology (Palo Alto, USA, 26–27 April 2007). Persuasive'07. Springer, Berlin, Heidelberg, pp 92–95

Rogers Y, Brignull H (2002) Subtle Ice-Breaking: encouraging socializing and Interaction around a large public display. In Proceedings of computer supported cooperative work 2002 workshop, New Orleans, USA, 16−20 Nov 2002

Steg L, Vlek C (2009) Encouraging pro-environmental behaviour: an integrative review and research agenda. J Environ Psychol 29(3):309–317

World Wildlife Fund (2008) Living planet report 2008. Available via http://assets.panda.org/downloads/living_planet_report_2008.pdf. Accessed 6 Sept 2013

Designing Mobile and Ubiquitous Games and Playful Interactions

Paul Coulton

Abstract Marshall McLuhan famously said in his 1964 book The Medium is the Message, *We become what we behold. We shape our tools, and thereafter our tools shape us*. In a relatively short time, video games have become a major feature of our cultural landscape that extend beyond the games themselves such that their aesthetic, iconography and operation are being reflected in the other more established forms of media: film, books and television. This chapter explores the rapidly growing field of mobile and the often associated field of ubiquitous games which are contributing significantly to the cultural spread of games by: opening up new markets, facilitating new player demographics and creating exciting new forms of game play, which will undoubtedly have a significant impact on future society.

Keywords Game · Design · Play · Interaction · Mobile · Ubiquitous · Sensors

1 Introduction

Since their appearance in the early 1990s mobile phones have become a ubiquitous consumer device with over six Billion phone subscriptions worldwide; this equates to 75 % of the worlds population having access to a mobile phone (World Bank 2012). This rapidly growing user base, combined with the increasing availability and range of data services on mobile networks, has resulted in the mobile phone becoming the main computing device for the majority of the world's population (World Bank 2012). As mobile phone functionality has increased, so has the evolution of operating systems (OS) to facilitate the creation and installation of applications from third party developers (Coulton et al. 2005) such as games.

P. Coulton (✉)
ImaginationLancaster, Lancaster Institute for the Contemporary Arts LICA Building, Lancaster University, Bailrigg, Lancaster LA1 4YW, UK
e-mail: p.coulton@lancaster.ac.uk

A. Nijholt (ed.), *Playful User Interfaces*, Gaming Media and Social Effects, DOI: 10.1007/978-981-4560-96-2_4,

Games on phones have come a long way since Taneli Armanto programmed a version of Snake for Nokia phones in 1997, and now constitute a significant proportion of the overall games market. For example, of the 43,000,000 h a day spent gaming in the UK 15 % of this is on mobile phones (http://newzoo.com). Whilst the early days of mobile gaming were largely characterised by recreating many of the games developed for the first home computers, i.e. BBC micro and Sinclair Spectrum or the early consoles such as the Atari 2600, we have seen them become the dominant platform for 'ubiquitous games' research that seek to offer new gaming experiences that go beyond the traditional console experience. These new forms of games often take advantage of not only the wide range of connectivity options, such as WiFi and Bluetooth, alongside commonly available sensors, such as multiple cameras and the microphone, but also an increasingly sophisticated range of on-board phone sensors, such as Global Positioning System (GPS), accelerometers, magnetometers (digital compass), Gyroscopes and Near Field Communications (NFC). These sensors have facilitated one of the primary drivers for mobile and ubiquitous games in the research community which is related to the inclusion of context awareness (Abowd et al. 1999), although such context aware games have yet to make any significant in-roads in the commercial environment. Other research games are more concerned with exploring new game design techniques and generally seek to expand the viewable game space on the screen into our lived reality and are the primary focus of this chapter. Whilst these games are often described as ubiquitous games, terms such as mixed reality, augmented reality, alternate reality, pervasive games, location-based games, big games and urban Live Action Role Play (LARP) (to name but a few) are often used to differentiate certain aspects of such games. In the following section, I will consider some of the most commonly used terms to provide further insight into the field.

1.1 Ubiquitous and Pervasive Games

The notion of 'Ubiquitous Games' has its roots within the ubiquitous computing movement that grew from Mark Weiser's vision paper in which he proposed the disappearance of computing into the background of everyday objects (Weiser 1993). Since the concept offers such a broad range of possibilities, we have seen an array of terms to describe activities such as ambient, physical, wearable, for example, although the majority would still be encompassed within the general field of ubiquitous computing.

Whilst games have often been created as part of the ubiquitous computing research agenda (Magerkurth et al. 2005) these have largely been used to illustrate particular hardware and software research platforms or the interactions with these platforms, rather than those that are explicitly considering these new forms of game play (Nieuwdorp 2007). The alternative view to this technological perspective is to consider these games from a cultural standpoint (Nieuwdorp 2007) which has led to the field of pervasive games. Although this term was also originally used from a technological stance, and almost interchangeably with ubiquitous games, it has emerged as a term used to describe a range of games *that extend and blur the spatial, temporal and social borders of the game spaces* (Montola et al. 2009). One of the most vocal commentators in this area is Jane McGonigal, argues that a great many of these games are, in essence, a rhetorical medium for furthering hardware and software research agendas (McGonigal 2006). For McGonigal, pervasive games are disruptive experiments that challenge the concepts of games, play and everyday life (McGonigal 2006). Although she views technology as an oft used enabler for this disruption, for many of the artists currently developing such games the technological element is lost and the games produced are predominantly performance based and more readily resemble urban LARPs. As this chapter is primarily concerned with the technological aspects of these games and the definition previously provided by Montola et al. arguably falls into the trap of the magic circle myth (Coulton and Wilson 2012), in that it assumes the boundaries of most games are impermeable to elements outside the game space, I would align with Nieuwdrop in that we should first decide *what makes a game pervasive?* (Nieuwdorp 2007) before making this distinction. Therefore, I have chosen to use ubiquitous games as a general descriptor for the many of games within this chapter.

1.2 Mobile Gaming

Mobile, in many respects, is an overused term and the term mobile game is most commonly used to describe games that run on a mobile phone. However, the International Telecommunications Union (ITU) definition, states, *the term mobile can be distinguished as applying to those systems designed to support terminals that are in motion when being used.* In other words, unless a game on a mobile phone uses a connection as part of the game, there is a strong argument that it is essentially a hand-held, portable or nomadic game. In this chapter, we will use the term in relation to games that run on a device which is capable of operating as a mobile phone as the term is commonly used in both academia and industry and as all the ubiquitous games presented in this chapter use mobile as their primary platform we could use these terms almost interchangeably.

1.3 Location-Based Games

A specific class of mobile games results from the personalisation of mobile phone applications using location has become commonplace; whether it be for mapping and routing applications, or used as a filter to provide the most relevant information based on the users current location. Games were one of the first systems to demonstrate the role location can play in a mobile service and a Location-Based

Game (LBG) is one in which the game play evolves or progresses relative to players' changing location (Rashid et al. 2006; Coulton et al. 2010; Lund et al. 2010). Thus, LBGs almost always support some kind of location sensing and the accuracy of this sensing is highly dependent upon the technology used.

Although we often consider providing the location of a mobile user as a recent problem, in fact, every generation of mobile phone system has effectively tracked a user's whereabouts at the cellular level. Each cell site has a unique Cell-ID that enables the system to locate a mobile user so that it can route calls to the correct cell. To obtain higher degrees of accuracy other techniques treat location finding as a relative exercise. In other words, the location of the mobile user is estimated against a known framework. This framework could be the locations of the base stations of a mobile phone network or the satellites of the (GPS). An alternative approach is to ascertain location from the user's interaction with objects whose location is known, thus, the location of the users can be implied. The interaction could be proximity within a physical area using communication technologies such as WiFi or Bluetooth or down to object level using one of the various forms of two-dimensional (2D) bar codes, such as QR codes or Radio Frequency Identification (RFID) tags. Over the years all of theses techniques have been used within location-based games (LBG) (Rashid et al. 2006).

2 Developing and Distributing Mobile Games

With the increase in mobile phone subscriptions, the ever-decreasing costs of hardware and mobile data connectivity, coupled with the lower barriers to entry for developers, the mobile phone makes a very appealing platform for games. However, this still leaves a variety of development options to be considered. These options primarily relate to restrictions imposed by different modes of app development and, whilst the purpose of this chapter is not to provide an overview of these alternatives, it is useful to consider the main differences. A 'native' app refers to an application built using the tools and languages that are compiled using the system native machine code and generally this offers the best performance for graphically intensive games. A 'mobile web' app sits on a server and is accessed via the phones web browser and is essentially independent of the phones OS. An 'interpreted' app is where the code is interpreted at runtime, whilst a 'hybrid' app is an application that is neither native nor web. These hybrid apps use a mobile WebKit to render the user interface (UI) through the phones native browser but they also have access to some device features unlike purely mobile web apps.

Notably, in the years since it was introduced, the iPhone has had a significant impact on the way we buy and play games. Although there have been earlier examples of what we define today as an 'App Store', the iPhone undoubtedly was the first commercially successful version. In this section, I will look at how app stores operate and what this means for researchers who may wish to use them as an experimental platform.

2.1 App Stores

Initially distribution channels were limited for mobile games developers, due to restricted availability through operator portals or aggregators using systems that had their roots in the ring-tone and wallpaper markets built around Short Message Service (SMS) payments and Wireless Application Portal (WAP) push (Garner et al. 2006). As these channels generally required a commercial working relationship with the companies operating portals they were effectively closed to many small mobile development companies. This often led to many researchers having to make applications available through their own websites and hoping for a 'viral' uptake to occur. Things improved somewhat with the emergence of aggregator sites such as GetJar™ (initially targeted at Java Micro Edition (JME) applications but now one of the few stores offering Apps for the majority of platforms), which allowed free Apps to be shared easily by researchers and small developers. In the early days the main difficulty with such services was the collation of data relating to applications. This was often marred by others users of the site 'high-jacking' an application rising in popularity by re-posting it on the store in an attempt to drive traffic to some particular web service (Chehimi et al. 2008).

When the Nokia WidSets™ platform appeared in 2006 it was the first adopter of what would now be considered as 'the App Store approach', made famous by Apple for its iPhone and more recently the Android Market. These characteristics being the simple search, installation, and rating of applications for and by the users and the ability for developers to easily push new versions of the applications to those users (Coulton and Bamford 2011).

2.2 Apps and App Store Life Cycles

In February 2009 the mobile analytics company Pinch Media released various metrics relating to the download of 30,000,000+ iPhone applications. They found that by appearing on the Top 100 list, applications would receive 2.3 times more downloads on average and often an order of magnitude higher for the Top 25 and Top 10 list. This highlights that while platforms such as iPhone make it possible to easily search or find any application within the library, the important aspect in gaining new users is through curation of the apps by content managers whose recommendations and suggestions greatly affect an apps likelihood of reaching the Top 'n' lists for popularity and ratings (Coulton and Bamford 2011).

Significantly, the research also showed that most mobile applications have a relatively short *shelf life*, with on average less than 25 % of users returning to the app 1 day after download, dropping to around 5 % after 30 days (Coulton and Bamford 2011). The category to which an application belongs also seems to have a strong effect on return rate—applications that are typically more dynamic, e.g. sports (results, league tables, etc.) and entertainment apps fared better than games, utility and lifestyle applications.

This suggests that the range and depth of content on platforms is as important as personalisation. In other words, users want to experience a wide range of content, but only a relatively few applications become part of their daily activity (Coulton and Bamford 2011). This represents a significant design consideration for researchers creating mobile applications (Coulton and Bamford 2011) if they are not simply testing a one off interaction but are trying to evaluate longitudinal use.

3 Mobile and Ubiquitous Game Play

Although the demographic of mobile gamers is different to that of console gamers, they are still often simply divided into the two traditional broad categories used by the games industry, that of casual gamer and hardcore gamer. It is worthwhile considering the popular profiles of hardcore and casual gamers, as they are undoubtedly influencing industry expectations about the mobile gamer (Bamford et al. 2006).

Hardcore Gamers:

- Purchase and play many games;
- Enjoy longer play sessions and regularly play games for long periods;
- Are excited by the challenge presented in the game;
- Will tolerate high levels of functionality in the user interface and often enjoy mastering the complexity;
- Often play games as a lifestyle preference or priority.

Casual Gamers:

- Buy fewer games, buy popular games or play games recommended to them;
- Enjoy shorter play sessions—play in short bursts;
- Prefer having fun or immersing themselves in an atmospheric experience;
- Generally require a simple user interface (e.g. puzzle games);
- Consider game playing as another time-passing entertainment like TV or films.

It is often assumed within the games industry that casual gamers form the majority of mobile gamers although I believe this is an over simplification, as the nature of the mobile environment is a major contributor to the formation of gaming habits. There is a strong argument that the game industry must establish new definitions specifically for mobile. For example, in an interview for the Game Daily Biz in February 2006, Jason Ford, then General Manager for Sprint Entertainment, suggested two specific types of hardcore mobile gamer:

- *First there are the 'cardcore' mobile gamers. These are people who play casual games in a hardcore fashion. The type that might spend hours and hours trying to get a Bejewelled high score but don't own a gaming console.*

- *Second is the 'hard-offs'. These are your more typical hardcore gamers, who are playing off their normal platform. They're the type more likely to check out the mobile version of a hit console title, because they know and like the brand.*

3.1 Single Player

Single player games or games where the same platform was shared to produce a two-player game dominated the early arcades and consoles as connectivity between devices presented significant challenges. This started to change with the emergence of the internet and games quickly developed to support multiplayer gaming to such an extent that, at its height in 2010, Blizzards Entertainments Massively Multiplayer Online Role Playing Game (MMORPG) *World of Warcraft* had 12 million players. Whilst multiplayer games have proved increasingly popular with the gaming public the majority of players on mobile games still tend to be for a single player, due both to the physical constraints of the platform (e.g. battery life, screen size, processing power, etc.) and the networking issues highlighted in the following section.

3.2 Multiplayer

There are a number of issues relating to multiplayer gaming on mobile that primarily come from networking and the nature of mobile gaming:

- Mobile game playing tends to be spontaneous rather than planned and is often used to fill spare moments resulting in short game sessions. Therefore, this characteristic must be at the forefront in the design of the lobby systems where players are able to find opponents and join or create games.
- Another big issue is how do systems handle users dropping out of games due to either accidental loss of connectivity (such as loss of signal while going through a tunnel) or deliberate disconnection to perform another task (such as answering a call). Scoring systems in particular should not encourage deliberate dropouts.
- Network latency can be as high as a few seconds in mobile, which is too long for the majority of fast action multiplayer games.

The other significant possibility for networking multiplayer games is to utilise Bluetooth, although this requires the players to be in close proximity and the number of devices that can be supported is typically limited to around 4 or 5.

3.3 Social Games

Social games often refer to a game that is being played as primarily a means of social interaction rather than competition. Typically, these games have a simple user interface, are easy to understand and often utilise a player's social graph as a resource within the game to help spread awareness of the game. This linkage means that these games can spread remarkably quickly, for example, the game *Draw Something* saw, at its peak, 83 million active players only 8 months after launch. Although such social games are currently in vogue we do not yet know if they will maintain long-term engagement amongst players.

4 Game Interactions

Traditionally, games interaction on mobile phones was either limited to the number pads or in some cases a small joystick or track pad. In this section, we consider new forms of interaction made available through both the increasing number of sensors available on smart phones and an increasing array of Bluetooth-enabled sensing peripherals that can easily be coupled to such devices.

4.1 Touch

When playing games that depend upon touch, movements on tablets or phones is the main design consideration, in that the touch points are within easy reach of the player and that in doing so the user does not obscure the screen for any extended period of time. Specifically, touch points are reported to the developer as (x, y) coordinates from which the game logic is used to determine what action should be taken. Note in some operating systems orientation and size of contact is also available. The number and size of the touch points that are detected depend upon the actual devices being used. Typically, the size range begins at a minimum of 5 mm and, in the case of iPhones/iPads, they can detect changes in size of approximately 1 mm increments. The number of concurrent touch points that can be tracked at any one time is also limited by the specific device, in terms of iPhones they are limited to five concurrent touch points, iPads are limited to 11. Note, that due to fragmentation of Android device capabilities, that range varies significantly from a minimum of two touch points upward.

Whilst capacitive touch screen phones and tablets are increasingly becoming one of the main forms of gaming platform, the nature of the touch interface and the lack of physical feedback are seen as limitations and in the section on Tangibles I present opportunities to address these issues.

4.2 Accelerometers

Although accelerometers are now commonly used in games on smartphones (one of first devices to contain such technology the Nokia 5500 sports phone), they were mainly used as a pedometer and speed/distance tracker for various exercising purposes (Vajk et al. 2008). The Nokia 5500 phone, in fact, provided the platform for some of the earliest studies in creating accelerometer-based games on mobile phones (Gilbertson et al. 2008). Typically, phones utilise 6 g accelerometers, that is, it can detect acceleration forces with a magnitude of up to six times that of earth's gravity. Typically, the accelerometer outputs three 12-bit signed data values that correspond to three phone axes (x, y, z). The limitation of 3-axis accelerometers is that they cannot detect rotational movement around the vertical axis (Vajk et al. 2008) (as this does not produce a change in the affects of gravity) and one of the reasons why magnetometers and gyroscopes have appeared on smartphones to enable the detection of a greater number of degrees of freedom including rotation.

4.3 Global Positioning System

Before GPS was integrated into commercial devices it was originally developed by the US military for defence purposes and is based on a system of 24 satellites that orbit the earth. A GPS receiver can triangulate its position as long as at least three satellites are visible. Typically, accuracy is around 2–10 m, although higher resolutions can be achieved. As the cost of GPS hardware continues to fall and GPS positioning remains a free service to consumers, integrating GPS hardware technology into smart phones is now commonplace, such that its almost considered a standard feature of the Smartphone. However, GPS has a number of operational limitations, such as poor operation in urban environments and inside buildings due to the requirement that mobile users be in *view* of the satellites, and a slow location acquisition time in the region of 10–60 s. To reduce the impact of these constraints, a hybrid solution 'Assisted GPS' (A-GPS) using WiFi access points or Cell-ID is often provided as a backup (albeit with lower accuracy).

GPS is now incorporated into the mobile's OS allowing for easy incorporation of photo geo-tagging, navigation, geo-tagged social networking updates (i.e. tweets and check-ins) into a variety of applications. Check-in services like *four-square* use GPS and A-GPS to allow users to indicate their presence at venues although the inaccuracies of this approach allow the system to be easily abused (Nandwani et al. 2011a).

One way to deal with the inaccuracies of the positioning technology used when designing LBGs is to adopt a seamful design approach (Broll and Benford 2005), whereby the inaccuracies become a feature of the game, as for example in the Tron Light Cycles inspired game *MarshOTron* (Lochrie et al. 2011). The primary

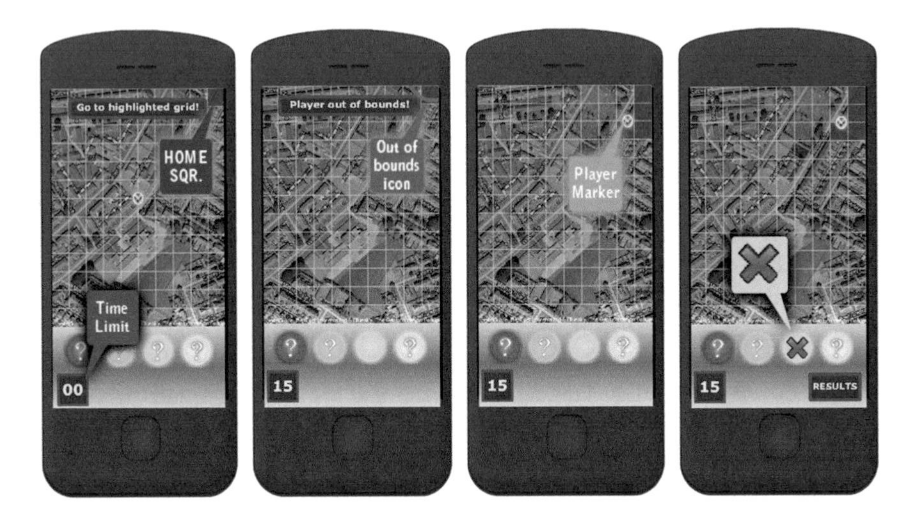

Fig. 1 MarshOTron GPS Game Interface

concept of *MarshOTron* is for players to create the longest light trail across the map using their light cycles (player's physical movement). This is achieved by augmenting their own physical position as an individual player marker on top of a fixed, pre-defined, futuristic styled map, overlaid by rows and columns made up of individual squares making as shown in Fig. 1. The game starts when all four players elect a different coloured light cycle (red, green, yellow or blue) and they physically move to their 'home' square. Game play involves player movement to initiate the light trails represented as captured coloured squares in the players colour on their mobile phone. In order to light up squares a player must occupy that square for 15 s. If a player re-enters a square they have previously captured, or is occupied by the light trail of another player, they are eliminated from the game. The game is finally complete when there are no active players left in the game and is won by the player who lights up the most squares. The seamful aspect is the use of squares that are an area rather than a specific location allowing for any inaccuracies in the GPS Fig. 2.

4.4 Compass (Magnetometer)

Ever since Egenhofer (1999) envisioned various forms of spatial information appliances they have generated considerable interest due to their natural mapping of a technological solution to existing human practices. Although Egenhofer described a number of possible devices, it is the geo-wand (a device to allow users to identify nearby points of interest (POI) by simply pointing at them) that captured the imagination of many. Such devices are now readily realisable with the

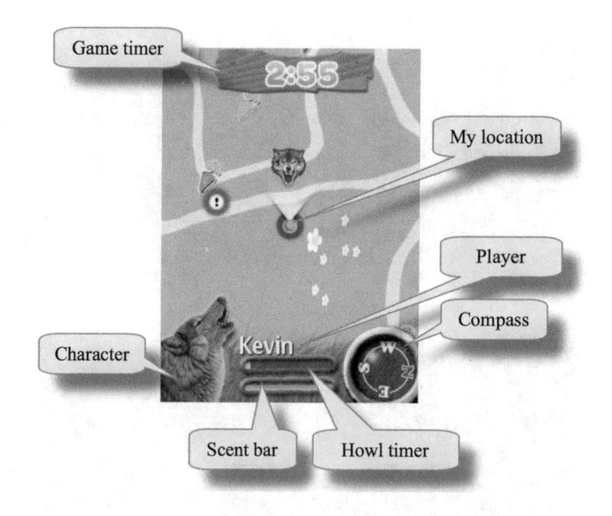

Fig. 2 They howl: GPS and compass game interface

recent appearance of commercial mobile phones with inbuilt GPS, 3D accelerometers and magnetometer (generally described as a digital compass) (Zhang and Coulton 2009). Similar to accelerometers, a magnetometer will typically provide directional information on three axes (x, y and z) and although direction can be derived from these three axes, phones often also supply a single reading of angle from North. This more natural interface offers interesting possibilities, particularly for LBGs. For example, in the game They Howl, direction had to be used collaboratively by the wolf pack to try to ascertain the location of the rabbit (Zhang and Coulton 2011) as shown in Fig. 2.

4.5 Camera

It seems rather quaint now to look back to the time when we used to specifically refer to camera phones as though they were special devices, now that a camera is arguably a standard feature on all mobile phones. The usage of the phone camera technology and photography within game play has had varying success and here we consider two distinct approaches in mobile games; using the camera to take photos and using the camera's viewfinder to augment objects over the physical world.

4.5.1 Photos

In terms of mobile gaming we are seeing photography integrated as a method of game play in a variety of ways. Some games have used photos as a method of collaborative user-generated content (Lund et al. 2011) or as a method of confirming

Fig. 3 Player photographs from big game huntr

a player's actions (Tuulos et al. 2007; Lochrie et al. 2010). For example, *Big Game Huntr* requires players to capture photographs of specified objects or actions in specific zones to gain the optimum number of points (Lochrie et al. 2010). The game itself is played using a mobile phone and is accompanied by an online web interface that allows players to design their own game and interact with the community. The following Fig. 3 shows the different interpretations of four clues during a game played on 18th March 2010 (Lochrie et al. 2010).

4.5.2 Augmented Reality

Augmented Reality (AR) is a real-time view of a real-world environment which is augmented by virtual objects that are contextually situated within the scene. In terms of mobile AR it is normally classified through the method used to obtain a reference plane in the real world from which the camera pose can be obtained, thus, allowing the virtual objects to be rendered onto the scene. These methods can be sub-divided into sensor- and vision-based approaches.

Sensor-based approaches take advantage of increasing numbers of sensors on mobile phones such as GPS, accelerometers, magnetometers and, recently, gyroscopes on commercial mobile phones. Combining the readings obtained from such sensors allow the position and orientation of the phone camera to be estimated. Such systems are relatively easy to implement and are the basis of many of the AR browsers such as Layer (www.layar.com) and Wikitude (www.wikitude.com). The main issue with sensor-based techniques is that accuracy is relatively crude which limits the contextual sensitivity of applications developed using this method.

The simplest and most widespread of the vision techniques involves the use of 2D fiducial markers to provide a pose estimate relative to the environment (Henrysson et al. 2005). The problem with this methodology is that the low quality and narrow field-of-view of many mobile phone cameras cause significant issues when tracking markers in a real-world environment and, when coupled with the practicality of covering real-world environment with such markers, limits the possibility for widespread use.

The alternate approach is natural feature tracking which remedies this problem by using features within the environment to produce the reference plane, although this comes at the expense of high CPU processing which is always an issue when implementing mobile AR. Whilst there have been a number of systems developed using natural feature tracking on mobile devices (Wagner et al. 2008), they require a priori knowledge of the environment. This a priori knowledge allows for a reference map of the environment to be created offline from normal use. Recent research has shown that Simultaneous Localisation and Mapping (SLAM) techniques, originally developed for robot navigation, can be successfully applied to natural feature tracking and positional estimation (Davidson et al. 2007). This work was extended (Klein and Murray 2007) to demonstrate that localisation and mapping could be considered independent, parallel processes, providing significant increases in both the accuracy and robustness of hand-held AR. Whilst any further developments for mobile markerless AR is on-going a significant amount of research is required in adapting SLAM techniques to mobile phones.

As an example of the use of AR, consider the following application of the 'scARecrow Time Machine'; created to represent the engagement of the Lancaster University with residents of a small rural village that has become a 'living lab' over the last 10 years. The application provides a playful way for families to explore the village in both space and time through its main cultural event, the annual Scarecrow Festival.

Fig. 4 a–f: Screen shots from augmented reality application scarecrow time machine

As the application created in this research is to be used outdoors in the village where the buildings are generally well spaced and not above two stories in height, a sensor-based system was deemed most appropriate. In terms of the application design, it was decided that only an AR view would be provided and not combined with a traditional map view as in many AR POI applications. This was done in order to evaluate the design challenges that must be considered if AR navigation becomes commonplace for pedestrians through either phones or AR glasses. One of the design challenges is to provide a sense of depth (distance) within the AR view. To this end, a number of common art techniques for adding perspective to 2D visualisations have been utilised as shown in Fig. 4a and together these go beyond the primary use of size and scale in the majority of geo-spatial AR applications. The grid provides linear perspective by creating a vanishing point, which is enhanced by adjusting the size and scale of the scarecrow icons that also utilise atmospheric perspective by adjusting their transparency. To avoid clutter within the AR view the scarecrow icons traverse along three horizontal lines that represent distance ranges as the phone is rotated. The only distance shown is that from the user to the closest scarecrow, which is shown by a different coloured icon. When the user gets within 5 m of the position the icon changes to a fully fledged scarecrow and the photo button is activated as shown in Fig. 4c. The photo view is also an AR view, shown in Fig. 4d, and shows the name of the scarecrow, its builder, the year it was made, alongside a photo of the scarecrow. The image is deliberately set to one side to allow the user to take a photo that could include themselves, their family or friends, or indeed the current scarecrow alongside the image. These users' photos are stored in the list view, shown in Fig. 4e, which is accessible from the main screen and provides a history of the users' interaction shown in Fig. 4f. As an important aspect of the project is to, 'tell the story' of the project to a wide audience, the application can also be used outside the village of Wray. The application uses a geo-fence around the village so that if a user is inside the fence the scarecrows are shown in the actual location where they appeared, otherwise they are randomly distributed around the users' location.

4.6 Audio

Sound is generally acknowledged as an important factor within games not only in providing feedback on player's actions but also contributing towards layers of immersion within the game. The majority of audio games thus far have been developed as 'blind accessible' games, for example, Multi User Dungeons and versions of *Quake* and *Doom* (Friberg and Gärdenfors 2004). However, more and more game developers are actively seeking to adopt sound as their central medium, for example, rhythm action games such as Konami's *Dance Dance Revolution* and the Nintendo DS title, *Rhythm Heaven*. Despite the fact that mobile phones provide extremely good sound facilities and limited displays, we have seen relatively few that have utilised sound as the main driver, but of these, many have linked it with movement such as *Songs of the North* (Rashid et al. 2006) which used GPS, and *Mobslingers* (Clemson et al. 2006) which used Bluetooth, although, I would note that a number of outdoor audio use mobile and GPS to trigger audio descriptions at defined locations. However, we have recently seen audio-driven horror themed games appear on iOS, with *Pape Sangre* and *Nightjar* developed by Somethin'Else, using binaural sound to create immersion. Given that many smart phones now provide the hardware functionality required to produce sensor-based AR there would seem significant opportunity to link binaural sound with movement through 3D space to produce innovative gaming experiences.

4.6.1 Speech

Speech control for computing devices has always been attractive as a form of user input as it releases us from the abstraction of other forms of interface. The most significant use within the games world has come through the Kinect from Microsoft, which originally centred on control of the Xbox console but has now seen integration in games such as *Skyrim*. The use of speech recognition technologies for controlling mobile functionality has also received considerable interest of late with the introduction of the natural language user interface Siri on the Apple iOS and the more recent Google Now on Android. In relation to games, as the APIs have yet to be released for these products we have not seen any games employing this technology specifically, although the iOS game *Pah* is a novel attempt at a voice-controlled game. *Pah* is a voice activated game and requires you to hum to fly your spaceship higher and shout Pah! in order to shoot oncoming asteroids.

4.7 Haptics

Haptics can be defined as relating to, or based on, the sense of touch and is of interest in mobile as it can be used to replace elements of a Graphical User Interface (GUI) thereby reducing the cognitive load on the user (Heikkinen et al. 2009).

Whilst complex haptic modalities have been used in different ways to code information, e.g. temporal and spatial parameters (Heikkinen et al. 2009) it has primarily been considered in games as either a form of feedback or alert to relay information to players. Given the current proliferation of touch screen devices, there is considerable research being performed into how haptics can be incorporated within these devices to provide much needed feedback to the user. Such technology will undoubtedly provide beneficial for games, as one of the major criticisms of touch-based control is that without haptics all feedback must be displayed visually which on a small-sized screen is problematic.

4.8 NFC

NFC is an interface and protocol that sits on top of Radio-Frequency Identification (RFID) and allows easy transfer of data from one device to another. Even though NFC has been around for some time it is only recently that we are seeing it integrated within a wide range of mobile phones. In mobile phones, its uses are primarily defined around making it simpler to perform transactions, device pairing and exchanging digital content. Aside from phones, Nintendo has recently incorporated NFC into the new Wii U game pad and they have highlighted potential uses such as incorporating play cards and figures as NFC-enabled objects within games.

Mixed reality mobile gaming using NFC as a method for defining location and making physical contact with objects was explored in the early days of mobile NFC by Rashid et al. (2006) who created the novel game *PAC-LAN*, which was a physical mobile LBG adaptation of the arcade classic Pac Man (Fig. 5). This work was extended by Nandwani et al. (2011b) to consider direct player-to-player interaction using NFC-enabled phones.

Thus far, there has been few commercial uses of NFC for games, no doubt due to the lack of consumer devices that incorporates the technology, although, a version of *Angry Birds* was created for the launch of the Nokia C7 device which allowed special levels to be unlocked by two of these phones together.

4.9 Tangibles

In terms of Tangible User Interface (TUI) for games, this is primarily achieved through physical game pieces or specialised controllers. These TUIs can be defined as providing a physical form of digital information and facilitates the direct manipulation of the associated bits and should posses the following attributes (Ishii 2008):

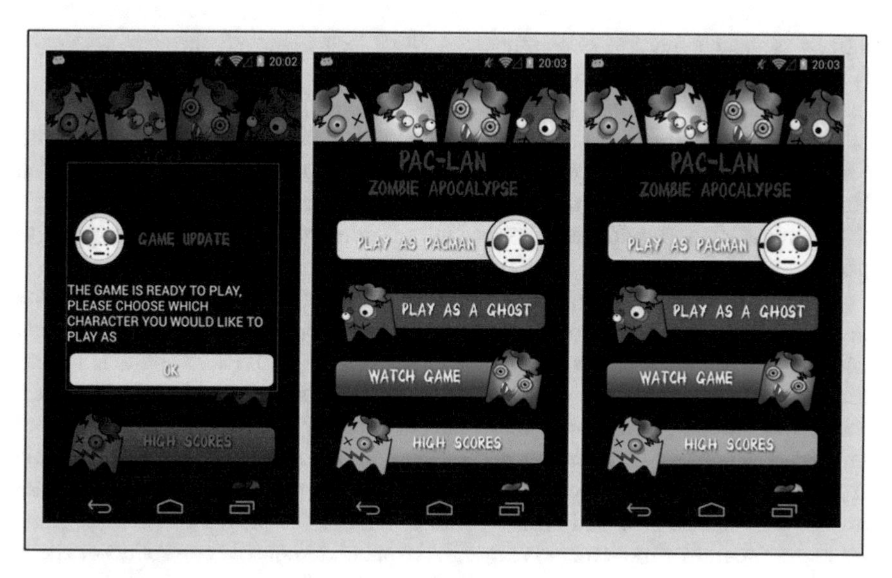

Fig. 5 Screenshots of 2013 version of NFC game Pac-Lan

- The form of objects should encourage and support spatial manipulation.
- Object affordances should match the physical constraints of the object to the requirements of the task.

The original concept of affordance was conceived by Gibson (1977) to define the actionable properties between the world and a person and was most notably developed by Norman (2002) who divided affordance between real and perceived. In particular, he used this as a means of distinguishing between the properties of an object that are controllable by a designer. In the case of real objects, both the real and perceived affordances are controllable, whereas for screen-based interaction generally only the perceived affordances are under the control of the designer as the computer system comes with built-in physical affordances (Norman 2002).

An additional consideration for games is what Juul (2009) describes as mimetic interfaces in relation to casual games, like *Wii Sports* and *Guitar Hero*. These games require players to perform actions that closely resemble the physical activity required by the avatar on the screen. These interfaces make games easier for players less familiar with the more 'traditional' configuration of game controllers utilising buttons and joysticks to pick up and play casual games. As Juul describes, the requirement for players to strum the guitar controller in *Guitar Hero* requires no explanation as most would be familiar with the image of a music artist playing a guitar (Juul 2009).

Whilst mimetic interfaces might suggest they are the same as Natural User Interfaces (NUI) the definition offered by Wigdor and Wixon contradicts this assumption in that they suggest, "natural refers to the user's behaviour and feeling during the experience rather than being the product of some organic process", and indeed suggest a natural experience "is NOT best achieved through mimicry" (Wigdor and Wixon 2011).

An interesting extension to this classification debate is the emergence of game interfaces using real-world artefacts such as the game *Rocksmith* from Ubisoft. The game comes with a cable that allows players to plug their own guitar into an Xbox 360®, PlayStation®3 or PC. In addition to the expected song tutor activity, the systems also provide a 'Guitarcade' of mini games that are designed to practice-specific techniques.

In terms of using physical game pieces to augment virtual games, much of the research has been concerned with augmenting tabletop games. However, a recent commercial game, *Skylanders: Spyro's Adventure* by Activision, augments a traditional role-play adventure game for all the main consoles. A player is provided with a 'portal', an RFID reader that connects to the console via USB or Bluetooth (a special Bluetooth Low Energy (BLE) version is now available for iPhone and iPad), and RFID-enabled game pieces. The game pieces are either character avatars that can be swapped on the portal to change the player's current in-game character (character movement is through the standard console controller) or spells and potions to enhance character performance (Coulton 2012).

In terms of physical games pieces for touch-based phones and tablets, there are two main approaches; one using active games pieces (Yu et al. 2011) that require their own power supply, or passive designs that use the natural conductance of the human body (i.e. Burnett et al. 2012) as shown in the Fig. 6. Burnett et al. created a game based on the arcade game Air Hockey and the classic Atari video game Pong. In both games the player controls a mallet and competes against another player controlling a second mallet (in Pong this could be a computer-controlled opponent) on the opposing side of the game area. Players use the mallets to hit a puck back and forth aiming to get the puck into their opponent's goal and earn points. The game ends when a player reaches a pre-determined number of points. The game created in this research, dubbed Pong+, is designed to be played with the iPad horizontal and using physical game pieces as mallets, as per Air Hockey, but with the virtual assets and game-play akin to Pong. Placing this particular implementation into context, it can be regarded as a fully embodied dynamic spatial tangible interface (Ullmer and Ishii 2000) and overcomes the limitations previously discussed regarding research on active objects (Yu et al. 2011) that only supports static interfaces.

4.10 Physiological Interactions

Physiological interaction goes beyond the physical interfaces whereby a computer has to use the real-time physiological data from the user. This could include heart rate, skin galvanometry, pulse oximetry, brain waves or, indeed, any other physiological system for which a sensor is available. As many of these sensors are now wireless for ease of deployment, it offers the possibility of using such data within mobile and ubiquitous games.

Fig. 6 Physical game pieces on touch-based tablet

4.10.1 Electrocardiogram (Heart Rate)

With obesity rates rising and with many people reluctant to engage in traditional forms of exercise, there has been a growth in interest of so-called 'exergaming' activities to provide novel and 'fun' ways of taking exercise. Exergaming is a term used to describe video games that also provide exercise (Yang et al. 2008), and is not to be confused with biometric gaming that has been linked to emotional gaming in that they try to incorporate player's emotional responses, such as stress or happiness, to influence game play. "Exergames" can be sub-divided into two main groups, those with a game specifically designed to use an exercise input device and those implementations using a particular genre, or a generic game to provide exercise (Wylie and Coulton 2008a).

The Wii Fit from Nintendo is probably the most well known of the former, utilising the Wii Balance Board peripheral to provide a series of games, the result of which are used to provide players with an estimate of their current fitness level and improve it. In terms of the second category, Dance Dance Revolution by Konami and Guitar Hero from Activision would exemplify a game requiring high levels of physical interaction thus providing a degree of exercise (Wylie and Coulton 2008a). With the emergence of personal electrocardiogram (ECG), heart rate monitors aimed at dedicated amateur runners that provide Bluetooth connectivity and can thus be coupled to mobile phones, there is now a possibility of enabling Mobile Exergaming incorporating real-time physiological data. Health Defender (Wylie and Coulton 2008b) was an early example of such a game which made use of the phone's on-board accelerometers connected with an Alive

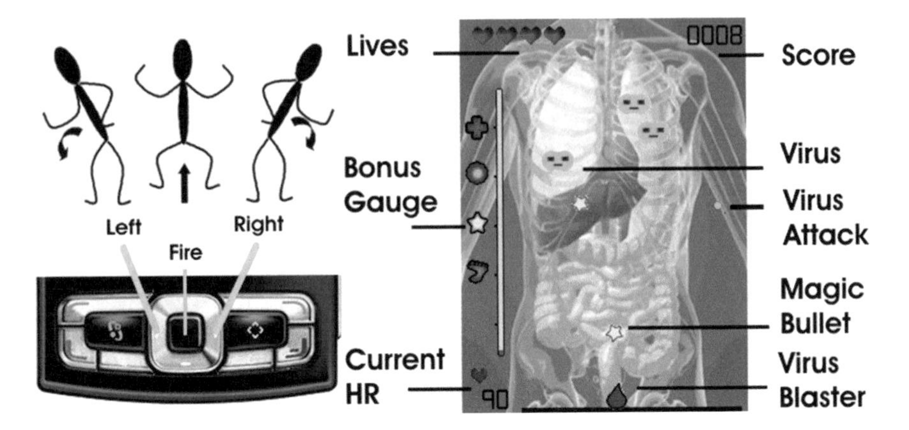

Fig. 7 Interactions for ECG game health defender

Technologies Heart Rate Monitor (HRM) via Bluetooth to both inform players on their current heart rate exertion during game play, and events such as triggered bonuses (see Fig. 7).

4.10.2 Electroencephalographic (Brain Waves)

The possibility of allowing our brains to interact with the environment without the normal physical intermediaries of nerves and muscle has fascinated scholars since Hans Berger introduced electroencephalographic activity (EEG) in 1929. Whilst the resulting prospect of Brain–Computer Interaction (BCI) has naturally received considerable interest for use with physically disabled users, the prospect of an exciting new control modality also has relevance for 'healthy users' and, with the inevitable reduction in size of the required technology, in applications applicable to mobile users to facilitate Brain–Mobile Interaction (BMI) (Coulton et al. 2011).

One of the drivers in the production of unobtrusive, low power, wireless, lightweight and wearable EEG devices has been the potential adoption by the game industry as peripheral sales already generate considerable revenue in addition to games. One such headset, NeuroSky MindSet, enabled the creation of the marble labyrinth game, named *Brain Maze* (Coulton et al. 2011). Alongside the traditional rolling marble action using tilt data derived from the phones accelerometers, the game was designed so that the attentive and meditative states available from the headset are used to unlock 'mind gates' which provide shortcuts or access to specific areas of the game (Coulton et al. 2011). To aid understanding of the attentive and meditative information of individual players during the course of a game Coulton et al. created 'Mind Maps' as shown in Fig. 8. These mind maps plot the individual intensity values of the attentive and meditative states at the (x y) position of the marble as the game was played. It is important to note that for clarity the points are plotted at the 1 s interval that new values where available from the

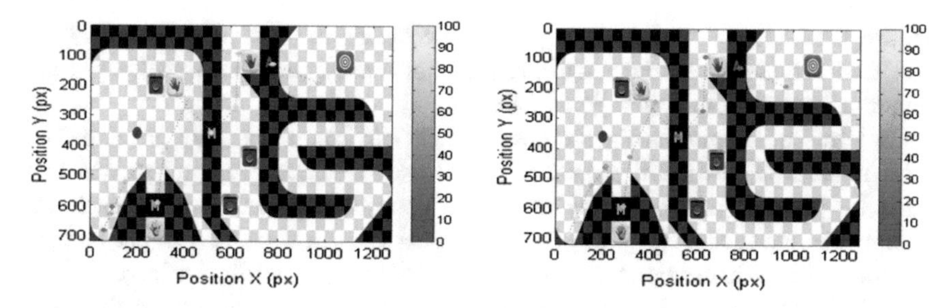

Fig. 8 Mind maps from playing brain-controlled mobile game brain maze

headset, rather than at the frame rate of the game (which is much higher, to produce smooth graphics and accurate feedback to the tilting input of the player). The figure is the representation of a single user's mind map whilst playing level one of the game. Note that the attentive readings are displayed in the left mind map and the meditative readings on the right mind map. The dotted line is provided to allow the actual route taken by the players 'marble' to be ascertained. As can be seen, this is a user who had a high initial meditative state allowing them to negotiate the first two meditation gates rather easily, then he deliberately decided to proceed via the attention gate in preference to taking the long way round and having to manoeuvre past the wormholes which put the players ball back to the start.

The possibility of controlling computers with our brains is obviously an exciting concept in general and no doubt its use within games will help drive the development of this technology at a faster rate. For mobile, in particular, it offers the enticing possibility of freeing ourselves from current interface conventions.

4.11 Further Bluetooth Add-ons

There are a wide range of Bluetooth peripherals appearing, such as, the Blobo, created by Finnish developers BallIT, to provide a six axis accelerometer and pressure sensor encased in a rubberised ball. In contrast, the Sphero robotic ball from Gosphero is fully steerable from a mobile phone and can be used as part of an AR game. These two products are merely examples of an expanding array of devices and toys that can be linked to mobile phones and the emergence of Arduino BLE boards will undoubtedly offer new possibilities for mobile and ubiquitous games.

5 Conclusions

Given the near 6.5 million hours a day already spent on mobile gaming in the UK alone, there is no doubt this is already a significant and growing part of the games industry. What this chapter clearly illustrates is that unlike the console games industry, which tends to work in predictable cycles of evolution of models, the world of mobile games is much more dynamic presenting both challenges and opportunities for researchers in this area.

In terms of mobile and game development environments there are already a wide number of possibilities open to games and interaction researchers and they must be prepared to adapt quickly to the opportunities provided by each and choose the one most relevant to the task, rather than the one they might be most familiar with. Whilst app stores present the exciting possibility for large scale in the wild evaluations they put a greater requirement on the quality of the game produced that would be normally be expected by researchers doing purely lab-based experiments. Whilst players testing a game in a lab-based environment will readily accept lower quality this is not the case on app stores and significant numbers of players will not be enticed to play without developing to a standard commensurate with commercial games if they are to overcome the participation inequality. The choice of app store is also highly significant in terms of both the number of users the researchers are likely to attract and the demographic of those users. There are also issues in regard to the availability of cheap data tariffs in certain countries as well as possible language localisation that must be considered as part research process as these again will affect adoption rates. Finally, there are ethical considerations in how you obtain informed consent from users that are much more difficult than would have been experienced during lab-based evaluations.

Mobile games offer the potential to reach very different player demographics than have been seen for console games and whilst many researchers adapt console inspired designs to evaluate a specific mode of interaction there is also a need to specifically expand game design to incorporate new and more challenging themes, such as death or sexuality, that will likely require new forms of interaction. There is a also a more general research requirement to better understand how people play mobile games beyond the cliché of standing at a bus stop. For example, the increasing use of mobile a second screen for live television programmes offers new opportunities for real time massively multiplayer games.

The main discussion of this chapter is the interaction modalities offered by the ever-increasing number of sensors both internal to the phones and those that can be easily connected using simple wireless technologies such as Bluetooth. The combination of these sensors with a device that has become a must carry item for the majority of the world's population. This presents opportunities for not only researching these new modalities as they emerge, and potentially in vast array of combinations, but also using it to provide opportunities for mobile and ubiquitous games researchers to expand context aware games to incorporate much more complex variables than simply location.

Although the main focus of this chapter has undoubtedly been on games it is important to acknowledge that their aesthetic, iconography and operation of games is apparent in many of the non-game applications we see emerging everyday and as such we can utilise many of the aspects contained within this chapter in a wide variety of scenarios.

References

Abowd GD, Dey AK, Brown PJ, Davies N, Smith M, Steggles P (1999) Towards a Better Understanding of Context and Context-Awareness. In: Gellersen H. -W. (Ed) Proceedings of the 1st international symposium on Handheld and Ubiquitous Computing (HUC '99), Springer-Verlag, London, 304–307

Bamford W, Coulton P, Edwards R (2006) A surrealist inspired mobile multiplayer game: Fact or Fish? In: 1st World conference for fun 'n games, June 26–28, 2006, Preston, UK, p 5–12

Broll G, Benford S (2005) Seamful design for location-based mobile games. In: Kishino F, Kitamura Y, Kato H, Nagata N (Eds). Proceedings of the 4th international conference on Entertainment Computing (ICEC'05), Springer-Verlag, Berlin, Heidelberg, p 155–166

Burnett D, Coulton P, Lewis A (2012) Providing both physical and perceived affordances using physical games pieces on touch based tablets. In: Proceedings of the 8th Australasian conference on interactive entertainment: playing the system (IE '12). ACM, New York, Article 8, 7

Clemson H, Coulton P, Edwards R (2006) A serendipitous mobile game. In: Proceedings of the fourth annual international conference in computer game design and technology, Liverpool, UK

Chehimi F, Coulton P, Edwards R (2008) 3D Motion control of connected augmented virtuality on mobile phones. In: Proceedings of International Symposium on Ubiquitous Virtual Reality (ISUVR), 67–70, 10–13 July 2008

Coulton P, Lund K, Wilson A (2010) Harnessing player creativity to broaden the appeal of location based games. In: Proceedings of The 24th BCS conference on human computer interaction—HCI2010—Play is a serious business, Dundee, Scotland, 6–10th September 2010

Coulton P, Rashid O, Edwards R, Thompson R (2005) Creating entertainment applications for cellular phones. Comput Entertain 3:33–3

Coulton P, Bamford W (2011) Experimenting through mobile 'apps' and 'app stores'. Int J Mob Hum Comput Int (IJMHCI) 3(4):55–70

Coulton P, Wylie C, Bamford W (2011) Brain interaction for mobile games. In: Proceedings of the 15th international academic mindtrek conference: envisioning future media environments (MindTrek '11). ACM, New York, 37–44

Coulton P, Wilson A. (2012) The Peoples Republic of Monsters. Performing Monstrosity in the City Symposium, Queen Mary, University of London, 1st September 2012

Coulton P (2012) SKYLANDERS: near field in your living room now. Ubiquity J Pervasive Media 1(1):136–138, 3p

Davidson A Reid, Moulton IN, Stasse O (2007) MonoSLAM: real-time single camera SLAM. IEEE Trans Patterns Anal Mach Int 29(6):1052–1067

Egenhofer M (1999) Spatial information appliances: a next generation of geographic information systems. In: 1st Brazilian workshop on geoinformatics, Campinas, Brazil

Friberg J, Gärdenfors D (2004) Audio games: new perspectives on game audio. In: Proceedings of the 2004 ACM SIGCHI international conference on advances in computer entertainment technology (ACE '04). ACM, New York, 148–154

Garner P, Coulton P, Edwards R (2006) XEPS—Enabling card-based payment for mobile terminals. In: The 10th IEEE international symposium on consumer electronics, St.Petersburg, Russia, June 29–July 1, 2006, pp 375–380

Gibson JJ (1977) The theory of affordances. In: Shaw RE, Bransford J (eds) Perceiving, acting, and knowing. Lawrence Erlbaum associates, Hillsdale, NJ

Gilbertson P, Coulton P, Chehimi F, Vajk T (2008) Using "tilt" as an interface to control "no-button" 3-D mobile games. Comput Ent 6, 3, Article 38, p 13

Henrysson A. Billinghurst M, Ollila M (2005) Face to face collaborative AR on mobile phones. In: Proceedings of the 4th IEEE/ACM international symposium on mixed and augmented reality (ISMAR '05). IEEE Computer Society, Washington, pp 80–89

Heikkinen J, Olsson T, Väänänen-Vainio-Mattila K (2009) Expectations for user experience in haptic communication with mobile devices. In: Proceedings of the 11th international conference on human-computer interaction with mobile devices and services (MobileHCI '09). ACM, New York, Article 28, p 10

Ishii H (2008) Tangible bits: beyond pixels. In: Proceedings of the 2nd international conference on Tangible and embedded interaction (TEI '08). ACM, New York, xv–xxv

Juul J (2009) A casual revolution: reinventing video games and their players. MIT Press, Cambridge

Klein G, Murray D (2007) Parallel tracking and mapping for small AR workspaces. In: Proceedings of the 2007 6th IEEE and ACM international symposium on mixed and augmented reality, ISMAR. IEEE Computer Society, Washington, pp 1–10

Lochrie M, Coulton P, Wilson A (2011) Participatory game design to engage a digitally excluded community. In: Proceedings of the 5th DiGRA: think design play. Utrecht, 14–17th September 2011

Lochrie M, Lund K, Coulton P (2010) Community generated location based gaming. In: Proceedings of The 24th BCS conference on human computer interaction—HCI2010—Play is a Serious Business, Dundee, Scotland, 6–10th September 2010

Lund K, Lochrie M, Coulton P (2012) Designing scalable location based games that encourage emergent behaviour: special issue on ambient and social media business and application (Part I). Int J Ambient Comput Int (IJACI) 4(4)

Lund K, Lochrie M, Coulton P (2010) Enabling emergent behaviour in location based games. In: Proceedings of the 14th MindTrek 2010 on envisioning future media environments, Tampere, Finland, 6–8th October 2010

Lund K, Coulton P, Wilson A, (2011) Participation inequality in mobile location games. In: Romão T, Correia N, Inami M, Kato H, Prada R, Terada T, Dias E, Chambel T (Eds) Proceedings of the 8th international conference on advances in computer entertainment technology (ACE '11), ACM, New York, Article 27, p 8

Magerkurth C, Cheok A, Mandryk R, Nilsen T (2005) Pervasive games: bringing computer entertainment back to the real world. Comput Entertain 3(3):4–4

McGonigal J (2006) This might be a game: ubiquitous play and performance at the turn of the twenty-first century. Ph.D. Dissertation. University of California at Berkeley, Berkeley, AAI3253985

Mobile Content Communities, HIIT Publications, Online: http://pong.hiit.fi/dcc/papers/mc2_final_report.pdf

Montola M, Stenros J, Waern A (2009) Pervasive games: theory and design. Morgan Kaufmann Publishers Inc., San Francisco

Nandwani A, Coulton P, Edwards R (2011a) Using the physicality of NFC to combat grokking of the check-in mechanic. In: MindTrek 11 Proceedings of the 15th international academic mindtrek conference: envisioning future media environments, ACM, New York, pp 287–290, MindTrek 2011, Tampere, Finland, 28–30 September

Nandwani A, Coulton P, Edwards R (2011b) NFC mobile parlor games enabling direct player to player interaction. Near Field communication (NFC), 2011 3rd international workshop on. IEEE Computer Society, p 21–25, 5

Nieuwdorp E (2007) The pervasive discourse: an analysis. Comput Entertain 5(2):1–17

Norman DA (2002) The design of everyday things. Published by Basic Books Sept 17, 2002. ISBN-10: 0465067107

Rashid O, Bamford W, Coulton P, Edwards R, Scheibel J (2006) PAC-LAN: Mixed reality gaming with RFID enabled mobile phones. ACM Comput Entertain 4(4):1–17

Rashid O, Mullins I, Coulton P, Edwards R (2006) Extending cyberspace: location based games using cellular phones. Comput Entertain 4, 1, Article 4

Ullmer B, Ishii H (2000) Emerging frameworks for tangible user interfaces, IBM Syst J 39(3–4):915–931

Vajk T, Coulton P, Bamford B, Edwards R (2008) Using a mobile phone as a "Wii-like" controller for playing games on a large public display. Int J Comput Game Technol 2008:1–6

Tuulos V, Scheible J, Nyholm H (2007) Combining web, mobile phones and public displays in large-scale: Manhattan story mashup. In: Pervasive'07, Toronto, pp 37–54

Wagner D, Reitmayr G, Mulloni A, Drummond T, Schmalstieg D (2008) Pose tracking from natural features on mobile phones. In: Proceedings of the 7th IEEE/ACM international symposium on mixed and augmented reality ISMAR. IEEE Computer Society, Washington, 2008, p 125–134

Weiser M (1993) Some computer science issues in ubiquitous computing. Commun ACM 36(7):75–84

Wigdor D, Wixon D (2011) Brave NUI world: designing natural user interfaces for touch and gesture, 1st edn. Morgan Kaufmann, San Francisco

World Bank (2012) Information and communications for development 2012—Maximizing Mobile, http://goo.gl/t85Jo

Wylie C, Coulton P (2008a) Persuasive mobile health applications. In: Proceedings of 1st international conference on electronic healthcare for the 21st Century, London, Sept 8–9, 2008, ISBN 978-9639799325

Wylie C, Coulton P (2008b) Mobile Exergaming. In: Proceedings of the 2008 international conference on advances in computer entertainment technology (ACE '08). ACM, New York, p 338–341

Yang S, Smith B, Graham G (2008) Healthy video gaming: oxymoron or possibility? J Online Educ 4(4):51

Yu N, Chan L, Lau S, Tsai S, Hsiao I, Tsai D, Hsiao F, Cheng L, Chen M, Huang P, Hung Y (2011) TUIC: enabling tangible interaction on capacitive multi-touch displays. In: Proceedings of the 2011 annual conference on human factors in computing systems (CHI '11). ACM, New York, p 2995–3004

Zhang L, Coulton P (2009) A mobile geo-wand enabling gesture based POI search an user generated directional POI photography. In: Proceedings of the international conference on advances in computer entertainment technology (ACE '09). ACM, New York, p 392–395

Zhang L, Coulton P (2011) Using deliberate ambiguity of the information economy in the design of a mobile location based games. In: Proceedings of the 15th international academic MindTrek conference: envisioning future media environments (MindTrek '11). ACM, New York, p 33–36

Part II
Indoor and Outdoor Playgrounds

Interactive Playgrounds for Children

Ronald Poppe, Robby van Delden, Alejandro Moreno and Dennis Reidsma

Abstract Play is an important factor in the life of children. It plays a role in their cognitive, social, and physical development, and provides entertaining and fulfilling activities in itself. As with any field of human endeavor, interactive technology has a huge potential for transforming and enhancing play activities. In this chapter, we look at interactive playgrounds, goals and considerations in their design, and we present the directions in which interactive playgrounds can be made more engaging.

Keywords Interactive playground · Ambient space · Computer entertainment

1 Children, Play, and Playgrounds

Interactive playgrounds are installations that combine the benefits of traditional playgrounds with advances in technology. To provide rich and engaging game experiences, they are designed to sense, learn, and adapt to the players' behavior. As such, they are capable of motivating children to explore and interact with the environment, develop social and physical skills, or promote positive social interactions amongst them. We refer to *play* as a social, bodily activity that children

R. Poppe (✉) · R. van Delden · A. Moreno · D. Reidsma
Human Media Interaction Group, University of Twente, Enschede, The Netherlands
e-mail: r.w.poppe@utwente.nl

R. van Delden
e-mail: r.w.vandelden@utwente.nl

A. Moreno
e-mail: a.m.morenocelleri@utwente.nl

D. Reidsma
e-mail: d.reidsma@utwente.nl

A. Nijholt (ed.), *Playful User Interfaces*, Gaming Media and Social Effects,
DOI: 10.1007/978-981-4560-96-2_5,

engage in for the fun of it. Play may be more or less structured, depending on the number of affordances of the environment and objects, and on the number of (social) rules that govern the interactions. Play is important for the cognitive, social, and physical development of children (Canning 2007; Vygotsky 1978). Recently, technology has found its way into children's games and play via digital games and toys. This trend has caused children to adopt a more sedentary lifestyle with fewer social interactions compared to traditional play. We argue that introducing technology into playgrounds can counter this trend while at the same time aid in children's development (Bekker et al. 2010). In this chapter, we describe how the introduction of technology can help to support certain goals in these interactive playgrounds. To this end, we discuss the goals and considerations in the design, and present directions in which interactive playgrounds can be made more engaging.

1.1 Benefits of Play

Many researchers have emphasized the role of play in child development. Vygotsky (1978) states that play provides an important opportunity for personal development. The interactions with peers, rules, and imaginative role-play provide important lessons which help to control impulses and make the conceptual distinction between thought, actions, and objects. Piaget (1951) also stresses the importance of play, as he considered personal experience to be a strong learning mechanism. Moreover, different skills are acquired depending on the child's development.

Cognitive skills such as creativity are typically mastered through the creation and adaptation of game rules, characters, and an imaginative world (Canning 2007; Vygotsky 1978). The playground is a safe environment, separate from the daily life, where children can immerse themselves and experiment with different roles while remaining in control (Huizinga 1950). This sense of control allows children to feel empowered and gain self-esteem. Through play and the reenactment of 'adult' situations, children can find out what and whom they like and dislike. Moreover, they practice and develop values and incorporate the norms and beliefs of their families and carers (Canning 2007).

Play teaches children to make inferences about others and themselves, about peer inclusion and the participation in social groups. Even aspects of play with negative connotations such as rough-and-tumble, play-fight, and dealing with conflict contain important factors for child development. Studies have shown that children who are deprived of free play time are prone to become violent, regardless of demographics (Hughes 2003).

Besides cognitive and social development, the bodily actions in play contribute to a child's development of physical skills. Play provides children the possibility to explore the capabilities of their own bodies, developing and refining fine and gross motor skills. Activities such as running and jumping help to develop and maintain muscular fitness and flexibility. Moreover, the introduction of objects such as balls and monkey bars can aid in the development of hand–eye coordination.

1.2 Open-Ended Play

Play can consist of games and open-ended play. The former have predefined rules, goals, and common ending conditions. In contrast, open-ended play is characterized by the lack thereof. Children invent games by introducing and adapting rules in an ad hoc fashion, including the goals and ways to meet these (Nijholt et al. 2009). The creation of a set of rules and the use of imagination are important parts of personal development for children and function as a stimulation of social interaction. Vygotsky (1978) stated that the change in child's play from overt fantasy with covert rules toward play with covert fantasy elements with overt rules outlines children's play evolution.

Often, the elements of open-ended play are borrowed and adapted from games, and might include characteristics such as cooperation and competition. Also, children frequently introduce roles and characters from their daily lives, television, or stories into the playground. Children might introduce and adhere to a theme, for example when searching for a treasure while some of the players are pirates, aiming to catch the treasure hunters. A similar observation can be made for objects and dedicated locations. Playgrounds can sometimes be adapted to provide a sense of immersion. While these ideas originate and are regulated by the children playing in the playgrounds, they depend strongly on the affordances of the physical playground. In the next section, we will discuss interactive playgrounds that can contain interactive elements to better support and stimulate children in their play.

1.3 Interactive Playgrounds

Interactive playgrounds combine traditional play with interactive elements in order to encourage the social and physical benefits of play (Bekker et al. 2010). In this chapter, we focus on interactive playgrounds in which *multiple players* play *co-located* using *natural interactions* as input, such as chasing, jumping, and making noise. We discuss mainly those playgrounds that support open-ended play, where no goals are imposed by the system. The feedback of these systems to the players is based on simple or more advanced adaptive game mechanics and can include lights and speakers, as well as large projections and screens. Preferably, these playgrounds are openly accessible without specific requirements on clothing or preparation. Interactive playgrounds can therefore support a free play experience in which children can enter and leave at any moment (Tiemstra et al. 2011). Such interactive playgrounds are suitable for a wide range of contexts and age groups. In this paper, we mainly focus on able children, but will also discuss user groups with cognitive or physical disabilities in Sect. 4.2.

The added use of technology in the playground allows for the enrichment of play when the following three properties are taken into account in their design: (1) context-awareness, (2) personalization, and (3) adaptiveness (Schouten et al.

2011). Context-awareness is about recognition of the situational context, including the locations, and behavior of the players, and their interactions with other players and the environment (Moreno et al. 2011). Context-awareness is achieved through sensors and their subsequent processing (see next section). Personalization considers the tailoring and configuration of the play experience to the players. Skills, experience, and interests of the players can be taken into account when regulating the game mechanics. Moreover, the interest of children can be sparked through the use of feedback, which can be adapted to their personal taste. Finally, adaptiveness relates to the manner in which the game mechanics are adapted to the current state of play, and is closely related to the goals of the playground (see also Sect. 2).

We will discuss design considerations of interactive playgrounds using the aforementioned three characteristics. It allows us to develop playgrounds that support open-ended play while keeping the players engaged. However, the mere introduction of technology into playgrounds is not a guarantee for an improved experience of the players. Of key importance is that the (group) behavior of the children can be understood and subsequently influenced when needed. For example, roles in games can be detected and indicated, to facilitate game play (Moreno and Poppe 2013). Alternatively, novel elements can be introduced when the interest of the children decreases. Novel (projected) elements can have different affordances, which have to be explored or defined by the children. This stimulates their imagination and their development of social skills such as negotiating and leadership. In this process, we can differentiate between players, if their experience and skills require to do so. Interactive playgrounds thus have similar benefits as traditional play, but with the added benefit that the experiences and interactions of the players can be steered to a certain extent. We discuss these considerations in Sect. 3.

We consider interactive playgrounds that are room-sized, such as classrooms, halls, or alleys (see Fig. 1 for a schematic example). Spaces of this size are small enough to encourage social encounters, yet large enough to encourage physical activity. This interplay between face-to-face interactions and the possibility of physically active play is important for the promotion and development of social and physical skills. Moreover, equipping these spaces with lights and sounds is possible, without having to rely on individual feedback, for example through the use of headphones and mobile screens. Some researchers have targeted remote, mediated interaction to support playing over a distance. For instance, Mueller et al. (2003) introduced playing soccer via a large interactive wall, with the remote player shown as shadow to increase the feeling of being co-located. While these distributed interactive playgrounds share many characteristics with co-located playgrounds, including their use of sensors and game mechanics, we do not discuss them in this chapter.

We focus on the use of natural interaction as opposed to the controller-based interaction that is currently common for playing digital games. It has been found that the use of body movements can increase the engagement and immersion of play (Bianchi-Berthouze et al. 2007). Moreover, we aim at promoting physical

Fig. 1 Impression of a room-sized interactive playground with sensors and actuators

health through the use of active play (see also Sect. 2). The advantage of using natural interaction is that we can detect and measure it unobtrusively without the players being required to wear and use specific clothing or devices. The interactive playgrounds that are the main focus of this paper rely strongly on the use of cameras that can be used to localize players, detect their body poses and movements, and give an interpretation to their bodily behavior. Besides the detection of body movements such as running and ducking, natural interaction includes the use of specific body poses and movements, such as dancing or posing. Interactions with the environment, such as leaning against a wall or climbing a rope can also be measured unobtrusively using appropriate sensors. Another source of behavior observation comes from the voice. Yelling, singing, and using commands such as "stop!" and "go!" could be detected by the playground without having the players to wear close-up microphones.

2 Goals in Interactive Playgrounds

Interactive playgrounds are typically designed to provide a fun and engaging game experience but can support other goals at the same time as well. These goals can be related to encouraging positive, healthy behavior or discouraging negative aspects of children's play. The feedback capabilities enable the use of interactive playgrounds for education purposes. In addition, the automatic sensing of behavior allows these playgrounds to be used as diagnostic tools to identify developmental problems in children at an early stage.

2.1 Engagement and Fun

Interactive playgrounds can elicit happiness in the players by providing a fun experience that keeps them engaged in play. Retaining a child's attention is not a trivial task. Children are active, curious, and their play is chaotic by nature (Tieben et al. 2011). They tend to get distracted, or become too focused and oblivious to their surroundings. However, this is where the technological aspect of interactive playgrounds can contribute. Presenting children with novel means of interaction and visualizations appeals to their curiosity and serves as a hook to keep them engaged. For instance, Lahey et al. (2008) showed people in the age range of 19–50 years were more hesitant to interact with robotic technology whereas children accepted it readily.

Fun experiences can be achieved in many different ways, both in games and free play sessions. Elements in the playground afford interactions, which appeal to the children's creativity. For instance, when colored shapes are projected on the floor, children are drawn to chase and stomp them (Tetteroo et al. 2011). When given a slide, children will go up and down until exhausted (Soler-Adillon and Parés 2009). When presented with interactive objects, they will use them as instructed but also in novel, creative ways (Bekker et al. 2010). The playground as a whole can also become a stage where fun activities can take place. Children can dance to the beat of the music or try to swim on a floor where colored lights become a stream of water (Moreno et al. 2012). The playground merely provides an environment where fun activities can take place. This can be seen, for instance, in the case of staged play where children have fun by pretending to be fictional characters and engage in fantasy play.

A child's level of engagement is partly determined by the amount of perceived autonomy and empowerment in a particular situation (Canning 2007). This might be affected by conditions such as a child's ability to understand the game, socialize with others, access the game, skills, interests, and mood. This further emphasizes that designing interactive playgrounds to provide engaging experiences is not a straightforward task, and careful considerations need to be made based on the desired goal and the target audience.

2.2 Physical Activity

Playgrounds have traditionally been used to promote physical activity. The same holds for interactive playgrounds, which have originally been created to counter the growing concern for the lack of physical activity in children and their increasing sedentary behavior due to playing video games (Vandewater et al. 2004). Encouraging physical activity can be achieved by getting children to physically exert themselves in any possible way.

Interactive playgrounds can support traditional games such as tag games or soccer, but can also add elements to further motivate children to explore and interact in a physically active way. For instance, Soler-Adillon and Parés (2009) designed a big, inflatable interactive slide designed to allow several children to slide down while a game was projected on its surface. The children were observed using cameras and could interact with the projected elements. Within one game, the children had to slide down, climb, and slide down again several times. Due to the game element, the children are likely to slide more, thus being more physically engaged while sliding and climbing. The Swingscape playground (see Fig. 2) is another example of a playground that stimulates exploration. Tetteroo et al. (2011) also implemented an interactive playground to encourage physical activity. A top-down projector displayed colored shapes onto the floor. Competition was encouraged by making children compete for the shapes scattered in the playing area. When a child moved within the vicinity of a projected shape, it attached itself to him and began following him. When another child stepped on a following resource, he effectively stole it, adding it to his own resource pool. Physical exertion was achieved by moving the shapes around, and by adding and removing shapes. Children were further encouraged to play together with balls equipped with motion sensors. The balls also changed the interactions the children could have with the projected shapes.

Sensing in interactive playgrounds allows for the measurement of players' skills. Moreover, the feedback can be adapted in such a way that the activity level required to achieve certain goals is tailored to each player. Derakhshan et al. (2006) presented an interactive playground consisting of tiles that children could step on and interact with through force sensors and LEDs. Machine learning techniques were used to learn and model different types of game styles. These styles were subsequently used as a basis to vary the amount of physical activity that the children had to engage in during the game.

Exergames are a type of game where players are required to undergo intense physical exertion (Mueller et al. 2003). They are not necessarily co-located or room-sized, but some can be considered interactive playgrounds. (Mueller et al. 2012) designed the game "Hanging off a Bar,'' where a flowing river with floating rafts was projected on the floor. Players were to hang from a bar when there was a river beneath them. Eventually, a raft would float by, which allowed them to come down and stand on the ground to take a rest. However, they had to hold onto the bar again once the raft had drifted down the river. The game was over when the player stood on the floor when there was no raft.

Often, an element of competition between players is used to stimulate players to be more physically active. "TacTower" was designed to support athletes (Ludvigsen et al. 2010). A "TacTower" consists of eight plastic balls stacked on top of one another and held in place by a steel structure that runs through them. The balls contain LEDs and sensors. When a ball lights up, the player can hit the ball horizontally, which would transfer the light signal to a ball on an adjacent tower, or vertically to make a ball above or below light up. Two players compete against each other on who manages to hit the colored balls. Even though originally meant

Fig. 2 Swingscape: an interactive playground by the Alexandra Institute, Denmark, aiming to solicit physical explorative behavior (Grønbæk et al. 2012)

for athletes, the concept is applicable in children's games as well. Exergames, through competition, can also facilitate social bonding amongst the players. For instance, Toprak et al. (2012) designed the "Bubble Popper," a game where two players play against each other to pop colored bubbles projected on a wall. The bubbles constantly move, so the players have to move as well to score more points than the opponent. This inevitably leads to physical contact when the players compete for the physical space.

2.3 Behavior Change

Playgrounds are places where children play together, and develop cognitive, social, and physical skills. Traditionally, they have been used to encourage positive behavior and discourage negative behavior. A teacher or trainer can perform this role, but it can also come from the interactions between children themselves. Cheating and aggressive behavior are typically not appreciated and often discouraged by other children. Interactive playgrounds present the opportunity to attempt to influence the children's behavior, in groups or individually.

Competition in play is important to teach children how to stand up for themselves. Often, this competition is achieved by striving for conflicting goals, such as competing for a limited number of resources, as discussed in the previous section. Besides motivating children to engage physically, competition in play also stimulates cognitive development. Metaxas et al. (2005) designed "Scorpiodrome," an augmented reality racing game that encouraged competition. In this playground, children were able to remote control toy cars that could pick up virtual rockets and use them against their opponents. Besides competition, a playground could also be designed to persuade children to cooperate. This can aid in children's development of negotiation skills. The "Scorpiodrome" playground also encouraged cooperation by having the children assemble the track and landscape together. Parés et al. (2005) designed an interactive installation that focused on cooperation by encouraging people to communicate and work toward a common goal. "Water Games'' consisted of several water fountains where each could be activated by forming a closed ring of people around it. Once the ring is formed and closed, players have to move in unison in one direction for the fountain to become active.

Competition and cooperation are group-level behaviors. Interactive playgrounds, however, can also be aimed at the individual by influencing personality traits. For instance, Hendrix et al. (2009) developed the "Playground Architect," an interactive board game aimed at helping shy children to confront their social incompetency. It is a multiplayer game where children have to build structures using readily available materials. There are two roles in the game: the architect and the builder. Shy children are always assigned the architect role as this requires coordination and leadership, traits that are typically underdeveloped in shy children. As both players complete the game successfully, self-esteem of the shy children is increased.

Care must be taken when attempting to change the behavior of the players in interactive playgrounds. The behavioral change must be designed into the game, and happen as a result of the game mechanics. More importantly, the change cannot be forced on the child but must be established through persuasion.

2.4 Education and Learning

Play is a fundamental part of learning, especially at a young age. At playgrounds, children can learn in a playful manner which encourages and empowers them (Henderson and Atencio 2007). Interactive playgrounds can be adapted to support certain themes and learning goals. For example, Charoenying et al. (2012) developed a body-centric game called the "Bar Graph Bouncer.'' They aimed at supporting children's ability to conceptualize numbers and interpret graphs. Children are presented with an animated scene that responds to jumping. As children jump, their corresponding bar grows in the animation, facilitating the understanding of correlation between the jumps and the bar. This gameplay is rather simple but it is easy to imagine how the positions and movements of children can be used to participate in quizzes or how specific objects can be used in interactive animations to explain topics in physics and biology.

Playgrounds also provide a social context where children can talk, observe, and ask. It allows them to participate and practice things they have learned with those around them. Besides schools, such playgrounds can be located in museums and other public places. For example, Carreras and Parés (2009) created the "Connexions'' playground for Barcelona's Science Museum. It is a full-body interactive playground with a floor projection of a virtual environment. The projected world resembles that of a network, where different nodes (concepts) are displayed around a hidden object. When children stand on a node, the node starts to glow if it was related to the hidden object. Children need to activate the different nodes and make them connect to each other. They do this by connecting to other children, for example by holding hands. The goal of "Connexions" was to facilitate the abstract understanding of science being a network of knowledge, composed of different concepts in different domains. The "Wisdom Well'' is another example, developed by Grønbæk et al. (2007). This playground supports learning through kinesthetic interaction. The game allows children to communicate and cooperate while interacting with simulations about geometry, physics, and geography.

2.5 Diagnosis

The automatic sensing of behavior in interactive playgrounds opens up possibilities of analyzing whether the playground or the objects therein are suitable and safe. This has been attempted by Ouchi et al. (2010) in the context of interactive playgrounds. A rock-climbing playground was developed with the aim to design safer playground equipment. To this end, it was analyzed how children played, specifically how and why children injured themselves while playing in playgrounds. The rock-climbing wall was equipped with sensors to measure children's behavior. They used machine learning techniques to model the changes in children's postures as they climbed the wall, and were able to predict how a child of a certain height would climb. Afterwards, simulation software was developed using the rock-climbing model that could be used to test new playground equipment for safety.

In contrast to the diagnosis of the playground itself, the behavior of players in an interactive playground can also be used to analyze whether the children are behaving normally, both physically and socially. While such diagnoses should ideally be made by professionals that can also provide the proper care, interactive playgrounds can assume the role of teachers, parents and others that supervise children in traditional playgrounds. As shown previously, play is very important in children's proper development. Studies have shown that children who fail to properly engage in social interactions during play can be linked to several mental or social disorders such as autism or mental retardation (Guralnick et al. 2006). Diagnosis of social deficiencies through the observation of play was shown to be possible, albeit with manual annotation (Gibson et al. 2011). Current practice in such studies involves either the analysis and annotation of video recordings of play sessions, or observational studies of live play sessions. Both are lengthy, cumbersome processes, prone to inaccuracies and a subjective bias of the observers. Interactive playgrounds could help by automating and consequently accelerating these processes, thus giving researchers proper tools to base their findings on (Moreno et al. 2012).

3 Developing Interactive Playgrounds

In order to meet the goals mentioned in the previous section, there are some aspects of interactive playgrounds that require careful design. Now, we discuss these design challenges, which can be linked to the dimensions context-awareness, personalization, and adaptiveness. We also discuss the issue of evaluating the playground's performance in terms of its goals.

3.1 Physical Setup

The rise of interactive playgrounds in the research and public domains is possible due to the increased availability of sensors, feedback elements, and the integrated combinations thereof. Sensors and feedback elements have become increasingly sophisticated and affordable. We briefly discuss some of the more common elements used in interactive playgrounds.

The most common and versatile sensor is the camera, with all its variations. Sensors are relatively cheap and there is an abundance of computer vision software available to analyze the footage in real-time. The players' locations, poses, and movements can be estimated relatively robustly and accurately from camera images (Moreno et al. 2013). Moreover, they can accommodate a space sufficiently large to observe players while they make unconstrained movements. Recently, Microsoft introduced its Kinect sensor to make a three-dimensional representation of the scene. This allows to robustly detect players and analyze their bodily configurations under a larger range of viewpoints and lighting conditions. In situations where cameras cannot be used, pressure sensors can be embedded in the floor to track people (Derakhshan et al. 2006). For the detection of sound, microphones can be used. These allow for the analysis of noise levels, patterns such as in singing, or to coarsely understand what has been said, for example by focusing on specific commands. When multiple microphones are employed and synchronized, players can be located based on the sounds they produce.

Feedback elements are increasingly versatile in modality, size, and how they can be embedded into the playground. Lights, for example rotating stagelights, can be placed virtually everywhere (Wakkary et al. 2008). LEDs can be embedded into the environment to mark locations or provide location-based feedback. Probably the most popular means of feedback are projections, for which any surface can be used as long as the projector is powerful enough. Moreover, projections can provide detailed feedback. When aimed at the floor or walls, they are very suitable for location-based feedback. Sound, especially when directed, can be used in a variety of interactive installations (Tieben et al. 2011). For a more complete overview of feedback elements in interactive playgrounds, we refer the reader to Schouten et al. (2011). Next, we turn to the goals that can be achieved by employing these sensor and feedback elements in playgrounds.

Depending on the playground's type and goals (Sect. 2), different sensors and feedback elements can be used. The choice for these elements depends on the required robustness and accuracy, and the physical properties of the space. Cameras using visible or infrared light are suitable to determine the position and movement of players in a relatively large space. In outdoor settings, variations in lighting might pose serious challenges in localizing the people. A similar challenge is faced when strong projections are used. Instead of relying on a single camera, cameras that use stereo vision, structured light, or time-of-flight can deal with these settings as they make a three-dimensional estimate of the field of view. Such a representation can be used to robustly localize objects in the foreground. Once objects in the foreground have been identified, they can be tracked to associate the movement to players. A top-down view is preferable, given that most of the players' movement take place on the ground plane (Tetteroo et al. 2011). For those playgrounds where such a setup is not possible, notably outdoors, cameras can be placed at a height on the corners of the space. Wide-angle lenses ensure that a larger area can be viewed. A similar setup with Kinect sensors allows for the quick and robust localization of players. This is ideal when the location of players and their interactions are sufficient as input for the playground, which is typically true for play aimed at engagement and physical activity.

For play that is focused on diagnosis or education, a closer analysis of players' body poses, actions, and interaction with other players might be required. Players' limbs are typically better visible from the side. However, this introduces challenges in dealing with occlusions due to other players and objects in the environment. For the analysis of direct interactions with the environment, pressure sensors are most convenient as these require little processing and can be robustly detected. They can be embedded in floor tiles, buttons, monkey bars, and other objects in the environment (Derakhshan et al. 2006; Mueller et al. 2012).

Sound can be informative of the level of engagement. When multiple microphones are used, players can be localized when they make noise. In addition, communication patterns such as yelling and responding to other players can be analyzed. This is useful to determine when players are not actively taking part in play. In addition, sound can be used as a direct means of communication, for example when the players have to shout certain words, sing along with the music, or have to make noise in a certain rhythm.

Appropriate types of feedback also depend on the specific playground. Floor projections are ideal to stimulate the players to move as the projections can be aligned to the positions of the players. They can be used to increase engagement and physical activity. Moreover, the projections can be used to indicate a player's role or status, or make explicit relations between players. For example, in a tagging game, the tagger could be marked, and the player that was tagged the lowest number of times could be highlighted. This way, the behavior of the players could be steered (Moreno and Poppe 2013). The drawback of projections is that players might be between the projector and the surface, thus casting shadows. However, this has not been found to be a major issue (Tetteroo et al. 2011).

The use of other visible forms of feedback such as lights or screens can be used to display the state of play. Information or narrative can be shown on screens (Bobick et al. 1999). Lights can be used to highlight certain areas in the playground, or as a way to visualize activity or observed engagement. Flashing lights are a convenient form of direct feedback, for example when a certain action has been performed. Sounds, especially when multiple sources are used, can support narrative or to give hints or updates on the current state of play, especially when used in addition to large screens and projections (Grønbæk et al. 2007). This is especially useful for educational play. Spoken encouragements can aid in increasing the engagement and can lead to higher physical activity.

3.2 User Adaptation

Since interactive playgrounds combine sensing, game mechanics, and feedback (Fig. 3), different rules or feedback can be given to different players. Such feedback can include roles such as tagger and runners, or leader and followers. For example, children with lesser-developed communication skills might be assigned the role of team leader, as in Hendrix et al. (2009). However, there can also be distinction made between the players' skills or expertise, similar to a handicap in golf. For example, play can be adapted such that slower children have to cover shorter distances within the same type of play. By adapting the game mechanics and feedback for each child, the chances of a child giving up because the play is too challenging or too simple are reduced. Instead, the play could be steered to be challenging yet rewarding for each child, which helps to ensure that the engagement is high.

Another aspect that can be tailored to the user is that of novel experiences. In interactive playgrounds, the feedback can be chosen such that it is not always the same for each child. For example, when using projections and sound, achieving a goal should not always lead to the same sound being played for a given player. It can be stored which player received which feedback for a certain action or for achieving a certain goal. On the other hand, when introducing elements in the playground, it can be ensured that each child will be presented with different elements. This will keep them engaged, especially since they might be curious whether there are more possibilities or rewards. This mechanism is especially relevant for interactive playgrounds that have an educational function. Ensuring that children will not be presented with the same information or narrative twice not only reduces frustration but might also keep children more engaged in play, thus allowing them to learn more in the same amount of time.

Fig. 3 Processing loop applied in interactive playgrounds

3.3 Interaction Mechanics

The interaction mechanics determine how the playground responds to the behavior of the players. It is the coupling of the playground's sensing with the feedback. These components together with the behavior of the players form a loop, see Fig. 3. Changes in the feedback, determined by the game mechanics typically lead to changes in the displayed behavior. The mechanics are closely tied to the goal of playground. For example, when the aim is to promote physical activity, the observation that the activity decreases should result in appropriate feedback to engage the players to move more.

Interactions can be specified at the level of inputs, such that system actions are triggered by the detection of certain behaviors. For example, a sound can be played when pushing a button, or a light could flash when a certain pose is assumed. Alternatively, feedback can be given based on a certain state of the players' behavior, for example when they move a lot or when their movements are in synchrony with the rhythm of music in the playground. Often, the interactions are conditional on the state of play. Actions performed by the players result in different ways of feedback from the playground depending on the roles and behavior of the players. As such, the behavior of the players can be steered, for example when the projections light up brighter when the players are close together.

When the feedback rules are more complex and take into account player roles, games such as tag games or "Simon Says" could be implemented. Feedback to the players is important as the observed game state not necessarily has to correspond with the one perceived by the players. For example, the playground could observe that a tagger hit a runner, whereas there was no physical contact. In this case, the internal state of the playground needs to be properly reflected in the feedback to resolve ambiguities.

Interactions can change over time. This stimulates players to explore the interactions, which could lead to a larger diversity in the displayed behavior. By varying the interactions or by making them less deterministic, the open-ended nature of play can be stimulated. Children are encouraged to explore the interactions and to come up with new rules and goals that use the changing types of feedback.

3.4 *Evaluation of Goals*

To evaluate whether a playground functions as intended, players can be asked about their experiences. Questionnaires could be used to find out whether children enjoyed the play, or whether they have learned something. Such evaluations are difficult as the goals of the playground are typically not explicit and might be achieved unconsciously (Poppe et al. 2007). Alternatively, one could have observers annotate the play session and reason about whether the playground's goals have been met. In such cases, the number of interactions, occurrence of certain actions, amount of movement or cooperation could be measured by the observers.

However, the automatic sensing in interactive playgrounds can also be used to obtain objective measures to evaluate the playground. For example, the effect of certain ways of feedback on the amount of movement and interactions of the players can be analyzed automatically. Such findings can give rise to the adaptation of the game mechanics during play, but also might reveal insights on how play can be made more enjoyable in a broader sense. Given the tremendous amount of possibilities of interactions between players and playground, it is useful to be able to automatically have a measure of which type of interactions are appreciated by the players.

4 Interactive Playgrounds of the Future

While interactive playgrounds are becoming more common, there are many ways in which they can be made more engaging, versatile and popular. The availability of novel and more affordable sensor and feedback technology presents opportunities to even better address the different goals of interactive playgrounds. We discuss some promising avenues along the previously identified dimensions of context-awareness, personalization, and adaptiveness.

4.1 *Socially Aware Interactive Playgrounds*

Sensing in interactive playgrounds is currently often limited to localization of players, analysis of movement, and the recognition of gross action categories. Additional context-aware information, such as social behavior, is usually obtained manually through observation or annotation of videos since its automatic analysis has proven challenging or nonviable. Advances in sensor technology and computer vision algorithms allow for more accurate automatic analysis of players' behavior. This allows researchers to analyze behavior that has been so far neglected or overlooked. This is true for both the interpretation of the bodily behavior of individuals, conscious or unconscious, as well as the social behavior of groups (Moreno et al. 2013).

The latter is especially useful since social behavior is an extremely important component of play. The study of social behavior could be achieved by looking toward social signal processing, an emergent field of research that deals with the automatic analysis of social signals (signals that people use to convey social information (Vinciarelli et al. 2012)). Being able to detect and understand social interactions during play, such as cooperation or competition, presents great opportunities to design new types of interactions, as well as novel evaluation methods.

As an example, imagine a playground that supports different types of tag games (Moreno and Poppe 2013). Roles (runner or tagger) could be indicated by a differently colored circles projected around each player. The playground could detect that children in a playground are playing together with the exception of one child who is excluded from the game. This may be due to personality traits, such as the kid being shy. The playground detects the exclusion and modifies the mechanics to integrate him, such as giving him a key role in the feedback, for example by persuading the other children to tag him by presenting directed projections. Also, differences in skills could be detected, for example when a child does not manage to tag anyone. In this case, the playground could decide to switch roles and appoint another child as the tagger. Another application could be the detection of boredom from the players. By sensing general playground features, such as lack of movement or low sound levels, along with specific player cues, such as having the arms crossed or not interacting with other players, the playground could radically change the mechanics to engage the players again. In the case of tag games, there could be more taggers, or there could be a "ghost tagger," a colored circle that assumes the role of the tagger.

4.2 Tailoring to User Groups

We have discussed playgrounds without focusing on players with specific age, interests or abilities. With the automatic sensing and the possibilities of adapting the game mechanics and feedback, we can tailor playgrounds to individuals but also to groups of users. Children that prefer physical exertion can be presented with more active play, whereas children that value novelty more might be presented with more narrative. This personalization is likely to present opportunities to further engage children into the play.

A user group that deserves special attention is that of players with physical or intellectual disabilities. Given that interactive playgrounds provide a safe yet controlled place for exploration and experiencing makes them the ideal tool to support these users. There is some initial work in this direction. Wyeth et al. (2011) designed the Stomp platform to encourage social interaction between players with intellectual disabilities, by using pressure sensors embedded in floor mats and projections.

However, there are challenges in the development of interactive playgrounds for these users (Wyeth et al. 2011). Classical principles of user-centered design and standard methods of evaluation will not always work. People with disabilities may have limited abilities to express themselves, or might express their states and intentions differently, which can lead to misunderstandings. There are also challenges faced in the design of the feedback, as some disabilities can affect the player's hearing, vision or cognition. Also, care should be taken in presenting feedback to the players as there is the risk of overstimulation, which can have severe consequences (Noordzij et al. 2012). Both for the design and practical use of playgrounds for these more vulnerable groups of users, care-takers will take a prominent role. The research into appropriate feedback might eventually also be useful outside the playground.

4.3 Novel Interactive Elements

The introduction of novel interactive elements into playgrounds can enhance their appeal, especially given the curiosity that children have for novel gadgets and technology. For example, interactive playgrounds could be equipped with interactive tiles (Derakhshan et al. 2006) or tangible objects that can make sound, emit light or vibrate in response to certain actions and interactions.

Moreover, we expect that stronger narrative, either through a voice-over or through projections, can increase the engagement of the players. We specifically foresee a more interactive way of narrative, for example, where a virtual character responds to the players' actions and guides them to explore certain types of play. When the interactive playground supports themes such as pirates or jungle, an increased feeling of immersion can also be achieved. Moreover, these characters can eventually become a play buddy.

In some cases, being able to play over a distance will be beneficial for both physical activity and social interactions. Playing with friends can lead to more engagement and more creative play as players might be less inhibited in their creativity. While distant players can be represented using projections and sound (Mueller et al. 2003), they could also control elements in the playground such as lights. Also, they might be represented as robots or virtual characters to be more versatile in their expression. Children have been found to enjoy robots (Lahey et al. 2008). Instead of having other players participate in the play through robots, they can also take on the role of narrator or automatically controlled team mate or opponent. This can increase the sense of competition or cooperation between players.

5 Conclusion

In this chapter, we have discussed interactive playgrounds and their goals. Promising avenues for further research, exploration, and development have been outlined. We are confident that future playgrounds will better meet children's needs of engagement, fun, and physical exertion. Moreover, there are opportunities to achieve behavior change, and to stimulate children to develop cognitive, social, and physical skills.

Acknowledgments This project has been made possible by support from the Dutch national program COMMIT.

References

Bekker T, Hopma E, Sturm J (2010) Creating opportunities for play: the influence of multimodal feedback on open-ended play. Int J Arts Technol 30(4):0325–0340

Bianchi-Berthouze N, Kim WW, Patel D (2007) Does body movement engage you more in digital game play? and why? In: Proceedings of the international conference on affective computing and intelligent interaction (ACII). Lisbon, Portugal, 2007, pp 102–113

Bobick AF, Intille SS, Davis JW, Baird F, Pinhanez CL, Campbell LW, Ivanov YA, Schütte A, Wilson A. (1999) The kidsroom: a perceptually-based interactive and immersive story environment. Presence: Teleop Virtual Environ 80(4):367–391

Canning N (2007) Children's empowerment in play. Eur Early Child Educ Res J 150(2):227–236

Carreras A, Parés N (2009) Designing an interactive installation for children to experience abstract concepts. In: Macías JA, Granollers Saltiveri A, Latorre PM (eds) New trends on human-computer interaction. Springer, New York, pp 33–42

Charoenying T, Gaysinsky A, Riyokai K (2012) The choreography of conceptual development in computer supported instructional environments. In: Proceedings of the international conference on interaction design and children (IDC). Bremen, Germany, pp 162–167

Derakhshan A, Hammer F, Lund H (2006) Adapting playgrounds for children's play using ambient playware. In: Proceedings of the international conference on intelligent robots and systems (IROS). Beijing, China, pp 5625–5630

Gibson J, Hussain J, Holsgrove S, Adams C, Green J (2011) Quantifying peer interactions for research and clinical use: the Manchester inventory for playground observation. Res Dev Disabil 320(6):2458–2566

Grønbæk K, Kortbek KJ, Møller C, Nielsen J, Stenfeldt L (2012) Designing playful interactive installations for urban environments–the swingscape experience. In: Proceedings of the conference on advances in computer entertainment (ACE). Kathmandu, Nepal, pp 230–245

Grønbæk K, Iversen OS, Kortbek KJ, Nielsen KR, Aagaard L (2007) Interactive floor support for kinesthetic interaction in children learning environments. In: Proceedings of INTERACT. Rio de Janeiro, Brazil, pp 361–375

Guralnick MJ, Hammond MA, Connor RT, Neville B (2006) Stability, change, and correlates of the peer relationships of young children with mild developmental delays. Child Dev 770(2):312–324

Henderson TZ, Atencio DJ (2007) Integration of play, learning, and experience: what museums afford young visitors. Early Child Edu 350(3):245–251

Hendrix K, van Herk R, Verhaegh J, Markopoulos P (2009) Increasing children's social competence through games, an exploratory study. In: Proceedings of the international conference on interaction design and children (IDC). Como, Italy, pp 182–185

Hughes B (2003) Play deprivation, play bias and playwork practice. In: Brown F (ed) Playwork–theory and practice. Open University Press, Philadelphia, pp 66–80

Huizinga J (1950) Homo ludens: a study of the play element in culture. Roy Publishers, New York

Lahey B, Burleson W, Jensen CN, Freed N, Lu P (2008) Integrating video games and robotic play in physical environments. In: Proceedings of the ACM SIGGRAPH symposium on video games, Los Angeles, CA, pp 107–114

Ludvigsen M, Fogtmann MH, Grønbæk K (2010) TacTowers: an interactive training equipment for elite athletes. In: Proceedings of the conference on designing interactive systems (DIS). Aarhus, Denmark, pp 412–415

Metaxas G, Metin B, Schneider J, Shapiro G, Zhou W, Markopoulos P (2005) Scorpiodrome: an exploration in mixed reality social gaming for children. In: Proceedings of the conference on advances in computer entertainment (ACE). Valencia, Spain, pp 229–232

Moreno A, Poppe R (2013) "You're it!": role identification using pairwise interactions in tag games. In: Proceedings of the conference on computer vision and pattern recognition (CVPR) workshops. Portland, OR, 2013, pp 657–662

Moreno A, van Delden R, Poppe R, Reidsma D (2013) Socially aware interactive playgrounds. IEEE Pervasive Comput 120(3):40–47

Moreno A, van Delden R, Reidsma D, Poppe R, Heylen DD (2012) An annotation scheme for social interaction in digital playgrounds. In: Proceedings of the international conference on entertainment computing (ICEC). Bremen, Germany, pp 85–99

Mueller F, Agamanolis S, Picard R (2003) Exertion interfaces: sports over a distance for social bonding and fun. In: Proceedings of the international conference on human factors in computing systems (CHI). Fort Lauderdale, FL, pp 561–568

Mueller F,Toprak C, Graether E,Walmink W, Bongers B, van den Hoven E (2012) Hanging off a bar. In: Proceedings of the international conference on human factors in computing systems (CHI). Austin, TX, pp 1055–1058

Nijholt A, Reidsma D, Poppe R (2009) Games and entertainment in ambient intelligence environments. In: Aghajan H, López-Cózar Delgado R, Augusto JC (eds) Human-centric interfaces for ambient intelligence. Academic Press, MA, pp 393–413

Noordzij ML, Scholten P, Laroy-Noordzij ME (2012) Measuring electrodermal activity of both individuals with severe mental disabilities and their caretakers during episodes of challenging behavior. In: Proceedings of the international conference on measuring behavior (MB). Utrecht, The Netherlands, pp 201–205

Ouchi H, Nishida Y,Kim I, Motomura Y, Mizoguchi H (2010) Detecting and modeling play behavior using sensor-embedded rock-climbing equipment. In: Proceedings of the international conference on interaction design and children (IDC). Barcelona, Spain, pp 118–127

Parés N, Durany J, Carreras A (2005) Massive flux design for an interactive water installation: Water Games. In: Proceedings of the international conference on advances in computer entertainment technology (ACE). Valencia, Spain, pp 266–269

Piaget J (1951) Play, dreams and imitation in childhood. In: Explanation of play. Routledge, pp 147–168

Poppe R, Rienks R, van Dijk B (2007) Evaluating the future of HCI: challenges for the evaluation of emerging applications. In: Huang TS, Nijholt A, Pantic M, Pentland A (eds) Artificial intelligence for human computing. Springer, Berlin, pp 234–250

Schouten BAM, Tieben R, van de Ven A, Schouten DW (2011) Human behavior analysis in ambient gaming and playful interaction. In: Salah AA, Gevers T (eds) Computer analysis of human behavior. Springer, Heidelberg, pp 387–403

Soler-Adillon J, Parés N (2009) Interactive slide: an interactive playground to promote physical activity and socialization of children. In: Proceedings of the international conference on human factors in computing systems (CHI). Boston, MA, pp 2407–2416

Tetteroo D, Reidsma D, van Dijk B, Nijholt A (2011) Design of an interactive playground based on traditional children's play. In: Proceedings of the international conference on intelligent technologies for interactive entertainment (INTETAIN). Genova, Italy, pp 129–138

Tieben R, Bekker T, Schouten BAM (2011) Curiosity and interaction: making people curious through interactive systems. In: Proceedings of the British computer society conference on human–computer interaction (BCSHCI). Newcastle-upon-Tyne, UK, pp 361–370

Tiemstra G, van den Berg RM, Bekker T, de Graaf MJ (2011) Guidelines to design interactive open-ended play installations for children placed in a free play environment. In: Proceedings of the international conference of the digital research association (DIGRA). Utrecht, The Netherlands, pp 1–17

Toprak C, Platt J, Mueller F (2012) Bubble popper: considering body contact in games. In: Proceedings of the international conference on fun and games. Toulouse, France, pp 97–100

Vandewater EA, Shim MS, Caplovitz AG (2004) Linking obesity and activity level with children's television and video game use. J Adolescence 270(1):71–85

Vinciarelli A, Pantic M, Heylen D, Pelachaud C, Poggi I, D'Errico F, Schröder M (2012) Bridging the gap between social animal and unsocial machine: a survey of social signal processing. IEEE Trans Affect Comput 30(1):69–87

Vygotsky LS (1978) The role of play in development. In: Mind in society. Harvard University Press, Cambridge, pp 92–104

Wakkary R, Hatala M, Jiang Y, Droumeva M, Hosseini M (2008) Making sense of group interaction in an ambient intelligent environment for physical play. In: Proceedings of the international conference on tangible and embedded interaction (TEI). Bonn, Germany, pp 179–186

Wyeth P, Summerville J, Adkins B (2011) Stomp: an interactive platform for people with intellectual disabilities. In: Proceedings of the conference on advances in computer entertainment (ACE). Lisbon, Portugal, p 51

Designing Interactive Outdoor Games for Children

Iris Soute and Panos Markopoulos

Abstract Mobile outdoor games for groups of children have emerged recently as a credible technological proposition and as an area of research and development that promises substantial benefits for children regarding a more active lifestyle and the development of social skills. This chapter examines specifically the design of Head Up Games, which are outdoor games that support embodied interaction and where players are collocated, e.g., in a playground, alley, park; the traditional loci of children's play over centuries. Designing such games and the emerging gaming experience presents its own set of challenges, such as designing the interaction of a group, ensuring pace in the game, and fairness for different contexts and groups of players. Not least, the added value of enhancing outdoor play and games with technology needs to be ensured. We describe some of the lessons learned from the design of a few of these games, how different design methods may contribute to the design process, and methodological issues concerning the early design, the prototyping, and the evaluation of these games.

Keywords Games for change · Serious games · Sustainability · Behavior change · Procedural rhetoric · Emergent dialogue · Persuasion · Design framework · Design guidelines

I. Soute (✉) · P. Markopoulos
Department of Industrial Design, University of Technology Eindhoven,
Den Dolech 2, 5612 AZ Eindhoven, The Netherlands
e-mail: i.a.c.soute@tue.nl

P. Markopoulos
e-mail: p.markopoulos@tue.nl

A. Nijholt (ed.), *Playful User Interfaces*, Gaming Media and Social Effects,
DOI: 10.1007/978-981-4560-96-2_6,

1 Introduction

Advances in technology are contributing to the increasing portability and ubiquity of mobile devices. As a consequence, interesting venues are opening up for outdoor, social gaming, supported by interactive technology—an area that is of growing interest to the research community. Our particular research interest is in Head Up Games (Soute et al. 2009b): interactive, outdoor games for children that resemble play behavior of "traditional" outdoor games (such as tag and hide-and-seek), i.e., games that are played collocated, encourage physical activity, and support social interaction. The concept of Head Up Games stems from the observation that outdoor pervasive games for children (e.g., Savannah (Benford et al. 2005); Ambient Wood (Rogers et al. 2004)) heavily rely on screen interaction. As a consequence, children are playing these games "head down" and we argue that this interferes with how children naturally play outdoors: running around, while engaging in rich face-to-face social interaction. Therefore, we proposed the concept of Head Up Games—to emphasize that these games are played with the players' heads up, and we aim to include technology to support, instead of interfere, with play behaviors as seen in traditional outdoor games.

Over the last few years we have designed several of such games and in this chapter we will reflect on the lessons learned regarding the design process of these games.

A generally accepted design process in HCI is the User Centered Design (UCD) process. This process advocates the involvement of users in all stages of the design process to ensure that the end product is valuable in terms of usage and experience for the user. Typical of a UCD process is that it is *iterative*, i.e., the product is iteratively created, tested, improved, and refined. It generally starts with a user requirements phase, in which users are interviewed or observed to gather requirements. Initial concepts are typically generated in brainstorm sessions. Next, a low-fidelity version, e.g., a paper prototype, of the intended design is created, which is evaluated with users. Results of such an evaluation are fed back into the design process, the concept is improved, and the process of creating and evaluating a new prototype is repeated. Typically, each cycle sees an improvement of the fidelity of the prototype, meaning that each time it increases in resemblance to the intended end product.

The literature in the field of (computer) Game Design advocates a similar approach: to adopt an iterative design process and to playtest often and early (Salen and Zimmerman 2003; Fullerton et al. 2004; Lundgren 2008). In this process, in contrast to the UCD process, game designers appear to be less concerned with involving their end users in all stages of the process; there is less emphasis on getting to know the user and gathering users' requirements. Instead, the concept generation phase is mostly attributed to the game designer, relying on his/her experience in this field. However, Game Design literature does emphasize the importance of iteratively designing the game, in combination with play-testing: as Salen and Zimmerman (2003) put it: "the act of play becomes the act of design." The general opinion is that the resulting play experience of a game cannot be predicted at the 'drawing'-table. A game designer designs the rules and mechanics of a game, but the resulting game experience is ultimately generated by playing that game (Costikyan 2002). As such, to be able to properly judge the game design, the game *must* be played. The design process proposed is to rapidly prototype a playable version of the game, starting with low-fi paper prototypes and increasing fidelity in subsequent iterations.

Our experiences in designing Head Up Games have generated insights into the design process of these games and how existing methods can, or cannot, be deployed in this process. We will discuss the value of a brainstorming session within the context of game design. Furthermore, we discuss several lessons learned from involving children, and also adults, in the process. Finally, and arguably the most significant lesson learned, we offer insights into the use of prototypes in the design process, more specifically regarding the fidelity of these prototypes.[1] We argue that to properly design for and judge the added value of novel interaction styles in Head Up Games, designers need to create working, interactive prototypes, in early stages of the design process, so that users can truly experience the gameplay.

In the next sections, all lessons will be described and illustrated with experiences we gained during the design of several Head Up Games.

2 Lesson 1: Idea Generation

A game design process typically starts with an idea generation phase. Inspiration can come from anywhere, at anytime, and there are many methods and tools available for kick starting the designers' creativity. One widely known method is brainstorming, either alone or in a team. Though issues have been identified that can reduce the effectiveness of brainstorming (Stroebe et al. 2010), when prepared well, a brainstorm session can definitely assist in generating concepts; Rossiter and Lilien (1994) provide a set of general principles for conducting successful brainstorm sessions. More specifically for gaming, Fullerton et al. (2004) give pointers for brainstorming (computer) games. In related work, many research projects in game design mention some form of brainstorming, e.g., (Kern et al. 2006; Valk et al. 2012).

Below we describe two brainstorm sessions that were held during the development of a Head Up Game because they illustrate possible benefits and challenges when brainstorming for these types of games. For the setup of the brainstorms we

[1] The notion of prototype fidelity was introduced by Virzi (1989) as a measure of how authentic or realistic a prototype appears to the user when it is compared to the actual service. Paper mock-ups of an interactive system are a typical low-fidelity prototype that allows a user to experience a simulated interaction and to help identify areas of improvement. The notion of fidelity was enriched in later years (McCurdy et al. 2006) to distinguish between different dimensions along which prototypes may seek achieve higher or lower realism of the intended design intent, namely visual refinement, breadth of functionality, depth (detail) of functionality, richness of interactivity, and richness of data model.

followed the principles of Rossiter and Lilien (1994), which are: (a) brainstorming instructions are essential and should emphasize, paradoxically, number and not quality of ideas; (b) a specific and challenging target should be set for the number of ideas; (c) individuals, not groups, should generate the initial ideas; (d) groups should subsequently join and refine the ideas; (e) individuals should provide the final ratings to select the best ideas, which will increase commitment to the ideas selected; and, (f) the time required for successful brainstorming should be kept remarkably short.

2.1 First Brainstorm Session

In the first brainstorm session eight designers participated, with backgrounds ranging from industrial design to game design. The session was organized as follows: first, as the participants did not know each other, we played a few games to familiarize them with each other. As the participants were not familiar with the concept of Head Up Games, we introduced it to them. We asked the participants to individually think about the games they liked to play in their childhood. Next, the participants were divided in three groups and we asked them to discuss their childhood games and identify elements of these games that added to the appeal of the games. We then asked them to create a new outdoor game for children. The participants should at least provide details on how the game could be won, though more details, like specific game rules, how many players, where to play, etc., were also encouraged. For inspiration we provided them with a set of commercially available board and card games; the participants could use these, or any other game that they knew themselves, as inspiration. At the end of the brainstorm session each group presented their final game concept and each participant picked out his or her favorite concept.

The results from this brainstorm were game concepts on paper; most of them included elements of tag, hide-and-seek, capture the flag, or a combination of these. Furthermore, what we concluded from the game concepts generated in this brainstorm is that many games seemed to be fun already, even without technology. This insight prompted us to conduct a second brainstorm session, but change the setup; with the change we hoped to generate concepts that would more meaningfully include technology in the game.

2.2 Second Brainstorm Session

The setup of the second brainstorm session was similar to the first brainstorm session. However, instead of asking the participants to use childhood memories or existing games as inspiration, we gave them several possible technologies for outdoor games as inspiration. Based on earlier experiences designing Head Up Games we compiled a list of technologies that we deemed appropriate for outdoor use. They were: RFID, distance detection, accelerometer, and a rotation encoder. All participants of this workshop were industrial design students, who where familiar with these types of technology and also with participating in idea generation sessions.

The second brainstorm session we started, similar to the first brainstorm session, by giving the participants an individual task: all participants were seated around a round table, and in the middle of the table a set of papers was placed. Each paper was marked with one of the technologies. We asked the participants to randomly pick a paper from the table, quickly jot down a game idea on that paper, and put the paper back. These game concepts need not be very elaborate or detailed. Next they could take a new paper and repeat the process. If the paper already contained an idea of one of the other participants, the participant could either start a new idea, or add to the idea already on the paper. After approximately half an hour many ideas had been generated this way.

Next, we grouped the participants in pairs of two. We provided them with two of the papers with ideas from the previous exercise and asked them to discuss and take inspiration from the strongest ideas to create a new, detailed concept of a game. This exercise took around 10 min, next we regrouped the participants and provided them again with two of the papers of the previous session and repeated the process. Afterwards, the game concepts were presented, discussed, and rated.

2.3 Reflecting on Idea Generation

The brainstorm sessions rapidly generated many ideas and concepts. However, we observed that in the first session some of the ideas appeared fun enough by themselves, without needing to add technology. This is not really surprising, as we did not explicitly ask participants to consider technology. In the second session we did ask participants to take technology into account in their game design. It turned out that in this case some of the ideas suffered from a "technology push," in that the games would have been fun when taking the technology out and replacing it by a nontechnical counterpart. We conclude from this that it is difficult to meaningfully include technology in game designs and we attribute this to the fact that the participants had never experienced such a novel form of outdoor play before.

Furthermore, we observed that ideas generated in the brainstorms were very extensive with regard to the number of rules and details. From a game design perspective this is undesirable; from our experiences creating Head Up Games we know that games typically do not benefit from having many rules. However, from a brainstorm perspective it is a good outcome: apparently the context of the brainstorm sessions allowed the designers to continuously create and expand on concepts. We need to keep in mind that concepts are not games yet; they serve as inspiration for designers. In that process, the observation that participants too easily add rules to game concepts is important to acknowledge and take into account when further developing concepts into games.

3 Lesson 2: Involving Children Early in the Design Process

The User Centered Design (UCD) process focuses on the user's wants and needs for interactive technology. To gather the user's requirements in an early phase of the design process, many methods are available: for example, user surveys (questionnaires, interviews) or observations can be conducted. The role of the user in these methods is more or less passive, in that he or she only reacts to what the designer proposes. A more active participation of children in the design process is proposed by Druin (1999): the Cooperative Inquiry methodology is a set of techniques that put children in the role of co-researchers or co-designers. Scaife et al. (1997) put forward the notion of *Informant Design*. Although in this framework children are not seen as co-designers, they are acknowledged as valuable participants who contribute to the design process: children and adults can work together in design activities to generate input for the various stages of the design process.

For eliciting children's requirements we have mostly involved children as informants; here we describe three methods we have applied in various projects designing Head Up Games.

3.1 Mission from Mars

Dindler et al. (2005) developed the method "Mission from Mars" to gather user requirements specifically for the design of children's technology. The method aims to create a shared narrative space that allows researchers to get insights into the user requirements in an informal, fun setting for the children. First, the narrative is established: children are introduced to the story of a "Martian" who is eager to learn more about a specific subject; the Martian is ignorant on this subject because it is nonexisting on Mars. Naturally, this subject is chosen such that it generates useful information for the researchers. As the Martian thinks that children on Earth are more knowledgeable on the subject, the Martian wishes to have contact to discuss and learn from them. Second, supported by the researchers, the children prepare for the encounter with the Martian. Finally, the children have the encounter with the Martian: a setup is installed, where the children can hear the Martian only, though the Martian can both see and hear the children, so that they can show what they have prepared. Practically, this means that the children are facing a video camera during the encounter; the children talk to a video camera and get feedback from the Martian through a set of speakers (see Fig. 1). That signal is forwarded to the room of the Martian, where the researcher acting as the Martian can respond to the children. The voice of the Martian is distorted, to make him sound more "believable."

The main reason for engaging in such an elaborate setup is to place children in the *expert* position, in which they feel free to share many details. The setting

Fig. 1 The Mission from Mars setup. *Left* the interview room for the children. *Right* the room with equipment for the Martian

allows the researcher to ask 'stupid' questions about details that would have been impossible to ask in a conventional setting. For example, during the development of *Camelot* (Verhaegh et al. 2006), the Martian asked the children what a ball was, to which the children gave a serious and elaborate answer. In a post hoc interview one of the children mentioned that in a conventional interview setting she would not have provided this level of detail, because she assumed adults to know what a ball is.

Dindler et al. (2005) used the method to gather insights for the creation of 'eBag', an electronic school bag. They applied the method with children 10–11 years. During the development of *Camelot* we applied Mission from Mars to obtain information on what games children prefer to play. We applied the method with children aged 7–9. Similar to Dindler et al. we concluded that indeed a significant amount of information is gathered using this method. Furthermore, Dindler et al. reflected on the credibility of the Martian narrative. Some of the children did not believe the story about the Martian to be true; however, this did not have an effect on the outcome of the study because the children played along anyway. In contrast, we observed that children from a younger age group did believe the narrative, and though the majority of the children enjoyed participating and communicating with the Martian, some of the children were quite anxious about meeting the Martian. This indicates that one needs to consider for what age group this method is appropriate: for younger children the method could arguably be too intimidating or at least to design the whole method to be more reassuring and comfortable for children, e.g., making contact with a more comforting or familiar character like an animation film hero.

3.2 Collage Making: A Creative Exercise

Another early user requirements gathering method is KidReporter (Bekker et al. 2003). In the KidReporter method, children are asked to undertake various activities that result in creating a newspaper with children's ideas on a certain topic. For example, children could take pictures and describe why they took these

Fig. 2 Making collages as preparation for the encounter with the Martian

pictures and what is on them. Furthermore, children could interview each other, reporting on that, or independently write an article about a topic. The KidReporter method inspired us to do a similar activity during the Mission from Mars method: to inform the Martian about the games children liked, we asked the children to create a collage of their favorite games. As a preparing activity, we gave the children small cameras so they could first take pictures of their favorite games and subsequently use these pictures in their collages.

This idea worked out well: the children really made an effort to take photos of their favorite games; most children documented the games they were playing that afternoon at school, and some children went as far as to stage all their favorite games after school hours so they could take photos of them. What we did not anticipate was that the act of making the collages, which we did the next day at school (see Fig. 2), would generate a considerable amount of information. Each group was guided by an adult whose main role was to make sure that children were kept focused on the activity. It turned out that, while the children were engaged in making the collages, they were very open to elaborate verbally on details of their favorite games. It was very easy for us to unobtrusively pose many questions to which the children answered freely. We attributed this to the fact that for the children the main activity was to make the collage, which they enjoyed, and they did not feel as though they were being interviewed.

During the development of a series of Head Up Games (see later in this chapter) we decided to again create collages with children to gather information. The main difference with the session described above was that this time we did not include the photo making activity—instead we brought crafting materials. Furthermore, the context was different: instead of children at school, we engaged with children of a Scouting group in the Netherlands.

Based on our previous experience with making collages with children we expected it to be a good opportunity to simultaneously interview the children. However, totally unexpected, this time our experience was very different: in contrast to the school children, the scouts did not enjoy the activity, fooled around a bit and did not provide us with any information.

Reflecting on this we argue that the context in which we executed the activity has a significant influence on the proceedings. At school, children are normally required to behave calmly and an activity such as making a collage is a welcome deviation from the normal school routine and thus perceived as fun. In contrast, children go to a scout meeting to be playful, active, and engage in games and play outdoors; it is a venue for the children to release pent up energy from a week's worth of attending school. In that context, an activity as collage making, which required the children to sit at a table and be relatively calm, was *not* seen as fun.

Furthermore, our experiences show that information can be generated at unexpected times: while preparing the Mission from Mars, we had not expected that making the collages would give us this much information; we merely saw it as a means for the children to prepare for the session with the Martian.

3.3 *Observing Children at Play*

In most of the design processes for Head Up Games, we spent time observing children's free play in their natural context. In our experience this is a necessary activity, at least for inexperienced game designers. Though the methods described above will result in more and detailed information, they also take more time to prepare, execute, and analyze; time that is not always available. However, we argue that it is important for a designer to familiarize him- or herself with the target audience, and observing children at play is a way to gather insight into the types of games they play, the language they use and the context in which the games are played. Not only will it give valuable insights for the design process, it will also help the designer/researcher to better prepare for evaluations with children of prototypes of games later on.

3.4 *Reflecting on Involving Children Early in the Design Process*

We have shortly highlighted three methods for gathering insight. What method best suits a design process depends on several factors: the amount of time available in the process, the desired level of involvement of the children, and, from a practical point of view, the accessibility to children. A method like Mission from Mars requires a substantial effort in time and resources to execute and we have seen that the method's suitability depends on the age of the children. Then again, if the aim of the whole process is to involve children as design partners or informants it is worth the effort to spend time with the children to build up a relationship for subsequent encounters.

Similarly, we argue that an activity as collage making can also be deployed as a requirements gathering tool, and arguably as a relationship catalyzer; though our experiences suggest that the context in which the activity is conducted must be carefully considered. Spending time with the scouts in a shared activity that better matches the scouts' context arguably would have been more informing for the design process.

Based on our experiences with Mission from Mars and the collage making activity, we argue that spending time with children in a fun, creative activity, or a shared narrative can provide valuable insights for a designer. In general, it is advised to create a fun experience for children when involving them in a design process, e.g., (Markopoulos et al. 2008; Gielen 2008). We add to this observation that it is necessary to carefully select the right activity that matches the children's context.

Finally, we acknowledge the fact that given the time frame of a design process it is not always possible to actively involve children, or alternatively, it is not possible to find a venue that allows for such active cooperation. For example, we found that it is not always easy to find a school willing to cooperate given the involvement we ask from them. Regardless the (desired) involvement of children, we feel that, especially for inexperienced Head Up Game designers, an effort should be made to at least (passively) observe children at play.

4 Lesson 3: Playtesting with Adults

A key element of user centered design is to involve representatives of the target user group, in our case children, and have them test intermediate prototypes with the aim to converge iteratively at a successful design. Such an involvement of children takes time and effort to arrange, and to ensure that the children's time is well spent, it is important that as many as possible glitches in the games have been already been identified earlier. Therefore, during the development of many of the Head Up Games we designed, we asked adults to playtest intermediate prototypes. It is commonly acknowledged that we should not treat children as miniaturized adults, and as such evaluations with children will benefit from methods especially designed for children (Markopoulos et al. 2008). In doing so, it is worth considering to what extent we can treat adults as oversized children for the purpose of evaluating Head Up Game prototypes. Comparing our experiences of evaluating both with children as well as with adults we observed the following.

First, the behavior of adults before playing the game was different from children. Before the game children often behaved excited, eagerly anticipating the gameplay. In contrast, adults acted placidly and seemed less excited about playing a game, or at least did not express this. Furthermore, we observed that adults patiently listened to our explanation of the game rules and game details, while at least half of the children did not bother to listen to the details once they had grasped the main goal of the game.

However, the moment a game started, instantly the behavior of adults changed and closely matched the behavior of children during gameplay: both groups became physically active, there was social interaction (shouting, cheering etc.) and adults responded similar to breakdowns in a game as children did. For example, in evaluations of *F.A.R.M.* (Soute et al. 2013, see also Textbox 1) we did not set a rule for the starting distance between the one player and the rest of the players. Adults responded in a similar fashion to this as children did; both commented that "it was unfair" if the distance was too close and both groups resolved the issue within seconds (see also next section).

F.A.R.M. (Finding Animals while Running and Mooing) is an individual chasing and collecting game. At the start of the game, each player receives an assignment to collect a set of animals, e.g., a cow and two horses. The player who first completes his assignment wins the game. Players take turns in being the "farmer". At the start of a turn the farmer gets assigned an animal, which can be won by other players if they tag the farmer within 10 s. Players are allowed to trade animals to better match their assignment.

Textbox 1: F.A.R.M

After playing the game, when we asked players for feedback, there was again a noticeable difference between adults and children: adults were more fluent in providing feedback than children were, which is not really surprising. Children have not yet properly developed the ability to reflect on a meta-level and/or simply lack the vocabulary to do so (Markopoulos et al. 2008). Furthermore, there was a difference in the type of feedback given. Children mostly reflected on actual events of the playtest; though children generally did not have problems to detect and fix "broken" game rules, they did have trouble to give feedback on the game at a more abstract level. Adults did not have trouble doing this and also commented more on the tactics of a game; they readily provided many more rules that they thought would enhance the gameplay. Nevertheless, adding rules does not necessarily improve the game: perhaps due to the fast pace of many of the games. There simply is less time to consider all these rules while playing the games. Also, children seemed to appreciate other challenges in games than adults: adults put more emphasize on developing play tactics in the game, and also favored rules that would support this. After playtesting *F.A.R.M.* the adults suggested to add more tactics to the game, for example to allow players to trick other players into losing their animals. In contrast, children seemed to be less concerned by this; and were less inclined to listen to the game rules at the start of the playtest. Interestingly, this did not seem to have a big influence on the gameplay. Not understanding all the rules while playing a game inevitably did result in confusion for some of the children, but in general they would just continue, figuring out most rules while playing. This observation was reflected in the informal interviews: when asked what children favored most in the games, some children referred to the physical activity, other children mentioned the fact that they were playing this game with their best friend. So in contrast to the adults, children did not seem to have a need for more (tactical) challenges in the games. Similarly, Sellen et al. (2009) concluded that the reaction to the same concepts can differ due to differences in age of the user groups. Thus, the play experience of adults may not be representative of how children would experience a game. Arguably, this is be expected since playing has a different importance to each.

Summarizing, we observed that it is indeed beneficial (and practical) to have adults playtest the games; the behavior adults display during play is similar to children's behavior in terms of physical activity, social interaction, and reaction to the game devices. Observing adults playtest the game can therefore identify usability issues with the game devices (e.g., sounds being not clear enough) or issues concerning the rules (e.g., when situations occur in which the rules are inconclusive, or conflicting). However, we would certainly not advise testing with adults only; though the behavior of children and adults is similar, children experience and value games differently than adults, and this cannot be revealed by testing with adults only.

5 Lesson 4: Tapping into the Children's Tacit Knowledge on Well-Played Games

Game Design literature, e.g., (Salen and Zimmerman 2003; Fullerton et al. 2004), states that it is impossible to design all the rules and mechanics of a game and predict the emerging game experience without playtesting. Therefore, Game Design literature stresses the importance of an iterative design process, in which designers playtest the games; based on the observations designers can improve the gameplay.

DeKoven (2002) argues that players implicitly know when a game is played "well." He states that a "well-played game" is impossible to define, as it is dependent on too many variables. However, the *experience* of a well-played game is familiar to every player. Hughes (1983) makes a similar observation: children intrinsically aim to play "nice," e.g., it is implicitly agreed that players will not hurt each other. Furthermore, Hughes, and also Salen and Zimmerman (2003) suggest that some rules are implicit, ingrained by the social context in which children play the games. For example, a child playing a game like *F.A.R.M.* with his younger brother would allow the younger child more leeway than he would were he playing with his best friend, who is roughly of the same age.

A designer can make use of this implicit tendency to "play well." First, we acknowledge the fact that a designer is not able to predict the game experience beforehand, and therefore is not able to design a definitive set of rules for a game in a single iteration of the design. Thus, we propose to purposefully design a limited, basic set of rules only. It is possible that during playtesting situations will arise that will "break" the game, because the basic set of rules is insufficient. If such a situation arises, we propose to rely on the children's tacit knowledge of a well-played game and their ability to come up with a new or changed rule to fix the gameplay. If possible that rule will take effect immediately, which allows us to instantly reflect on the suitability of the rule.

During the design process of the Head Up Games we have encountered examples that this way of working is indeed useful for informing the design process. For example, while designing *F.A.R.M.* we did not explicitly state in the rules what the starting distance between the players should be. At the start of the game, it immediately became clear to the players that this distance had a big influence on the chances of winning for the player who was the *farmer*. Players commented on the unfairness of this situation and we discussed with the players how to improve this. The players suggested giving the *farmer* some leeway; they argued that this was common in other games as well, and this largely solved the issue as we experienced immediately during the subsequent playtest.

Another example occurred during the evaluation of *Save the Safe* (Soute et al. 2009a, b). In *Save the Safe* two teams compete to capture a key that opens a safe to win the game. At the start of the game the key is randomly assigned to one of the players. We compared two types of gameplay: one with a digital (virtual) key and one with a physical key (a ball). Unexpectedly, the game with the physical key, the ball, ended very rapidly, because the first player grabbed the ball and sprinted to the safe to end the game. Immediately, the opposing team started protesting that this was "unfair," since "you are not allowed to walk with the ball!" In fact, we had not imposed any rule stating such a thing, but many ball games indeed have such a rule: the player who has possession of the ball is forbidden to move. After a very short discussion—the winning team, at first reluctantly, agreed, since they saw too that there was no fun in playing a game that ended this abruptly—we agreed to impose the rule (not walking with the ball) for this game.

5.1 Reflecting on Tapping into Tacit Knowledge

Concluding, we argue that we can make use of the observation that children are in fact domain-experts to our advantage for informing the design process. However, we should keep in mind that children are domain-experts regarding gameplay, though not regarding technology. It is mostly impossible for children to comprehend in what ways technology can be used in the game; and this can result in either children not being able to imagine novel interaction styles in games, or alternatively, children imagining game interactions that are technically infeasible. By having children create rules and immediately play them, we are certain that these rules are playable. Still, the "blue sky" suggestions of children, combined with observations of children playing the game, can provide valuable hints to a designer on what direction to take in the game design process.

6 Lesson 5: Using Prototypes

The generally accepted approach for designing products in HCI and Game Design is to start with low-fidelity prototypes that, through subsequent iterations, gain in fidelity and start to resemble the intended product more closely. In the design process of early Head Up Games we adopted this approach. For example, during the development of *Camelot* (Verhaegh et al. 2006) we playtested three game concepts using simple paper cards and boxes that represented some of the game ideas. Similarly, during the development of *F.A.R.M.* (Soute et al. 2013) we playtested the game using paper prototypes. Although these evaluations were successful at first sight in that they provided a considerable amount of insight in the gameplay, the question arose whether or not the information gathered using paper prototypes was valuable for informing the design process of Head Up Games.

One of the challenges for the evaluation is the high pace of the games; often, the games are designed such that, at least in parts of the game, players are running or chasing each other. Using low-fidelity prototypes (e.g., paper prototypes) in a playtest can seriously disrupt the intended flow of the game, or at least alter the game mechanics to such an extent that it is no longer valid to compare the experience of playing with a noninteractive prototype to the experience of playing with an interactive prototype. As a consequence, the feedback generated with the nontechnical prototypes is reflecting on irrelevant game mechanics, which results in the design process optimizing towards a game that is playable as is, i.e., without interactive technology. Subsequently integrating technology degrades new interactive features to unconvincing post hoc add-ons that do not integrate well with the game.

Therefore, to evaluate the impact of interactive technology on the game experience of outdoor games, we argue that high-fi prototypes should be employed in an early phase of the process. Furthermore, the games we intend to design are to be played by multiple players in an outdoor context. As a consequence, not only the player technology interaction plays a role in the emerging game experience, but also the player–player interaction and player context interaction have a significant impact on the game experience (see Fig. 3). For that reason we argue that the games should be evaluated in situ, i.e., in a context where children normally play games, as opposed to a lab setting. This poses high demands on the robustness of the prototyping medium.

In similar games where the design process is disclosed in related literature (e.g., *Ambient Wood* by Rogers et al. (2004), and *StarCatcher* by Brynskov and

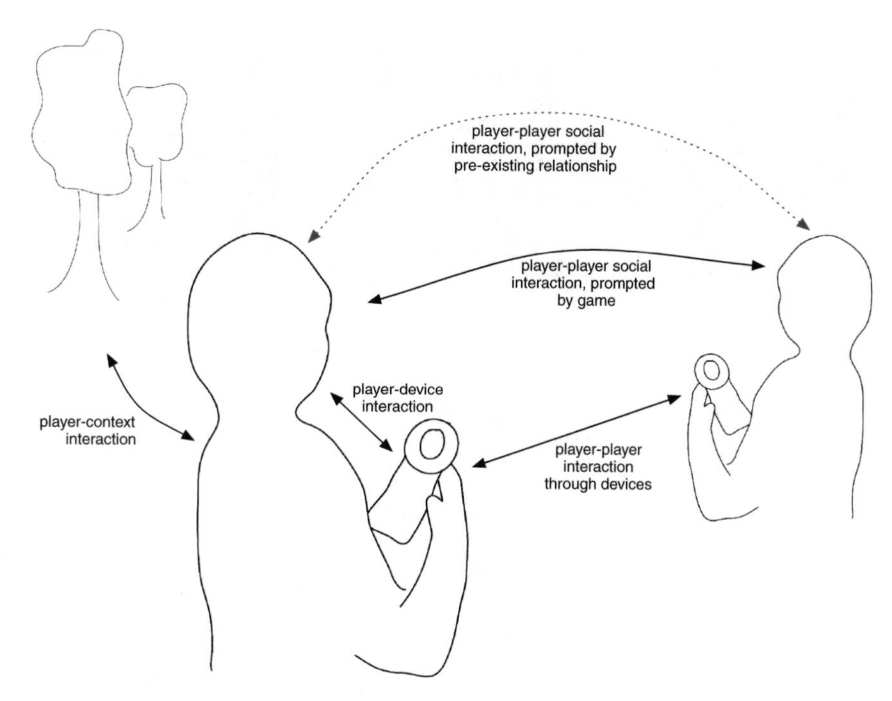

Fig. 3 Types of interaction in multiplayer, outdoor games

Ludvigsen (2006)), it is only reported in the last design iteration that a partially or fully working, playable prototype is created that covers a reasonable part of the game mechanics. Most commonly in this field, authors report only the final game design, how that was evaluated by users and suggested potential improvements of the 'final' game. None of the papers report or reflect on an iterative process for improving the game design using interactive prototypes.

To enable a rapid, iterative design process using interactive prototypes we developed the RaPIDO (Rapid prototyping of Physical Interaction Design for Outdoor games) platform. Early Head Up Games had been developed from scratch each time—a time-consuming activity. Thus, based on the earlier experiences of building prototypes we designed the RaPIDO platform to include useful technologies for outdoor games. The platform consists of a set of player devices (see Fig. 4) that offer several modes of interaction: e.g., visual, haptic, and auditory, see Table 1.

Besides offering the hardware, the platform includes software libraries to easily address all hardware components and program new interactive games. Furthermore, the exterior of the devices is robustly designed, to survive multiple outdoor evaluation sessions with children.

To evaluate the use of RaPIDO and its impact on an iterative game design process, we engaged in a study in which we iteratively designed several Head Up Games for children. Our approach shows similarities to the RITE method (Rapid Iterative Testing and Evaluation) by Medlock et al. (2002); the RITE method

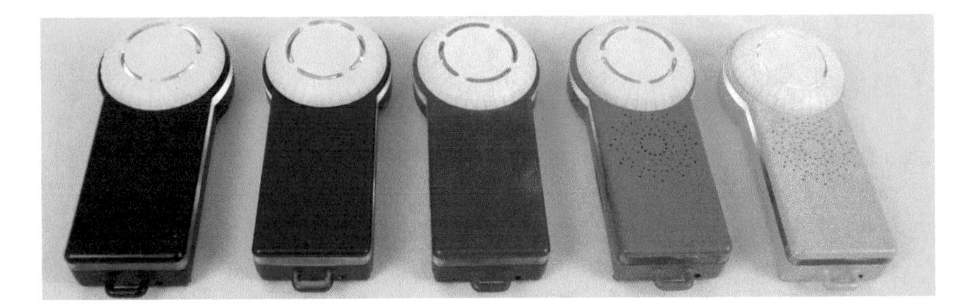

Fig. 4 RaPIDO player devices

Table 1 Main components of RaPIDO

Technology	Interaction style
4 RGB leds	Provide visual cues, e.g. by blinking, or changing color
Sound chip + speaker	Provide auditory cues, can read and playback .wav files from SD card
RFID module	Detect objects tagged with RFID-tags
XBEE module	Provides: (1) inter-device communication (2) distance measurement between devices
Vibration motor	Provides tactile feedback
Rotation encoder	Measures degree of rotation of wheel
Accelerometer	Measures movements

advocates an iterative approach in which iterations are executed at an extremely high pace. Problems identified in the interface (in Medlock's case a tutorial of a popular game) are immediately changed, and more importantly, immediately evaluated, sometimes even within hours of implementation.

The games were repeatedly playtested with children (7–10 years) of a Dutch scouting association in an outdoor context. Details of the study can be found in Soute and Markopoulos (2013). Reflections on the iterative design process are discussed next.

6.1 Reflections

The experience of creating and evaluating several games has generated insights into different levels, namely (1) on the rapid, iterative, design process (2) on evaluating with children in this particular setup, and (3) on what interaction mechanisms and technology are appropriate for interactive, outdoor games for children.

6.1.1 Design Process

Most design changes that we implemented based on our direct observations of the gameplay and comments of the children, concerned directly the play and interaction functionality. Seemingly small details, like the duration of some interactions, influenced heavily the emerging game experience, showing the inadequacy of evaluating a mock-up of the game interactions for example when playtesting functionality with Wizard of Oz interventions. For example, in *Save the Safe* (Soute et al. 2009a) a 'key' is represented virtually, using tactile feedback. It is easy to see that replacing the virtual element by a physical object would alter the game: a physical object is clearly visible to the other players, especially when passing it around between players, so the element of guessing which player actually possesses the key (as is present when the key is virtually represented) is taken out of the game completely. And particularly that feature of the game turned out to be the most fun part. Thus, we argue that instead of playtesting with paper prototypes, for developing Head Up Games it is best to immediately focus on the actual, working, interactions.

With regard to the time it earlier cost to develop a working prototype, we conclude that with the RaPIDO platform we were able to bring this time drastically down. Over the course of 6 weeks we were able to develop, evaluate, and improve on four games. Furthermore, because it was relatively fast and easy to create working prototypes, the platform allowed us to play around with the technology, and thus freely explore the design space.

In our design process, we decided to keep a high pace in developing new iterations of the games—typically we took 1–2 weeks to develop new iterations. As a consequence, little time remained to analyze the results of the evaluations (e.g., run a structured observation, or content analysis of the interviews). Further, the interviews did not yield as much information for improving the games as we had expected; directly observing the gameplay was much more effective. Nonetheless, the little information that was deduced from the interviews was useful for corroborating our findings from the direct observations.

Another benefit of rapidly iterating over small changes is that it becomes easier to observe the impact of a small feature change. We argue that this way the design process becomes a "self-steering" process: if based on an observation a wrong conclusion is drawn and subsequently a wrong decision regarding the game mechanics is implemented, the next session will immediately show the (negative) effect and the design decision can be undone quickly.

Testing early and often in the design process makes sure that as a designer you do not "fall in love" with your own (features of the) games. After only 1 week of implementing a game, it is much easier to toss a feature in favor of an improved version or abandon a game altogether. In contrast, if one has taken months to implement a game, it is much more difficult to part from it, if at an eventual user test it turns out that certain features do not work out as expected.

6.1.2 Iterative Testing with Children

We found that repeated evaluations with the same group of children have a few distinct advantages: first, we got to know the children, which made it easier to interpret observations. For example, a child that behaves in a certain way may or may not do that as a consequence of playing the game and it is relatively hard to tell the difference from a single observation only. However, when observing the same children over time, as a (game) designer it becomes easier to tell which behaviors can be attributed to a child and which might be the result of playing a certain game.

The second advantage is that the children got to know us and because of that gained confidence and were at ease in their interactions with us. An often-argued side effect of the children getting acquainted to researchers is that the power imbalance that might exist between a child and an adult (Hennessy and Heary 2005; Markopoulos et al. 2008) is lessened. In fact, we even experienced this to the extreme; the power imbalance was *reversed* in a sense that we had to try hard to assert ourselves on the participants, simply to get and hold their attention. We attribute this to two causes: first, as we indicated, the children became familiar with us; and second, because the children clearly outnumbered the researchers, children did not feel at all intimidated by the two researchers (in contrast to evaluations where only one or two children are present). In our case, we eventually had to claim a leading role in order to quiet the group down, and make sure they were all paying attention. This did not seem to have a negative impact, possibly because the children equated us to their scout leaders and they too addressed the children in this manner.

This brings us to another observation: when observing "in the wild" (i.e., in the user's natural context, e.g., observing children playing at their regular playground) it is important to adjust to the context of an evaluation (see also Rogers (2011)), and more specifically how an evaluator should interact within that context. In our case this meant that we positioned ourselves in the roles of scout leaders. Related to our observation above is the notion that an evaluation method cannot simply be transplanted from one context to another. Certain patterns of behavior have been established between the children and the scout leaders and as a researcher we argue that you should be aware of this and plan your evaluation accordingly. An example of this is the observation that using collages to elicit information from children as a method did not work well in this particular context, simply because the children were not used to sedentary activities within this context. In contrast: we have applied the same method earlier in a school context where it worked well.

The age of the children ranged from 7- to 10-year old. This is something we had not anticipated, but in the end had to adjust to: for some games it might occur that the challenge for a 7-year old to compete with a 10-year old becomes too high, resulting in a negative game experience for the 7-year old, and maybe even for the 10-year old, as the competition is too low for him. We observed this during the playtest of one of the games, and later adjusted for this by not randomly mixing the children but instead sorting them by age group. Then the chances for winning the game became more equal for all players, resulting in a better game experience.

6.1.3 Interaction in Outdoor Games

In the games we designed we used a variety of interaction styles and technologies, but the one technology most commonly used was radio communication. We used it for two purposes: for communicating game events between the RaPIDO devices, so they could appropriately respond to what was happening in the game with respect to other players. Further, we used radio technology for getting a rough estimation for distances between devices (and thus players). Both features contributed significantly to the novelty of gameplay, as it allowed us to introduce elements in the games that have no similarity to features in traditional outdoor games. An example is the virtual key in *Save the Safe*, which was transferred between players based on proximity.

Furthermore, for feedback to the players we often used auditory, visual, and tactile cues. We found all modalities appropriate for supporting outdoor games, though that does not automatically imply that every style of using it is appropriate in the context of outdoor gaming. To give an example: in one game we needed to convey to the players how many steps they could take in a turn. At first we implemented it by letting LEDs blink, each blink accounted for one step. However, this enforced the players to be paying attention to their device at a specific frame of time within the game. Also, the information is volatile, once it is shown it is gone. So, a moment of distraction, for example when team players are talking to one another (i.e., engaging in social interaction, which we want our games also to encourage!) would result in the loss of game information. Therefore, we redesigned that part of the game to have the LEDs continuously shine; the number of LEDs switched on corresponded to the number of steps. This was a more persistent way of showing the same information.

Another technology we made heavy use of was RFID technology. Each device is equipped with an RFID reader, which allowed us to program the devices to detect objects tagged with an RFID tag. Though we used it moderately in the games themselves, we employed the RFID tags mostly for setting up the games.

Based on our experiences we conclude that the process we followed is very suitable for games, where the emerging game experience is not only a result of interactive technology, but also of the context the game is played in, and other existing game rules; these games need to be really experienced and cannot be tested with lower fidelity prototypes. Arguably, a similar process might be valuable for other interactive systems that are novel to users and are designed to change behaviors in users. To valuably generate feedback from users on such systems, they have to be tested and experienced "in the wild," meaning in context, with actual users using a product under realistic conditions rather than those which are anticipated by designers or that are easy for designers to work with.

7 Recommendations for Designing Head Up Games

In this chapter we have focused on the design process of Head Up Games. Based on our experiences we can now present several recommendations that can inspire and inform first-time Head Up Game designers.

The first recommendation is the most radical, as it deviates from the generally accepted way of involving low-fi to high-fi prototypes in the design process. Instead, we emphasize the necessity of using high-fi prototypes from an early stage of the design process: these games really need to be played with working technology to assess the effect of the game design and technology on the game experience. Instead of gradually increasing the fidelity of the prototypes, we went ahead and immediately created high-fi (with respect to interactivity) prototypes. We argue that, for games involving physical activity, outdoor play, groups of players and embodied interaction, it is virtually impossible to test with paper prototypes as the lack of interactivity distorts the game dynamics intended by the designer and leads to very different play experiences. Moreover, we argue that, if children are involved, it becomes more apparent, as children might be less able to reflect on the impact of interactivity and the resulting game dynamics without actually experiencing it.

Furthermore, we suggest starting the design process with designing and implementing a limited set of game rules and rely on the players' ability to detect and fix a broken game.

We also discussed the process of engaging children in creative activities to gain insight into children's requirements for games. The amount of information gathered is dependent on the effort put in the activities. The Mission from Mars method is very time-consuming, but offers a considerable amount of feedback. However, the method relies on a narrative that might not be suitable for all children. We furthermore discussed making collages with children. This method was a success in a school context, leading us to repeat it during evaluations with scouts. Unexpectedly, in that context the method failed, suggesting that the context of the evaluation plays a large role in its success rate. Finally, we suggest that at least designers should make an effort to observe children at play, if time is too limited to execute the methods described above for gathering children's requirements.

Finally, we recommend to playtest with children as often as possible. Our experience suggests that some issues can also be identified by adults. Indeed, we would recommend letting adults playtest intermediate prototypes of the games to root out early usability issues.

In short:

- Use high-fi prototypes from an early stage in the design process. Head Up Games really need to be played with working prototypes, most notably with regard to the game interaction, to assess the effect of the game design and technology on the emergent game experience.
- Start with an incomplete set of game rules. Rely on the players' innate ability to detect and fix a broken game to fill the gaps.

- Engage with children in a fun, creative activity as a way to facilitate discussion. Though be aware that ‘fun’ is context dependent.
- Adopt an iterative process and playtest with children as often as possible. To prepare for these sessions, and/or to test intermediate designs, playtest with adults too.

References

Bekker M, Beusmans J, Keyson D, Lloyd P (2003) KidReporter: a user requirements gathering technique for designing with children. Interact Comput 15:187–202. doi:10.1016/S0953-5438(03)00007-9

Benford S, Rowland D, Flintham M et al. (2005) Life on the edge: supporting collaboration in location-based experiences. In: Proceedings Chi 05. ACM, Portland, Oregon, USA, pp 721–730

Brynskov M, Ludvigsen M (2006) Mock games: a new genre of pervasive play. In: Proceedings 6th conferences Designing Interaction systems ACM, University Park, PA, pp 169–178

Costikyan G (2002) I have no words and I must design: Toward a critical vocabulary for games. In: Proceedings of the computer games digital cult conferences, pp 9–33

DeKoven B (2002) The well-played game: a playful path to wholeness. Writers Club Press, San Jose

Dindler C, Eriksson E, Iversen OS et al (2005) Mission from mars: a method for exploring user requirements for children in a narrative space. In: Proceedings of 2005 Conference Designing Interaction for Children, pp 40–47

Druin A (1999) Cooperative inquiry: developing new technologies for children with children. In: Proceedings of SIGCHI conference on Human factors in computing systems. Chi Is Limit. ACM, Pittsburgh, Pennsylvania, United States, pp 592–599

Fullerton T, Swain C, Hoffman S (2004) Game design workshop. Focal Press, San Francisco

Gielen MA (2008) Exploring the child’s mind—contextmapping research with children. Digit Creat 19:174–184. doi:10.1080/14626260802312640

Hennessy E, Heary C (2005) Exploring children’s views through focus groups. In: Greene S, Hogan D, (eds) Researching children’s experience Approaches Methods. Sage Publications Ltd, Thousand Oaks, pp 236–252

Hughes L (1983) Beyond the rules of the game: why are Rooie Rules nice? In: Manning, F. (ed) The world of play Leisure press, New York

Kern D, Stringer M, Fitzpatrick G, Schmidt A (2006) Curball - a prototype tangible game for inter-generational play. IEEE Computer Society, Los Alamitos

Lundgren S (2008) Cover story - designing games: why and how. Interactions 15:6–12. doi: 10.1145/1409040.1409042

Markopoulos P, Read JC, Macfarlane S, Höysniemi J (2008) Evaluating Children’s Interactive Products: Principles and Practices for Interaction Designers. Morgan Kaufmann, San Francisco

McCurdy M, Connors C, Pyrzak G, et al. (2006) Breaking the fidelity barrier: an examination of our current characterization of prototypes and an example of a mixed-fidelity success. In: Proceeding of Sigchi In: Proceeding of Sigchi conference on human factors of computer system ACM, New York, pp 1233–1242

Medlock MC, Wixon D, Terrano M, et al. (2002) Using the RITE method to improve products: A definition and a case study. Usability Professionals’ Association

Rogers Y (2011) Interaction design gone wild: striving for wild theory. Interactions 18:58–62. doi: 10.1145/1978822.1978834

Rogers Y, Price S, Fitzpatrick G, et al. (2004) Ambient wood: designing new forms of digital augmentation for learning outdoors. In: Proceeding of IDC 04. ACM, Maryland, pp 3–10
Rossiter JR, Lilien GL (1994) New "brainstorming" principles. Aust J Manag 19:61–72
Salen K, Zimmerman E (2003) Rules of Play: Game Design Fundamentals, Illustrated edition. The MIT Press, Cambridge
Scaife M, Rogers Y, Aldrich F, Davies M (1997) Designing for or designing with? Informant design for interactive learning environments. In: Proceeding of Sigchi conference on human factors of computer system ACM, Atlanta, Georgia, United States, pp 343–350
Sellen KM, Massimi MA, Lottridge DM, et al. (2009) The people-prototype problem: understanding the interaction between prototype format and user group. In: Proceeding of Sigchi conference on human factors of computer system ACM, New York, pp 635–638
Soute I, Lagerstrom S, Markopoulos P (2013) Rapid prototyping of outdoor games for children in an iterative design process. In: Proceeding of 12th international conference IDC, ACM, New York, pp 74–83
Soute I, Kaptein M, Markopoulos P (2009a) Evaluating outdoor play for children: virtual vs. tangible game objects in pervasive games. In: Proceedings of the 8th International conference on interaction design and children. ACM, Como, Italy, pp 250–253
Soute I, Markopoulos P, Magielse R (2009b) Head up games: combining the best of both worlds by merging traditional and digital play. Pers Ubiquitous Comput 14:435–444. doi:10.1007/s00779-009-0265-0
Stroebe W, Nijstad BA, Rietzschel EF (2010) Chapter four—beyond productivity loss in brainstorming groups: the evolution of a question. In: Zanna MP, Olson JM (ed) Advance exports society psychology Academic Press, pp 157–203
Valk L de, Rijnbout P, Bekker T et al (2012) Designing for playful experiences in open-ended intelligent play environments. In: Proceedings Iadis international conference on game and entertainment technologies lisbon. Portugal, pp 3–10
Verhaegh J, Soute I, Kessels A, Markopoulos P (2006) On the design of camelot, an outdoor game for children. In: Proceedings 2006 conference on interactive designing child ACM, Tampere, Finland, pp 9–16
Virzi RA (1989) What can you learn from a low-fidelity prototype? In: Proceedings human factors ergonomics society annual meeting, pp 224–228

Smart Ball and a New Dynamic Form of Entertainment

Sachiko Kodama, Toshiki Sato and Hideki Koike

Abstract This chapter introduces a smart ball entertainment system, which we call "Bouncing Star." A smart ball is a ball that contains electronic devices such as sensors, LEDs, microprocessors, and wireless modules. It has the ordinary functions of a ball, but advanced functions are achieved by the combination of various electronic devices. In our system, the ball's state of motion (static, thrown, bounce, etc.) is recognized by the analysis of data received via a wireless module, the ball's position can be tracked using image processing techniques. Using this system, we created several applications that integrate real-time dynamic computer graphics with responsive sounds, which we have exhibited in museums where people can participate in ball play. The goal of our project was to establish a new dynamic form of entertainment based on the combination of a ball and digital technologies.

Keywords Ball interface · Wireless module · Sensing technology · Image processing · Augmented sports · Computer-supported cooperative play · Interactive art · Digital sports

1 Introduction

Since ancient times, many people throughout the world have used balls as entertainment equipment. The design of balls has become more sophisticated to facilitate their use as sports equipment in soccer, baseball, and many other sports enjoyed by people.

S. Kodama (✉) · T. Sato · H. Koike
University of Electro-Communications, Tokyo, Japan
e-mail: kodama@inf.uec.ac.jp

A. Nijholt (ed.), *Playful User Interfaces*, Gaming Media and Social Effects,
DOI: 10.1007/978-981-4560-96-2_7,

Recently, small electronic devices have been installed inside balls. Thus, we can buy toy balls that contain light-emitting diodes (LEDs), which used to be sold on night markets. These toys can be considered as the source of smart balls, although they are too simple to be referred to as smart. These balls simply contained a light but no other advanced functions. Since the introduction of LED balls, technical breakthroughs have been transforming smart balls. It has become possible to use various small electronic devices, which were originally designed for smartphones, have been installed inside these balls.

We use the term *smart ball* to refer to a spherical (or hemispherical) ball that people can use for sports or play, which contain small electronic devices that provide new functions. These devices affect the appearance of the ball (such as LEDs embedded in a ball) but they can also sense, process, and store data, and communicate information externally. These functions make balls "smart."

Our team has operated the "Bouncing Star" smart ball project since 2007 (Kodama et al. 2007). Kodama initiated this project as an artist, director, and inventor. Kodama and Izuta developed a new ball in 2007, after which we made improvements to the ball and the interactive system. This smart ball has been demonstrated at academic conferences and in museum exhibitions and it has been used for research purposes. Thus, we discuss the successes and problems encountered with our smart ball entertainment system based on experiments conducted during our six-year project.

2 Smart Ball and a New Dynamic Form of Entertainment

2.1 Background

A smart ball has the normal functions of an ordinary ball. It falls, rolls, spins, and can be grasped, thrown, kicked, hit, caught, bounced, turned around, etc. However, a smart ball can have additional functions based on the small electronic devices it contains. Smart balls have sensors that allow them to capture data from around/inside the ball and to sense the conditions of the ball. They can also be used as display devices to represent multimodal information. The design of the basic concept of the smart ball was guided by the definition of an "Organic User Interface" (OUI) (Vertegaal and Poupyrev 2008), and we consider that a smart ball is a type of OUI.

The addition of wireless communication technology to smart balls facilitates a new dynamic form of entertainment. This new dynamic form of entertainment is achieved by using a smart ball that communicates with a computer support system so the ball play is connected directly to the playing field. The field graphics and/or sound effects change according to the real-time state of the ball. The rules and processes of the game are connected directly with the ball's state. The people (in the playing field or stadium) and internet viewers throughout the world can participate in the game via the computer system. We consider that this new dynamic form of entertainment will facilitate artistic applications that merge sports and interactive art, thereby developing a new field of "digital sports." The goal of our project is to establish this new dynamic form of entertainment, which can be achieved by combining balls with digital technologies.

Pioneering research on ball play combined with digital technology produced PingPongPlus, which was developed by the Tangible Media Group at MIT (Ishii et al. 1999). This was a classical ping-pong game, which used an ordinary ping-pong ball and paddles. However, digital technology was embedded in the table to enhance the game. Their "reactive table" had eight microphones beneath the table that sensed the location of the ping-pong ball, which created a novel athletics-driven, tangible, computer-augmented interface that incorporated sensing, sound, and projection technology. A projector displayed patterns of light and shadows on the table top. The position of the bouncing ball was detected to leave images of rippling water. The rhythm of play was used to drive the accompanying music and visual effects. In PingPongPlus, the ball itself was an ordinary ping-pong ball, which had no electronic parts.

After PingPongPlus, several athletic-tangible interfaces were developed that used balls. Moeller and Agamanolis devised a system for playing "sports at a distance" via a life-size video conference screen using an unmodified soccer ball (Mueller et al. 2005). Rudorf and Brunnett developed a table tennis application, which allowed real-time tracking of the rapid movements of a ball (Rudorf and Brunnett 2005). Iyoda developed a VR application for pitching in baseball, which included a wireless accelerometer inside a "screen-shaped" split curtain equipped with IR sensors (Iyoda et al. 2006).

Adidas has been researching the construction of a new system for football games. In 2013, Adidas uses sensors and computers to communicate data between a ball and a smartphone.

As we mentioned above examples, there have been many attempts to develop new balls for sports, they are aimed mostly at enhancing traditional sports such as ping pong, baseball, soccer, etc. But also, we know several studies have tried to develop smart ball entertainment systems, which can be more flexible approach to create a new form of entertainment. The Japanese artist Kuwakubo produced the media artwork "heaven seed," which used an accelerator sensor and wireless module to generate sounds when a ball was thrown (Kuwakubo 2003). Sugano presented an augmented sports game called "Shootball," which used a ball equipped with a shock sensor and wireless module to conduct experiments with a novel, goal-based, sports game. Their system used multiple cameras and multiple projectors on the walls in the field (Sugano et al. 2007). At UPM, Torroja developed a ball device, which was an interface for musical expression that sent MIDI messages via a wireless connection to a computer. (In 2008, Kodama experienced Prof. Yago Torroja's ball at his laboratory at UPM while she was making a large ferrofluid installation work for the Reina Sofia Museum in Madrid). The American company Orbotix developed a robotic ball called "sphero," which people could control a motion of the ball on the floor by controlling a wheel mechanism inside the ball using their smartphone. Their product was released in 2011. Their function and playing style are different from our Bouncing Star ball.

2.2 Concept of the Smart Ball

Before describing the development of our system, we consider a ball's characteristics as a human tool, because balls are used in unique ways. We consider three important characteristics of a ball: (1) spherical shape, (2) predictability (in relation to its controllability), and (3) elasticity.

(1) The merit of the spherical shape is that it leads to instability in the position and movement of a ball. In general, it causes problems if a tool changes its position too easily. For a ball, however, this instability is beneficial for its use. Its 3-D symmetrical spherical shape makes it possible for a ball to move around (360°) because of gravity.
(2) A ball touches a surface via a very small surface area, so the friction is small enough to generate an unstable condition. However, if a strong force is applied to a ball, the ball's movement becomes predictable and controllable. These contradictory properties (random instability and controllability) merge and facilitate the popular ball games that we enjoy.
(3) The ball's elasticity is also an important aspect that causes feeling of joy when we play with a ball. The elastic texture elicits a pleasing sensation when we grasp a ball, which causes a feeling of unity between our body and the ball. This high elasticity allows us to dribble the ball and engage in rhythmic play by bouncing a ball on a wall or floor.

Figure 1 shows the generalized possible interactions with our hand and body using a ball. A smart ball works as an interface by acquiring real-time data from the ball, which can be used to generate data for control output devices and the ball itself when people operate the ball.

2.3 System Overview

We started the Bouncing Star project in 2007. We specified three necessary functions for our ball. They are the precise detection of the ball's bounce, the precise tracking of the ball's position, and the ball's durability to tolerate powerful shocks.

First, we developed a new ball, Bouncing Star (Hane-Boshi in Japanese), which contained electronic devices. At the same time, we developed an augmented sports system using Bouncing Star ball and computer programs to provide an interface between the digital and physical worlds. This program could recognize the ball's state of motion (static, rolled, thrown, bounce, etc.) by analyzing data received via a wireless module. The program also tracked the ball's position using image processing techniques. A high-speed camera was fixed to a point where it could capture the whole playing field and it acquired images of infrared (IR) lights inside the ball while it was on the playing field at a frame rate of 200 fps. The camera was equipped with an IR filter that only detected IR light. The corresponding computer graphics were generated by a PC and projected onto the floor (or ground).

Ball Control

Tap Touch
Turn Spin
Shake
Push
Grasp Hold
Straight move (with and without spin)
Ball throwing (with and without spin)
Ball catching (received)
Bounce (on the wall or floor)
Hit Kick

Natural Interaction

Ball and Playfield

Light
Sound
Music
Text
Image
Graphic
Haptic sensation
Shape change

Fig. 1 Generalized concept of the smart ball interface

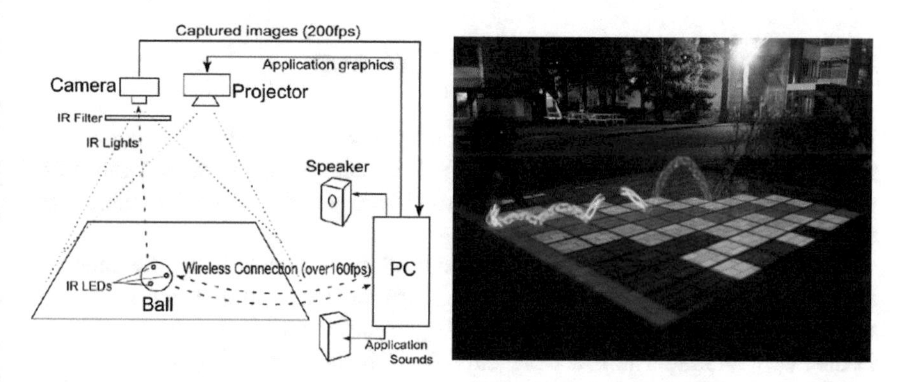

Fig. 2 Overview of the bouncing star system (*left*) and the playing field for an application developed using the system (*right*)

Sounds were generated by a separate application and played through two separate stereo speakers. Figure 2 (left) shows an overview of the system hardware, while the photograph on the right shows the application of the system.

Using this system, we can develop applications that integrate real-time dynamic computer graphics and responsive sounds, which are synchronized with the ball's motion characteristics. We also conducted experiments to investigate the full potential of this new entertainment system and tried to design new games, with new rules, and new styles of play.

3 Implementation Technology

3.1 Basic Design of the Ball

Bouncing Star is a simple ball that contains electronic components, which is strong enough to be thrown and bounced off walls and floors. It can recognize various states, such as thrown, bounced, and rolled, using built-in sensors. Furthermore, the ball is equipped with infrared LEDs, so the system can detect the position of the ball using a high-speed camera. The ball, thanks to its Multicolor LED, can change color based on speed and its status. We have produced several different ball-based game applications by exploiting these characteristics of Bouncing Star. In these applications, real-time CG and sounds are generated in the playing field, as well as the changes in the color and flicker speed of the light, which are linked directly to the movement of the ball. The player and audience can understand what is happening during the game by watching the ball itself.

The Bouncing Star ball comprises a core component with a cover. The core contains electric circuits, which are housed in a plastic (polycarbonate) sphere. The cover protects the core.

3.2 Core and Shock-Absorbing Mechanism

We decided that the ball should express itself visually, so we added color LEDs that responded to the conditions of the ball, and the color of the ball and flicker speed changed depending on the context of the game, the field graphics, and sounds based on the game rules (Fig. 3).

The core comprised a PIC micro-controller, a three-axis accelerometer, a microphone, a XBee wireless communications module, a lithium ion battery, six full color LEDs, and six infrared LEDs. A gyro sensor and photosensor were added in the 2013 version. The ball weighed 170 g. The accelerometer could determine accelerations between +6 and −6 G on three axes (x-axis, y-axis, and z-axis). The sound sensor could interpret sounds that occurred within the ball. The data obtained from the sensors were processed by a micro-controller (PIC16F88), approximately 160 times per second. The wireless module used a XBee platform, which allowed reliable connectivity to over 30 m. Wireless communications between environments, PCs, or with other Bouncing Stars used the serial communication protocol. The lithium-ion rechargeable battery was charged for 2 h by connecting a stereo mini-jack and the ball could be used for 1 h after it was fully charged.

The cover of the ball required a material, which was strong enough for use with applications during real ball games, but it had to have sufficient elasticity and transparency to transmit LED light through the outside of the ball. As reported in our 2010 paper (Izuta et al. 2010), we developed three different balls. The first was a transparent beach ball type. The second was Sepak Takraw ball type which is made by synthetic fiber. Sepak takraw is a sport native to the south Asian Maley-Thai peninsula; the ball is hand-woven using rattan or plastic stems (Fig. 4). The third was a rubber ball type (Fig. 5). We selected these three types because the ball had to express itself using full-color LEDs, while it also needed to be protected against strong shocks. The rubber ball type was the most suitable for our smart-ball system because of its transparency to express information, its round shape, and high elasticity (Table 1).

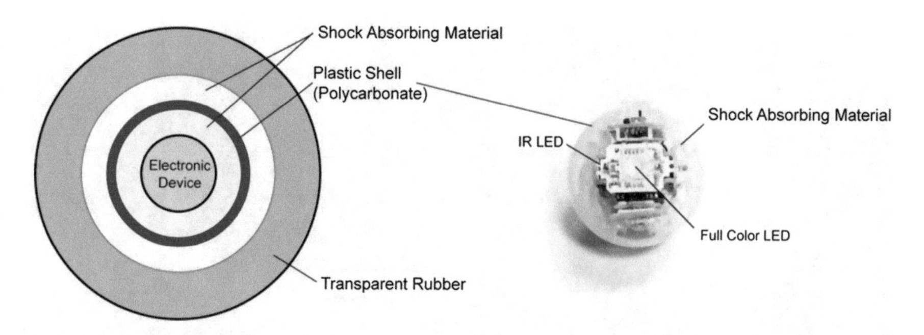

Fig. 3 Internal composition of the ball

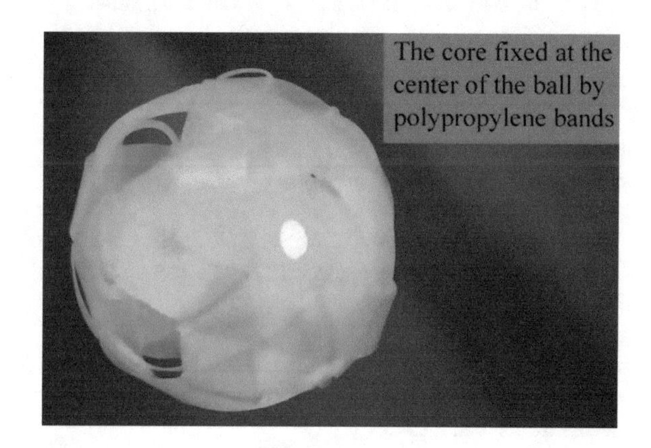

Fig. 4 Bouncing star sepak takraw ball

3.3 Bounce Detection Algorithm

We developed a new algorithm for recognizing the ball's state (static, rolled, thrown, caught, etc.) using information obtained from the acceleration and sound sensors. The ball spins irregularly during play so we calculated the acceleration values for the x-axis, y-axis, and z-axis, and used them to obtain the value of the total acceleration, $A\left(A = \sqrt{X^2 + Y^2 + Z^2}\right)$. When the ball collided with an object, the sound sensor detected the value of the sound within the ball, S. We compared the acceleration value A and the sound value S with constant thresholds for each, As and Ss. Combinations of these allowed us to recognize the state of the ball (Table 2).

Moreover, the three different states of the ball (static, rolling, in the air) were used to calculate the value of the acceleration due to gravity. When the ball was on the ground, we could detect the value A as the constant value of the acceleration due to gravity. The rolling state was detected by analyzing the acceleration due to gravity on the x-, y-, and z-axes. The in the air state was recognized when $A = 0$,

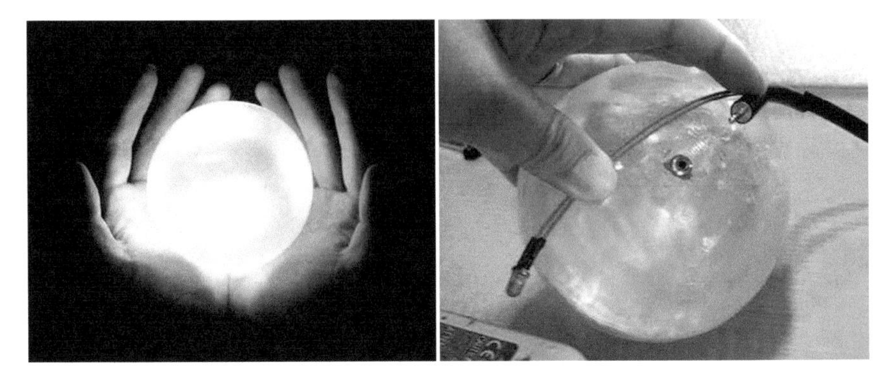

Fig. 5 Bouncing star rubber ball (*right* July 2012 version)

Table 1 Specifications of the three different types of ball

Ball type	Material	Diameter (mm)	Weight (g)	Reflection coefficient
Rubber ball	Silicone rubber	98	550	0.70
Beach ball	Vinyl	220	290	0.34
Sepak takraw ball	Polypropylene 135	135	250	0.43

Table 2 Algorithm used to recognize the ball's state

	A > As	A < As
S > Ss	Bounce	(Nothing)
S < Ss	Thrown	Static Rolling In the air

because the accelerometer could not detect the acceleration when the ball was in the air. These processes are performed by the micro-controller equipped in the ball and the state is transmitted to the application program via wireless communication. The threshold values As and Ss could be changed from the PC interface to modify the value, depending on the environment of the playing field or the specific application's demand for the ball. The changed threshold values were saved if someone switched the ball off. The threshold value was always saved in the memory after it was changed. If we could not use a projector and a high-speed camera with the system, the LED light emissions from the ball continued to be linked to the movements of ball, so we could play new sports with the ball alone. Five different modes of full color LED emission patterns were programmed for the ball in the micro-controller (Table 3).

Table 3 Light emission pattern modes inside the ball

Mode name	Interaction description
Gradual mode	The color changed gradually when the ball was rolled and the color changed slowly when the ball bounced
Pulse mode	The cycle of the light pulse depended on the acceleration value, A
Skip mode	The color changed quickly when the ball bounced or rolled
Burning mode	The ball's color changed from blue to green, to yellow, and to red as the acceleration value, A, increased
Vanishing mode	The light was switched off for a few seconds when the ball was thrown

3.4 Position Tracking Algorithm

We developed an image processing program to recognize the position of the ball with a camera using the IR lights inside the ball. In a real demonstration environment, we calibrated the field coordinates based on the camera coordinates using projective transformation. Next, the image captured by the camera was labeled by chain code algorithm after binarizing process with a constant threshold and it also recognized the positions of the IR lights. If several light sources were detected in the space closer than a specific distance threshold, the ball was considered to be located at the center of the sources of the lights. Six IR LEDs were present inside the ball so some light sources would always be detected from a ball.

4 Production of Applications

We produced interactive graphic and sound applications for the rubber-type ball.

The graphics were created based on the ball tracking information (location and timing of bounces). The applications were written in Visual C++ with Open GL or DirectX. We used an nVIDIA Geforce 6800GT graphics card. A projector was suspended 10 m above the playing field surface where the graphics were projected. We used a BenQ SP 870 projector for the experiments at the University of Electro-Communications and for public demonstrations at three different places.

The environmental sound effects were generated according to the timing of bounces. The sound effects were designed to match the contexts of the graphics and the player's motions such as shaking or bouncing the ball in a specific scene.

4.1 Simple Graphic Effects for Ball Play

Figure 6 shows the simple graphical effect we refer to as *spotlight effect*, which we produced at a very early stage of this project. A bright spotlight was projected continuously at the ball's position (x, y) on the floor. The players and the ball were illuminated.

Fig. 6 Simple graphical "spotlight effect" where a projector was placed in the balcony on the third floor of a building and a real-time spotlight image was projected onto the ball, the player, and the ground

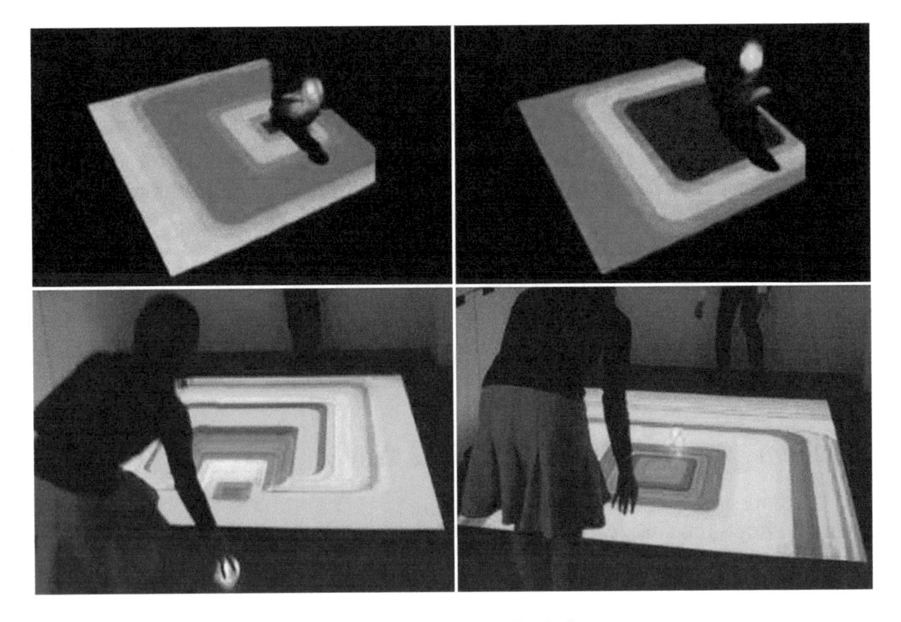

Fig. 7 A simple geometrical graphic spreading from the ball

Figure 7 also shows a simple graphical effect that spreads from a moving ball. In this program, the positions where the ball bounced on the floor (or the wall) were detected using data from the 3-D accelerometer and a sound sensor (microphone), with image processing using a high-speed camera. A 2-D graphical pattern spread rapidly from the point where the ball bounced.

4.2 Graphic Effects Using a Particle System

After generating a simple graphic effect, such as a spotlight or a spreading simple geometric pattern that followed the ball's position and movement, we used a particle system to generate stars that followed the ball's tracks. In conditions where there was no acceleration against the ball, small stars spread out from it. After the ball bounced, large stars appeared and spread our (Fig. 8).

We also produced fluid graphics effect using the Navier–Stokes equations to generate fluid ("smoke") graphics where the movements were generated by the ball's position and movement. People could produce swirls by turning the ball on the table (Fig. 9).

4.3 Collision with the Real Ball and Virtual 3-D Objects

Figure 10 shows our first attempt at a collision between a real ball and virtual 3-D objects projected onto the tabletop. The physical movements of the virtual objects after a collision were calculated in real time.

4.4 Augmented Sports Applications

After producing the simple graphic modes that allowed people to enjoy playing with the ball and interactive graphics on the floor, we created more complicated sports game applications where people could move dynamically to compete or collaborate in the context of a sports scenario.

4.4.1 Space Ball I

We developed an application called "Space Ball I" where our system had no wireless communication module or a microphone in the ball. Therefore, the ball information was obtained using two high-speed cameras. This was the only information that the application in the PC could acquire from the system. The information detected by the accelerometer was used only to change the light

Fig. 8 Graphical effect using a particle system (*small* and *large stars*)

Fig. 9 Graphical effect using a fluid equation

emission from the ball itself. A projected CGI (using Open GL) of 10 × 10 squares was spread across the field.

Figure 11 shows Space Ball I where the player could score points by hitting the ball at these panels. Two players could compete to hit these panels to score points. Our challenge was to recognize the bounce of the ball based on image processing alone using a second high-speed camera and displaying CGI effects, such as scattered stars, when the ball hit the boundary. Indeed, boundary identification using the second camera led to much false positive and/or false negative detection.

4.4.2 Space Ball II

In Space Ball II, we used a ball with an added sound sensor (microphone) and the wireless communication module inside. This application generated dynamic CGI effects on the playing field, which changed in synchronization with the ball's characteristic motion because the ball's state was detected by a image processing program. We considered three different ball states, i.e., bouncing, rolling, and in

Fig. 10 Collision between the real ball and virtual 3-D objects (a cube and a sphere in virtual space). The "spotlight effect" was applied to the two balls

Fig. 11 Space ball I (Laval Virtual 2008)

the air, which were detected by the program. The program used the position information obtained from the ball (with the high-speed camera) as parameters to determine the direction of the game. Table 4 shows how the direction of the game was determined using the information acquired from the ball. This application was designed as a multi-player cooperative game. There was a time limit of 60 s per game. A player could score by hitting the ball at a target projected as a CG spot (Fig. 12).

The targets with the same color were displayed on the playing field. Their color and placement changed if the player bounced the ball outside the field (the color of

Table 4 Direction of space ball II using ball information

Ball information	Direction in space ball II
Ball bounced outside the playing field	Change the color of the ball
	Change the target's placement
Ball bounced inside the playing field	Display the shock wave effect
	Hit surrounding targets
	Randomly generate several different colored new targets
Rolled	Extend the remaining time based on the number and interval of the hit targets, if several targets are hit during one throw
In the air	Cannot score points when the ball moves over the targets

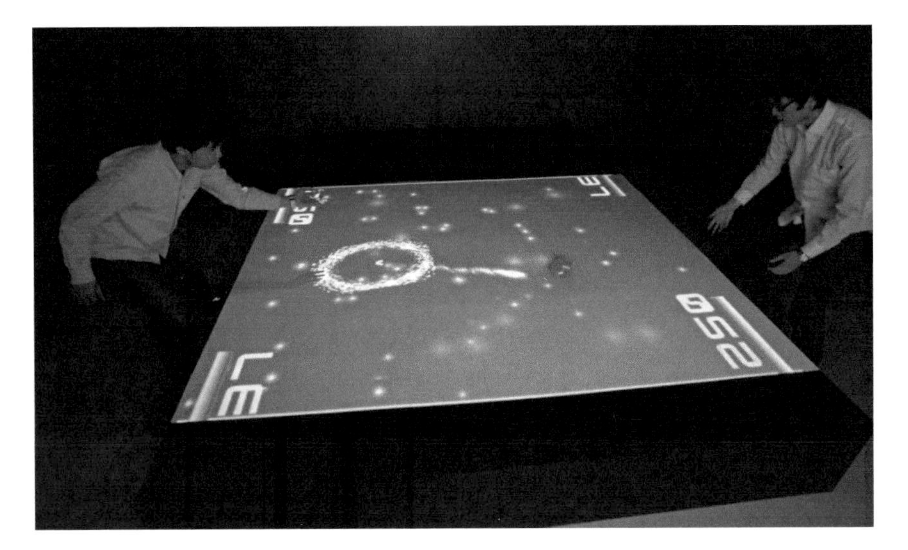

Fig. 12 Space Ball II, an augmented digital sports game (Miraikan 2010)

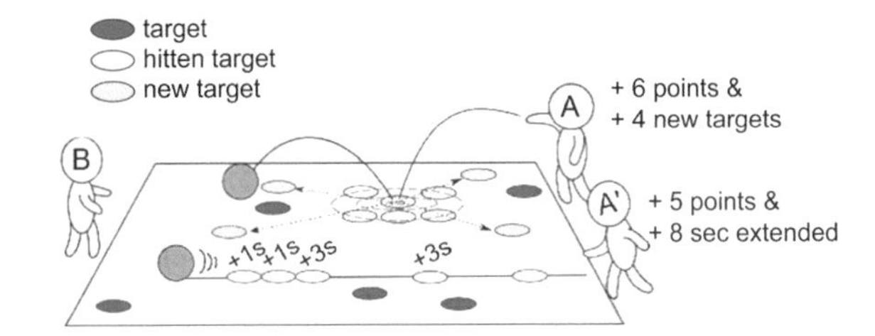

Fig. 13 Description of the rules for scoring points and making new targets in Space Ball II

the ball also changed at this moment). The players could choose their favorite placement from the target spots, which made it easier to obtain high scores by changing the ball's color if the player dribbled it on the floor with their hands. Hitting a target in one bounce or rolling the ball along a line of targets generated higher scores. Figure 13 shows the rules that determined how points were scored and how new targets were made, as well as how the time limit was extended.

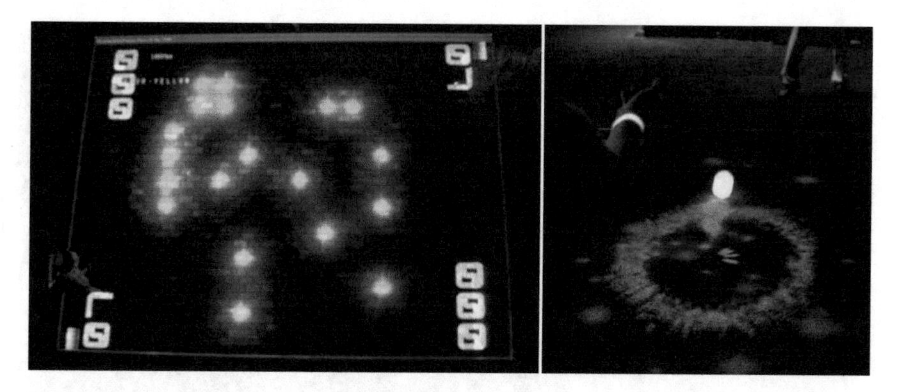

Fig. 14 Floor coordination (*left*) and shockwave effect after bouncing (*right*) in Space Ball II

Sound effects had an important role in Space Ball II. We used up-beat music as a basic BGM during the play. This up-beat music was aimed at making people excited during the game. During continuous BGM, we added four different sound effects based on the ball's bounce. The sounds differed depending on the context of the scene, which let players know what was happening in their game (e.g., they changed the target's coordinates, they scored points by hitting the target, or the ball simply bounced inside the playing field but failed to hit the target). Each sound was designed and recorded beforehand and played when a bounce occurred, with no delay (Fig. 14).

4.5 Ball Play with a Word Puzzle

Further applications are shown in Figs. 15 and 16. We already produced a 2-D game (Bouncing Star I), which resembled the game Othello where people could compete by making a square panel change color (red or blue) when they hit the surface with the ball. Later, we produced a panel on a 3-D cube so people could turn the virtual cube (like dominos) by throwing the ball. There were sound effects (which sounded like real wood blocks falling down) when the cubes turned.

In 2012, we added text/graphics on the face of the cube to produce an application called "Soradama", a type of 3-D crossword puzzle in a virtual space (Fig. 17) (Kodama et al. 2012).

5 Public Exhibits (2007–2013)

Table 5 shows the exhibition history of Bouncing Star. Our interactive applications were exhibited thirteen times (eleven times in Japan) and about 1,000 people participated in the experiment in Miraikan Museum during 2010.

Fig. 15 Children playing with a ball using 3-D virtual cubes like dominos where the surfaces had symbols from the Japanese alphabet (at Yumeminato Tower, Japan, 2012)

Fig. 16 In Soradama, a ball was used to solve 3-D crossword in a virtual space (photograph taken in Kodama's laboratory at UEC, 2012)

In 2008, we demonstrated our first system in Japan, France, and the USA (Izuta et al. 2008a, b, c). Subsequently, the ball and interactive contents were exhibited: four times at academic conferences, and ten at public media art exhibitions. The number of visitors to the seventh, eighth, and ninth exhibitions were about 70,000, 4,000, and 50,000, respectively. Not all of the people who visited the exhibition experienced the ball, but we consider that at least 10,000 people interacted with our system.

At the Miraikan museum show in 2010, the Kodama Laboratory conducted a questionnaire survey of 1,000 participants who played with the interactive applications of Bouncing Star system demonstrated at the museum. We conducted the survey to analyze sex and generation differences in the influence of sound/graphical effects in the ball play. The result was reported in our student thesis (Satake 2012).

Table 5 Exhibition history of Bouncing Star

No.	Location	Exhibit Dates
1	'Device Art' exhibit, Miraikan (Tokyo, Japan)	Sep/26–30/2007 Exhibit of ball only
2	The National Art Center (Tokyo, Japan)	Feb/6–17/2008 12 days demo
3	INTERACTION (Tokyo, Japan)	2008 March 3 and 4 2 days demo
4	Laval Virtual 2008 (Laval, France)	Apr/9–13/2008 5 days demo
5	SIGGRAPH 2008 New Tech Demo (Los Angeles, USA)	Aug/11–15/2008 5 days demo
6	'Invisible Garden' exhibition, (Miraikan, Tokyo, Japan)	Dec/1/2010–Mar/11/2011
7	'Art of Light' exhibition, (Malios, Morioka, Japan)	June/30–July/24/2011,
8	'Magical Art Museum' exhibition, (Takehara Museum, Hiroshima, Japan)	Aug/5–Sep/25/2011
9	'Magical Art Museum' exhibition, (Yumeminato Tower, Tottori, Japan)	Mar/17–May/6/2012
10	'Magical Art Museum' exhibition, (Matsuzakaya Museum, Nagoya, Japan)	July/28–Sep/2/2012
11	'Mugendai Bijutsukan' exhibition, (Hamada Children's Museum of Art, Shimane, Japan)	July/23–Sep/23/2012
12	1st workshop on Smart Material Interfaces in conjunction with ICMI2012 (Santa Monica, USA)	Oct/22–26/2012 1 day demo, workshop
13	'Magic Art Museum: Light Art Is Fun for Everyone' exhibition, (Contemporary Art Museum, Kumamoto, Japan)	July/6–Sep/8/2013
14	'Art in Wonderland' exhibition, (The Ueno Royal Museum, Tokyo, Japan)	Sep/6–Oct/6/2013

6 Discussion

During 2008 and 2009, we conducted many experiments in outdoor and indoor fields at the University of Electro-Communications campus. We also conducted many public demonstrations at three different places. Hundreds of people, including small children, experienced our applications and they enjoyed the practical interactive demonstrations. In some cases, we only gave people the Bouncing Star Ball without any projected images. In these cases, we found that the "vanishing ball mode" had a more favorable reception from many people. In this mode, the ball appeared to have disappeared. This mode is used in a dark place, so players feel a sense of thrill at catching the disappearing ball. This experience elicited a positive reaction in this mode. The ball was entertaining without any graphics applications because it had six different light modes. Thus, people could play catch in dark areas. We plan to consider the possibilities of using a smart ball without any graphical applications of a projector, by adding other components inside the ball.

Fig. 17 Children enjoying Soradama at the exhibition site. (Media art exhibition at the Contemporary Art Museum, Kumamoto, Japan, 2013)

Throughout all of our exhibitions, young people showed a strong interest in our ball and the graphical applications. As we expected, small children preferred playing with the simple graphical stars and fluid effects. At the *matsuzakaya* (department store) museum, a child aged 1.5 years played with the ball with the fluid graphics for 20 min with her mother. Children aged three to seven years preferred games with simple rules. In these games, scoring points was not the objective and there were no winners in these games. However, people could continue playing with the ball and graphics, while handling the ball freely on the table. People enjoyed cooperative play around the table. By contrast, a game such as Space Ball II has complicated rules like video games, so small children found it difficult to understand the rules immediately.

During the exhibitions, the projected image size was changed ($2 \times 2\ m^2$ to $4 \times 4\ m^2$), depending on the space requirements. We found that a large image made people exercise more. In the Space Ball II game, it was a challenge for people to score high points by collaborative play, which made people feel very excited and they played the game until they felt as tired as they would if they were playing normal sports.

The interactions between the ball and graphics/sounds were made smooth and natural using a high-speed camera (200 fps) and X-bee wireless communication. The ball interface was natural because it had no cables and no restrictions on movement, so people could move it freely.

Unfortunately, the weak point of our ball was its lack of toughness. We asked people not to hit the table strongly because there was a possibility that the ball might break. By asking them to be gentle, no repairs were needed during the four months of the Miraikan month exhibition, where the floor and table were made of wood. During exhibition nine, however, the floor was concrete and the ball sometimes broke when it fell to the floor.

There were small gaps between the ball's movement and the game CG scene in Space Ball I, so the detection in this game was relatively slow using the ball position information alone because at that time the application could not use the acceleration and sound information from the ball.

Therefore, we added the wireless communication module and a combination of the microphone and accelerometer to the ball for Space Ball II. This allowed us to synchronize various ball movements, thereby obtaining the ball's states and the ball's position information. This helped us to implement real-time bounce detection in the game. Based on the analysis of the acceleration information, we could identify the "rolling" and "in the air" states. The further development of this software allowed us to produce several unique methods for recognizing the state of the ball, which were not available in the previous project because it only used electronic devices.

However, we have yet to adapt the Bouncing Star system to all of the possible ball movements. Thus, we are now developing a new ball, which includes a gyro sensor and photo sensor, and we are conducting experiments to implement roll direction and grasp identification.

7 Conclusion

In this article, we summarized the smart-ball concept used in our Bouncing Star project and we reported the development of our interactive graphical applications for the Bouncing Star ball system.

We also reported our exhibition history. Given people's positive responses to the interactive contents with sound and graphics effects during the practical exhibitions, we conclude that the ball system is entering the product level phase (we will need to consider the production cost). People were often enthusiastic about the games. Players made full body movements during the games and they sometimes played continuously for 15 min, perspiring during their workout. The players and many audience members enjoyed the entire game scene around the playing field.

In this project, we mainly developed a rubber-type ball, but this ball needs further innovation in terms of its materials and methods. The durability needs to be improved to protect the electronic devices inside the balls against strong shocks.

We believe the smart ball and this new dynamic form of entertainment has great potential for the development of human communication and as an entertainment activity. The fusion of the ball and digital technology has just begun in this past decade, but we believe that the use of a ball interface for augmented sports has great potential as an entertainment activity.

Our next phase will include making tools to provide an interface for connection on a large scale (like a stadium) in the real physical world with virtual information resources in a very smooth and intuitive manner.

Acknowledgments We appreciate all of the people who have supported our project and activity. The development of the ball and system were supported by the project "Expressive Science and Technology for Device Art," which was funded by the Core Research for Evolutional Science and Technology (CREST) of the Japan Science and Technology Agency (JST).

References

Ishii H, Wisneski C, Orbanes J, Chuu B, Paradiso J (1999) PingPongPlus: design of an athletic-tangible interface for computer-supported cooperative play. In: Proceedings of CHI'99, pp 394–401

Iyoda A, Kimura H, Takei S, Kakiuchi Y, Du X, Fujii S, Masuda Y, Masuno D, Miyata K (2006) A VR application for pitching using acceleration sensor and strip screen. J Soc Art Sci 5(2):33–44

Izuta O, Nakamura J, Shibasaki K, Kodama S, Koike H (2008) Development of a rubber ball bouncing star for digital sports: a ball containing infrared and full color LEDs and an acceleration sensor. In: Proceedings of interaction'08, IPSJ Symposium Series vol.2008, no.4, Tokyo (in Japanese), pp 65–66

Izuta O, Nakamura J, Sato T, Kodama S, Koike H, Fukuchi K, Shibasaki K, Mamiya H (2008) Bouncing star, exhibited at Laval Virtial 2008, Salle Polyvalente, Laval (Apr/9–13/2008)

Izuta O, Nakamura J, Sato T, Kodama S, Koike H, Fukuchi K, Shibasaki K, Mamiya H (2008) Digital sports using bouncing star rubber ball comprising IR and full-color LEDs and acceleration sensor. In: Proceedings of ACM SIGGRAPH 2008, new tech demos, article no.13

Izuta O, Sato T, Kodama S, Koike H (2010) Bouncing star project: design and development of augmented sports application using a ball including electronic and wireless modules. In: ACM proceedings of the 1st augmented human international conference, article no.22: 1–7

Kodama S, Izuta O, Nakamura J (2007) IR rubber ball: bouncing star, exhibited at the device art exhibition, Miraikan, Tokyo (Sep/26/2007–Sep/30/2007)

Kodama S, Sato T, Koike H, Fujimoto A (2012) Bouncing star smart-ball project: focusing on the interaction of expressions and exhibitions. In: Proceedings of the 1st workshop on smart material interfaces: a material step to the future, article no. 6:1–4

Kuwakubo R.(2003) heavenSeed, exhibited in Digital Art Festival Tokyo exhibition, Panasonic Center (Aug/8-17/2003)

Mueller FF, Agamanolis S (2005) Sports over a distance. ACM Computers in Entertainment, 3(3):1–11

Rusdorf S, Brunnett G (2005) Real time tracking of high speed movements in the context of a table tennis application. In: Proceedings of ACM VRST, Monterey, pp 192–200

Satake T (2012) Development and evaluation of bouncing star entertainment System. Master's thesis, University of Electro-Communications, Tokyo, Feb 2012 (in Japanese)

Sugano Y, Mochizuki Y, Usui T, Okude N (2007) Shootball: the ball sport using dynamic goals. In: Proceedings of ACM ACE'07, pp 262–263

Sphero's home page. http://www.gosphero.com/

The Bouncing star ball project website. http://www.kodama.hc.uec.ac.jp/bouncingstar/index-e.html

Vertegaal R, Poupyrev I (2008) Introduction to organic user interfaces. Commun ACM 51(6): 26–30

Part III
Games for Change, Personalization, and Teaching

Games for Change: Looking at Models of Persuasion Through the Lens of Design

Alissa N. Antle, Theresa Jean Tanenbaum, Anna Macaranas and John Robinson

Abstract Games for Change are digital games that purport to change people's opinions, attitudes, or behaviors around specific issues. While thousands of games have been created, there is little evidence that such games do persuade or contribute to behavior change. To address this problem, address the research question: How do elements of the different models of persuasion and behavior change manifest within Games for Change? We identify and focus on three models: Information Deficit, Procedural Rhetoric, and a new model called Emergent Dialogue. To answer this question, we had to determine what "clues" there were in games that we could use to identify each model of persuasion. Using a collaborative version of a Close Reading methodology we analyzed ten Games for Change about sustainability. Based on our results we propose six categories of design markers. Each marker can be used to identify or implement specific design elements associated with a particular model of persuasion. In this chapter, we describe our methodology, present six categories of design markers, and describe the specific strategies for each marker associated with each of the three models of persuasion. We illustrate each model and its design markers through canonical examples including a new game called Youtopia that we have created to encode the Emergent Dialogue model into a digital game. We conclude with proposed guidelines for game design of Games for Change.

A. N. Antle (✉) · T. J. Tanenbaum · A. Macaranas
Simon Fraser University, Surrey, BC, Canada
e-mail: aantle@sfu.ca

T. J. Tanenbaum
e-mail: tess.tanen@gmail.com

A. Macaranas
e-mail: amacaran@sfu.ca

J. Robinson
University of British Columbia, Vancouver, BC, Canada
e-mail: john.robinson@ubc.ca

A. Nijholt (ed.), *Playful User Interfaces*, Gaming Media and Social Effects,
DOI: 10.1007/978-981-4560-96-2_8, © Springer Science+Business Media Singapore 2014

Keywords Games for change · Serious games · Sustainability · Behavior change · Procedural rhetoric · Emergent dialogue · Persuasion · Design framework · Design guidelines · Close reading

1 Introduction

Persuasive computing focuses on how interactive technologies and services can be designed to change people's attitudes and behaviors (Fogg 2003). The motivational power of playing video games has been leveraged by groups interested in social change for some time now (Games for Change Society 2014). Games for Change and Serious Games are increasingly being used as play-based tools for behavior change. Games have been created to address social issues such as sustainability practices, bullying, political lobbying, and personal health care. But do these games actually enable behavior change?

In this chapter, we explore how digital Games for Change can be designed to influence behaviors, thoughts, and feelings. Many Games for Change are developed based on implicit knowledge or assumptions about how external persuasion can motivate attitude and/or behavior change. One of the contributions of this chapter is to make these models of persuasion (and expected behavior change) explicit. We describe three models of persuasion: Information Deficit, Procedural Rhetoric, and Emergent Dialogue.

There are other models of persuasion—more than we can discuss in one chapter. We delimit our work by focusing on three models. The first, called the Information Deficit model, is one of the most common models (He et al. 2010). The Information Deficit model focuses on persuasion through information. Our analysis found evidence of this model in most Games for Change. The origin of this model in Games for Change is likely that "best practices" were mapped uncritically from educational games to Games for Change. However, if the purpose of Games for Change is not learning related but future behavior change, is this appropriate? The second, called Procedural Rhetoric, is a model that has recently emerged from the games studies community as an alternative, and possibly more effective approach to serious game design (Bogost 2007). The Procedural Rhetoric model focuses on persuasion through interaction. Lastly, we introduce a new model of persuasion that has recently emerged from the environmental studies community in response to the failure of existing models to elicit the change they purport to enable. We call this model, Emergent Dialogue (Robinson 2004). The Emergent Dialogue model focuses on persuasion through participation in discussion around personal values. We suggest that this model is similar to a model of behavior change recently appearing in the counseling psychology literature called Acceptance and Commitment Therapy (ACT) (Hayes et al. 1999).

There is no empirical evidence that any one model is more effective than the others, and each may have an important role to play in persuasion depending on the goals of those seeking change and the specifics of the social issue at stake. However, we think that the practice of developing Games for Change can be made more effective if designers are explicitly aware of the persuasion model they are using, its benefits and weaknesses, and the kinds of design decisions entailed by each model.

In order to relate design decisions to models of behavior change, we introduce a methodology for deriving design markers from game play experience. A design marker is an identifiable element or strategy that is either encoded in a game during development or emerges through game play and which indicates the implementation of a specific behavior change model. When designers create a serious Game for Change, they encode their own assumptions about persuasion and behavior change into the game system, sometimes without any awareness that this is happening. When we play and analyze a game we are able to identify evidence of one or more models of behavior change encoded into the system in the form of design markers. Markers may involve or be apparent based on different game elements such as game controls, content, visuals, interface features, rules, game mechanics, or rewards. In this chapter, we present six important design markers that reveal underlying behavior change models. We define, describe, and illustrate each marker type for all three persuasion models in order to make explicit the relationship between models of persuasion and game elements. We delimit our study by focusing on Games for Change related to social issues where the intent of the game is to produce some form of short- or long-term attitude or behavior change in the players. Specifically, in our games analysis phase, we looked at Games for Change around issues of sustainability and the environment. We suggest that our results will be applicable to other Games for Change where the core objective is behavior change. We conclude this chapter with best practices for designers of Games for Change.

2 Three Models of Persuasion

We focus on three different models of persuasion that can be observed within the design of Games for Change: the Information Deficit model, the Procedural Rhetoric model, and the Emergent Dialogue model. These models are described in detail in Tanenbaum et al. (2011, 2013). We provide a summary of each model here. The three models are not mutually exclusive. A single game may have evidence that more than one model is at work. They are not formal models in any sort of "framework creation" sense. Rather, each of these models of persuasion represents an intellectual commitment to specific ideas about how persuasion happens (or should happen) in games that are designed to promote attitude or behavior change. These models are therefore attitudes about persuasion in games.

2.1 The Information Deficit Model

The Information Deficit model assumes that providing correct knowledge about the phenomenon in question will lead to behavior change. Many current approaches to sustainability are based on this model of behavior change. The Information Deficit model posits that providing information changes values; value change drives changes in attitudes; attitude change drives changes in behaviors (He et al. 2010). For example, it is common for local governments and organizations to run community workshops and lectures intended to educate participants in the benefits of recycling, conservation, reuse, and other environmental friendly practices. These types of workshops are based on the model that unsustainable behaviors arise from a lack of education.

The Information Deficit model assumes a top-down model of sustainable behavior change where some entity or organization (such as a national government, NGO, educational institution or other authority) already has determined what the optimal behavior is for the individual to adopt. There are five common motivational models that conform to this approach: (1) Attitude, (2) Rational-Economic, (3) Information, (4) Positive Reinforcement, and (5) Elaboration Likelihood Model (He et al. 2010). The Attitude model assumes that changing an individual's attitudes will result in changes in behavior. The Rational-Economic model assumes that financial factors alone will motivate positive changes in resource use behavior. The Information model, similar to the Attitude model, assumes that providing information to people will encourage improved behavior, reasoning that, "once you know what to do, you will do it" (He et al. 2010). Positive Reinforcement encourages desired behaviors through positive feedback stimuli. Finally, the Elaboration Likelihood technique uses a more sophisticated approach, combining logical arguments and emotional persuasion to motivate behavior change. All of these models have been implemented using networked technology for a variety of applications. All of these models assume a top-down approach where an authority provides "correct information" about what to do, and possibly why to do it.

As such the key assumption behind all of these persuasive models depends on the intellectual commitment that what the public is largely lacking is information. These information-centric models assume that by using best new media practices to design and communicate the right information, behavior change will follow.

Quiz games are the quintessential form of Information Deficit-oriented design. These types of games have been around for decades and are still in use today. Consider the interactive section of NASA's website on Climate Change (Fig. 1a, b).[1] It includes a selection of quizzes on topics such as the impact of global warming, sea level rise, and the state of the Earth's glaciers and ice caps. These quiz games all operate on the premise that learning the facts about climate change will change how the player thinks and acts when confronted with questions of sustainable living. They

[1] http://climate.nasa.gov/interactives/quizzes

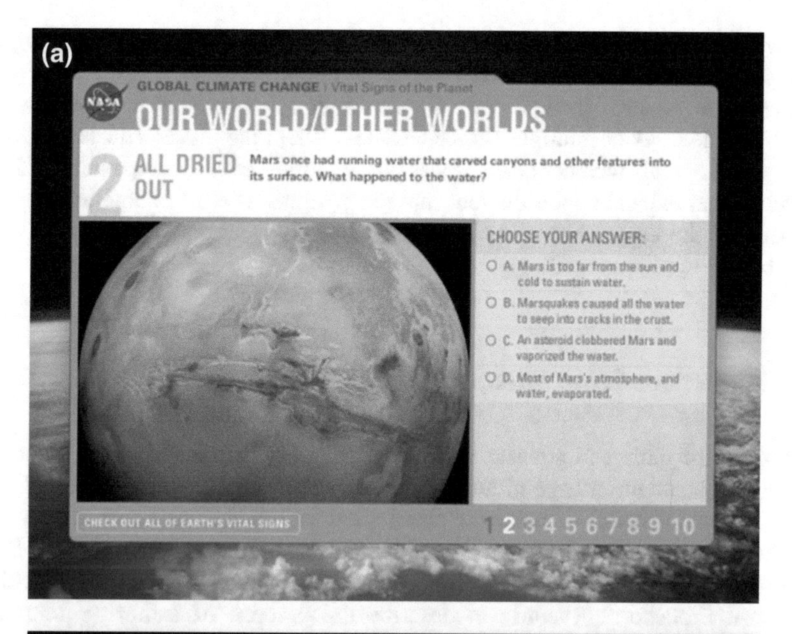

Fig. 1 **a** NASA quiz game based on information deficit model of persuasion in which information/facts are presented through the quiz questions and answers. **b** Players win by choosing the correct answer/fact and are judged based on their performance (e.g., expert or student)

are primarily interested in testing the knowledge of the player, and correcting any misconceptions that he or she may have about climate change.

As game systems there is very little to them. A player is presented with a question and a set of multiple choice answers. When the player selects an answer the system reveals whether or not it was correct, often with accompanying didactic material that expands upon the fact that the designer wished to communicate. At the end of the experience, the player is evaluated, scored, and either praised, or not, by the system accordingly.

2.2 The Procedural Rhetoric Model

The study of games to educate and persuade has led to the formulation of new models of behavior change grounded in the dynamics of simulations. The Procedural Rhetoric model of behavior change has emerged from game studies as a response to the criticism that many Games for Change, most based on the Information Deficit model, were either ineffective, unappealing, or both. Bogost coined the term Procedural Rhetoric to describe the practice of authoring arguments through interactive processes (Bogost 2007). In this model the argument of persuasion is not represented through information but through interaction. The game designer(s) create the rules of interaction in the game mechanics that are in line with their argument for attitude or behavior change. Arguments are represented through the dynamics of interaction rather than specific textual or visual forms. Instead of simply providing the player with the desired information, the player is given an opportunity to interact, observe, and reflect within a dynamic game system.

Games often involve a possibility space that represents a microworld or a simulation, in which players can explore and play in a simulated situation related to some particular social issue. The world enables players to understand how that world works. For example, most Games for Change about sustainability involve game play with a small world in which the goal is to create a sustainable situation for that world.

The set of rules or mechanics that constitute a game define the possibility space for that game. The game rules define what a player can and cannot do, and what happens when they do or do not make certain choices. As a media form, video games can represent cultural values, norms, and expectations. They can do so in the content they represent, such as text-based instructions, auditory dialogue, visual depiction of settings or characters, and background sounds. However, as Bogost points out in (Bogost 2008), the game rules or processes that define the possibility space may also represent cultural values. For example, in a microworld simulation style environmental Game for Change, using a lot of energy usually results in effects such as high energy prices, environmental degradation or energy shortages. The implicit message is often that these effects are bad or negative. Cultural values around sustainability and responsible energy use are communicated to the player through the game rules, triggered by their choices through interaction. This is what Bogost calls *Procedural Rhetoric* (Bogost 2008). Rhetoric refers to a claim or argument being made, here about a particular cultural value around sustainable energy use. Procedural refers to the rules programmed into a game through its rules, algorithms, and other forms of code. Through Procedural Rhetoric a particular claim or argument is programmed into the game rules.

An underlying assumption of Procedural Rhetoric used in Games for Change is that by creating a set of game rules (procedures) that enable players to experience—through their choices and interactions—particular events, they will modify their behavior in line with the claims of the argument being made. Both the Information Deficit model and the Procedural Rhetoric model employ a top-down approach to content and information. These two models are based on the assumption that the desired outcome is a known quantity that must be advanced through the delivery of either facts or processes. Unfortunately, there is little evidence that either model of behavior change works.

Environmental role playing games often employ the Procedural Rhetoric model of persuasion. The BBC's Climate Challenge role playing game (Fig. 2a, b) has elements of Information Deficit present in it.[2] However, the primary mode of communication is in the dynamics of the game's simulation, making it an excellent example of Procedural Rhetoric at work. In Climate Challenge the player is responsible for dictating the national, trade, industrial, local, and household policies of the European continent. The player is the "leader" of Europe, and alternates between setting policies, and engaging in negotiations with the leaders of other continents (Fig. 2a).

The game uses internal iterations of cause and effect, and interaction and reflection, to augment and communicate its core message. After each turn, the number of variables and options available increases slightly, so that decisions increase in complexity over time. At the same time, the core mechanics of selecting policy and experiencing consequences remain firmly in place throughout the game. The values of the game are apparent through play: carbon is bad, but without popular support, economic stability, and access to food, power, and water a government cannot operate (Fig. 2b). The core rhetoric of the game is about balancing this equation. In most of the game, the player is led to interpretations about the consequences of her actions via a variety of contextualizing factors, such as natural and civic disasters, and newspaper reports.

2.3 The Emergent Dialogue Model

The third model of behavior change that we consider is a relatively new one that has emerged from environmental studies, specifically from research into creating and running policy workshops with the general public around sustainability issues.

---

[2] http://www.bbc.co.uk/sn/hottopics/climatechange/climate_challenge

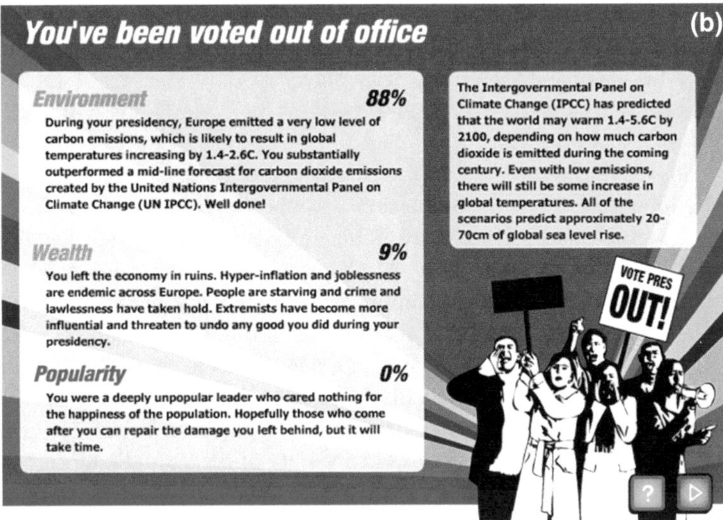

Fig. 2 **a** BBC climate challenge based on procedural rhetoric model of persuasion uses advisors to communicate facts to player. **b** Players experience the consequences of their actions represented by voter response

It has been primarily advanced in the work of our collaborator, John Robinson, who developed it in response to extensive critiques of the Information Deficit model. Robinson's group argues that the previous conception of a unidirectional flow from information to behaviors is incorrect: that people often bring their attitudes in line with their behaviors, rather than the other way around. The Emergent Dialogue model of behavior change has emerged from environmental studies as a response to the failures of the last few decades of sustainability education, which have not resulted in widespread behavioral change. John Robinson points to some reasons for this:

> Multiple conflicting views of sustainability exist [that] cannot be reconciled in terms of each other. In other words, no single approach will, or indeed should be, seen as the correct one. This is not a matter of finding out what the truth of sustainability is by more sophisticated applications of expert understanding… Instead we are inescapably involved in a world in which there exist multiple conflicting values, moral positions and belief systems that speak to the issue of sustainability (Robinson 2004).

Robinson contends that for behavior change to occur the critical element is not information but personally meaningful *participation* in discussions about information, decisions and personal values. Unlike the previous two models, this approach is bottom-up. The Emergent Dialogue model holds that behavior change occurs when participants become engaged stakeholders in the process of co-constructing their own narrative about a desired future. Unlike Information Deficit and Procedural Rhetoric, Emergent Dialogue is agnostic about desired outcomes, instead focusing on supporting a collective meaning making process of social change. Informational processes in an Emergent Dialogue model take the form of static representations of information or dynamic simulations that model potential consequences of actions—but both are ideally neutral, rather than top-down value-laden stances about desired outcomes.

An underlying assumption of Emergent Dialogue is that information in a participatory process is not determined ahead of time—instead it emerges from dialogue alongside personal values in an iterative ongoing process. This process is less about educating people about what behaviors are correct or incorrect. Instead it is about enabling people to generate their own understandings of how their behaviors are in line with their values, or not, and how their behaviors will shape the world they live in.

To date this model has only appeared in workshops and facilitated sessions (Robinson 2004). The challenge of the Emergent Dialogue approach is in finding ways to support it through design decisions about content, procedures, rules, and rewards in games. It does not readily lend itself to being encoded in software.

As we explored this model, we noted similarities to an emerging model in behavioral psychotherapy called ACT (Hayes et al. 1999). ACT has been recently proposed as an effective method of supporting experiential behavior change around issues including depression, anxiety, and pain management and to increase psychological flexibility. A key element of this approach is the explicit discussion of personal values and plans to take actions in accordance with those values.

We have just developed a sustainable land use activity for children (aged 9–12) on a Samsung Pixel Sense interactive tabletop (i.e., Microsoft Surface) that incorporates the Emergent Dialogue model of behavior change. In the activity, called Youtopia, we support players to explore information as they need it, to discuss their own values and to see how their decisions impact the world they create (Antle et al. 2013). To support emergent dialogue we created an open-ended activity. There is no explicit goal or "winning state" in the game. Instead players are asked to make a world they would like to live in. The interaction in Youtopia involves using physical stamps to designate land use types on an interactive map (Fig. 3a). Players can chose to work with either a small or large population. There are different types of shelter, food, and energy sources as well as nature reserves, each with different benefits and limitations. Players take on the roles of builder or natural resource manager. Creating the world involves decisions about using natural resources (land, trees, coal, water) and building developments for human needs (energy, food, shelter) as well as designating reserves (mountain, forest, river).

Building any development requires two stamps used sequentially. A natural resource stamp must be used to designate a resource for use, and then a builder stamp must be used to place a related development in a suitable location. When natural resource stamps are assigned to one player (or group of players), and builder stamps to another, a situation of positive interdependence between the two players (or groups) may result. For example, one player must use their lumber stamp to designate an area of forest as logged before the other player can use the shelter stamp to build housing using that lumber. We expect that players will need to engage in dialogue about the decisions they need to make. For example, they need to decide which natural resources to use, which to preserve, and which and where to put developments. This dialogue will in all likelihood include reflection about personal values.

At any given time players can see the amount of shelter, food, energy, and pollution that they have created. There is no reward for meeting the population's needs without over polluting the world. And conversely, not meeting the population's needs or creating pollution is not associated with right or wrong judgements or values. Information about the relationships between resources, development and pollution is available when the player wants. It is secondary, and the information the game contains is factual but not value laden. For example, an information card on coal plants simply explains the resources used, energy produced, and pollution created without indicating if any of these things are good or bad, right or wrong. Players can use the eraser tool to change any decisions and experiment with options. The activity ends when the players decide they are satisfied with their world.

The open-ended goal to create a world "you" (i.e., the players) want to live in–combined with system design for interdependent action, game mechanics that involve no winning state and value-free content–may better support players to discuss, reflect, and make decisions based on their personal values about what is important in balancing human and natural needs compared to games in which

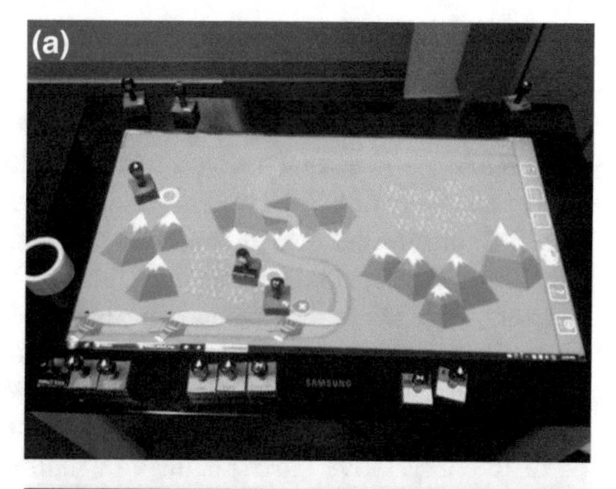

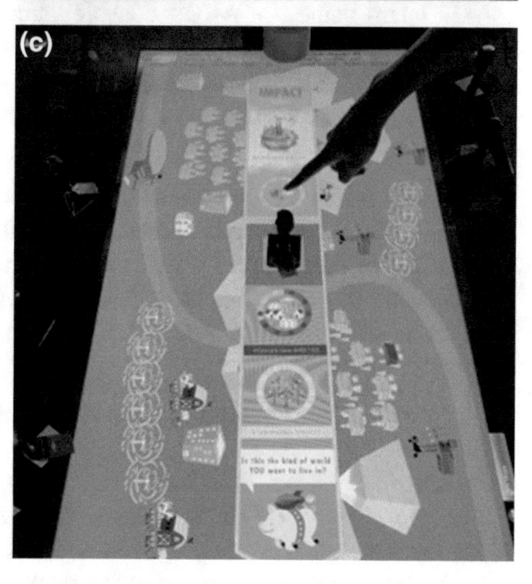

Fig. 3 **a** The Youtopia land use simulation based on Emergent Dialogue model of persuasion enables two players to create their own world. **b** Information is presented on demand when players place a stamp in the information ring and is value-free. **c** Impact stamp shows state of the world in terms of how many people have their needs met and pollution levels, and asks if this is a world the players want to live in. If not, the players can erase and make new decisions

players focus on getting the right answer or winning (Information Deficit) or avoiding the wrong action(s) (Procedural Rhetoric). We are currently conducting a study with 40 children (aged 10 and 11) to evaluate this proposition.

3 Methodology

In order to derive design markers of persuasion models, we used a variant of the close reading and textual analysis methods outlined by Bizzocchi and Tanenbaum (2011) and Carr (2009, Carr et al. (2003). The goal of using close reading was to gain insight into the design strategies of Games for Change in the realm of sustainability. We departed slightly from the pure humanities methods outlined in these chapters, extending the method to support the interpretive processes of three researchers looking at several digital Games for Change in parallel. Previous versions of this method primarily deal with a single researcher interacting with a single game. In our variation, we systematize the process so that multiple analysts may perform comparative close readings of multiple games. In this way we enhance validity with multiple analysts, and by analyzing multiple artifacts to look for commonalities in the ways that persuasion models are encoded into game interfaces, content, and mechanics. The purpose of our close readings was to identify common markers within a number of games in order to better understand how these models of persuasion have been encoded within Games for Change as persuasive artifacts.

To select games for this study we started with some broad criteria:

- *Topics.* Sustainability, environment, climate change, and urban development. *NOT* social justice, gender, speech, civil liberties, and other sustainable development issues.
- *Platform.* Downloadable, playable, Browser Based, PC, or iPhone/IPad games. *NOT* board games, or ARGs, console games or anything that is unplayable.
- *Audience.* All ages. If necessary to reduce numbers we will exclude games that are directed specifically at older teens and adults.
- *Quality.* The goal is to have the best representative sample, rather than the largest sample.

We first searched for games using the criteria of "topic," identifying 35 games that were worth considering, based on an initial survey of serious games on the Web. Eight of these games were excluded when it became apparent that they did not meet the playability or the platform criteria. All of the remaining 27 games were reviewed by two of the researchers independently and annotated for the following information:

- *Game Title.* The name of the game and a link to a playable version (or an install file).
- *Funders/developers.* Who had commissioned and developed the game.
- *Funding Type.* Whether the game was the result of a corporate interest, an academic research project, a governmental initiative, or some other source (such as the broadcast media, or an independent environmental organization).
- *Cost.* How much money it cost to play/access the game.
- *Genre/style.* What type of game was it?
- *Topic.* Which aspects of sustainability did the game address?

- *Audience.* What age/demographic was the apparent target audience of the game?
- *Platform.* What technology was needed to access/play the game.
- *First impressions of quality/importance/usefulness.* A more subjective assessment of the game after an initial play through, annotating whether the reviewer enjoyed the game, what the learning curve was like, and whether or not it appeared (on the surface) to utilize any of the models of persuasion under discussion.

This review revealed a number of interesting patterns among the 27 games under consideration, including several common subgenres and game types that recurred multiple times throughout. These included simulation games that emphasized sustainability policy making, or corporate/industrial management, turn-based strategy games, environmental simulations, non-game simulators, and arcade/reflex-oriented games. It also became apparent that games fell along a spectrum of difficulty, which rendered them more or less appropriate for audiences of different ages. Within specific subgenres, some games stood out as clearly higher quality than others, containing more sophisticated visual and audio assets, more comprehensive training systems, and game play that both reviewers found more engaging.

Based on this initial review we selected 10 games to perform more deep analysis on. This included a selection of the standout games from the most common genres, as well as several games that stood out for their inability to fit comfortably within any genre category. We also made certain to include games from a range of sources and platforms, including large-scale commercial games, and free-to-play independent games developed by NGOs, media organizations, independent game developers, and small family run foundations. The final list of ten popular Web and DVD Games for Change around the topic of sustainability included: Spore (EA), Futura (Tangibles Lab, SFU), Flower (That Game Company), Climate Challenge (BBC), CEO2-The Climate Business Game (WWWF/ Alianz), Energyville (The Economist Group/Chevron); Energy City (Filament Games/JASON Science), Precipice (Centre for Digital Media/Global EESE), and Rizk (Player Three/W. Garfield Weston Foundation).

An important part of close reading is the use of one or more analytical lenses. Analytical lenses enable us to focus on elements of interest during close readings (Bizzocchi and Tanenbaum 2011). In this project, we used two analytical lenses. The first was rooted in theoretical concepts from three models of persuasion. The second was our knowledge from game studies of key elements in game design. Based on these two lenses, we began our process of analyzing the ten Games for Change about sustainability. Through individual and subsequent close readings of these Games for Change, we iteratively refined our design marker category definitions and filled in details about how each model of behavior change presented in each marker category. Our process revealed aspects of the lenses that were inaccurate or insufficient to the task. During close reading of each game we documented and tracked insights into the operations of the three models of persuasion as well as defined our design marker categories. The main outcome of our close reading process was to identify strategies used by designers to persuade players.

In a close reading methodology, in addition to applying analytical lenses to the games, we need to account for the active presence of the researcher within the play experience (Bizzocchi and Tanenbaum 2011). A rigorous critical reading of games requires an ongoing act of oscillation between a distanced and critical state of mind, and an active and engaged process of play. Rather than assume a specific role (such as expert gamer), we chose to conduct a collective reading, in addition to the solo readings. This enables us to productively triangulate between our three different play experiences.

4 Design Markers

Based on our iterative close reading analysis, we propose that the dominant behavior change model instantiated in a game may be revealed by identifying the nature of a series of six design markers. The six markers are: content; interpretation; mode of communication; game goals; game motivation and rewards; and game path and outcomes (see Tables 1 and 2). It is possible to identify the specific characteristics of these six design markers because there are observable patterns in the game that reveal specific design decisions about game elements. In order to base a game on a model of behavior change, either implicitly or explicitly, design decisions are made. Markers are game elements that reveal these decisions. Note that it is not necessary for the game developers to be aware of the model of behavior change that they are using. It may often be the case that they have internalized a particularly model, most commonly the Information Deficient model, and simply proceed in game development using this model subconsciously.

Design markers then describe specific elements of the game that provide evidence of the designer's commitment to the different aspects of one or more models of persuasion. Markers in this case might be seen as "things in the game or play experience that indicate to us when a specific model of persuasion was in operation during the game's design." They are, to use a term from hermeneutics, "textual markers," or "things in the text (game) that indicate the author's (designer's) intent." We have distilled how elements of each model manifest within the design of the game.

The first three design markers are interrelated (Table 1). As such it is difficult to write about any one of them without discussing the other two. In part this is because our unit of analysis is the entire game, not a small component, so design strategies work together. One way to think about the distinctions between the first three design markers is through the following questions that can be used to identify the markers:

What is the game's message? (content)
Why do you think this is the game's message? (interpretation)
How do you know the message of the game? (mode of communication).

Table 1 Comparison of first three design markers for each model

	Content	Interpretation	Model of communication
Information deficit	Info, facts	Enforced	Didactic presentation
Procedural rhetoric	Values, judgements	Led to	Through interaction
Emergent dialogue	Our stories or world	Open to	Dialogue, exploration, co-creation

4.1 Content

The Content marker is about the information, meaning or "text" of the game or the message that the game is trying to communicate. It is one of the most important markers as it deals with the "what" of the game. What is the core message of the game? We distinguish between the three kinds of content based on the dominant meaning that a game system encodes and expresses. While each of the three models is theoretically able to communicate any form of content, in practice we are making a distinction between them based on the dominant meaning that a system attempts to encode or express.

The Information Deficit model deals primarily with content representing facts and information. Through information the content of the game has one or more clearly articulated positions about values and behaviors that are right and wrong. The information is known, defined, and encoded into the game at design time. When a player encounters this content it is already shaped. The player is simply accountable for consuming this content. For example, NASA's Climate Change quizzes present facts about topics related to climate change, which is clearly articulated as undesirable (Fig. 1a).

The Procedural Rhetoric model deals primarily with content that supports preset judgements and values about what is right and wrong. This content is encoded within the dynamics of the game as enacted by the player. It is not *present* in the game artifact. It is there to be *inferred* and *experienced* rather than to be consumed. The player experiences the effect of making his or her decisions about actions in the game based on his or her own judgements, or values about the content domain. The player's values and judgements are explored through and shaped by their actions within the game. For example, BBC's Climate Challenge presents information through advisors (Fig. 2a).

The Emergent Dialogue model deals with the player's personal narratives about the content domain, rather than an authored or encoded message or judgement. This content is not present in the game, rather the game provides opportunities to reflect and discuss personal meanings and values outside of the mechanics of the game. In this model the game artifact serves as a means of *eliciting* a player's perspective on the content domain by providing the player with a kit of reconfigurable expressive elements. Content in this model is the most socially situated of the three, involving a dialogue around the content domain between both system and player, and with a community of players. For example, in Youtopia, the system requires one player to stamp resources as usable before the other player can use them to build. This co-dependence mechanism supports negotiation and discussion about decisions.

4.2 Interpretation

The Interpretation marker is about how the designers of the game intend the core content to be interpreted by players. Do players reach their own interpretation of the core message as they experience the results of their actions in the game play? Or are they left to form their own interpretation of what the core message of the game was? Interpretation can fall anywhere on a continuum between "closed" or forced and "open" or unenforced.

Interpretation in the context of this analysis may be seen as synonymous with "authority." The question of "how much of/how the content is open to player interpretation?" might be also framed in terms of "whose authority shapes the interpretation of the content?"

The Information Deficit model assumes that the designer's perspective or authority is dominant. The player must bend to this perspective in order to succeed. Interpretations are fixed, and pre-encoded within the artifact. For example, in the NASA quiz, a player must get the right answer to win.

In the Procedural Rhetoric model, more authority/responsibility is given to the player. Rather than the system *telling* the player what the designers want him to know, the dynamics of the game constrain the player in ways that provide *designed experiences* that ideally lead the player to conclude the desired message. Consequently, this model is less rigid, especially as the complexity of a system increases. For example, BBC's Climate Challenge enables a player to take the role of a politician, make decisions, and see the consequences of their decisions through voter response. However, the information presented by the advisors is not open to interpretation. They explicitly advise how voters will response to various actions.

In the Emergent Dialogue model, there is a more even division of authority and interpretation between the experience designer and the player, who is invited to contribute her own interpretations and perspectives to the meaning making process. In Youtopia, players are invited to create a world they want to live in (Fig. 3c—near bottom pig). It is open to the players to decide the size of the population, how much of the population's needs they will support and how much clean water and air pollution they are willing to live with.

4.3 Mode of Communication

This marker deals with how content is communicated through the game to the player. Are players told or shown the core message through text, graphics or sound? The mode of communication is a comparatively simple marker, and is largely a function of the interaction of the previous two markers.

Table 2 Comparison of last three design markers for each model

	Game goals	Motivation/reward	Path/outcomes
Information deficit	To win	Extrinsic/external—getting facts right	Single unidirectional/Single predetermined
Procedural rhetoric	To complete, play or experience game	Either/internal—doing it right	Multiple unidirectional/multiple predetermined
Emergent dialogue	To create/express and/or to communicate	Intrinsic/internal—authentic participation	Multiple bidirectional/multiple undetermined

In the Information Deficit model content is communicated *didactically* through text, visuals or sounds. A canonical example of how content is encoded is an information screen (Fig. 1a), card or non-player dialogue.

In the Procedural Rhetoric model content is communicated through interaction with the game and subsequent game responses. Content is communicated *dynamically* by experiencing events in the game that result from player actions and choices. Content is encoded within the dynamics of the game play and the logic of the simulation. In Climate Challenge, players can choose to access their advisors to get information or not.

In the Emergent Dialogue model opportunities to reflect on content are created through the game interface or mechanics *indirectly* or content *directly* (e.g., through a question—What do you think about…?) Content emerges as a bidirectional phenomenon that happens via a process of player dialogue, exploration, and co-creation supported by game elements. For example, in Youtopia, players can choose to access information to inform their decisions by placing any land use stamp in the info ring which halts the game and displays value-free information. Figure 3b shows the apartment stamp in the info ring which prompts the display of content indicating that one apartment needs three lumbers and provides shelter for a medium group of people. Based on the players' own values about balancing human and natural needs, they can then decide how to proceed.

The last three markers can be described individually, but they may be integrated or entwined with other markers in a game (Table 2). For example, game goals and game rewards may involve content about facts and getting the "right" answer and may be communicated with value-laden content.

4.4 Game Goals

Most contemporary definitions of games include some notion of winning and losing: games are structured experiences with some sort of goal state that the player is struggling to attain (Jesper 2005). In Games for Change, variations in these goal states are often markers of different models of persuasion at work.

The Information Deficit model is primarily concerned with whether or not the player has acquired the desired knowledge, so goals in this model are essentially

"Grades" on a test: quantitative assessments of performance in the form of score cards and categorical rankings. In the NASA quiz game the goal is to get a near perfect score.

The Procedural Rhetoric model is more concerned with the process of the play, so the emphasis is less on winning or losing and more on providing value oriented feedback that moves the player toward a desired interpretation of the experience. Some games that use Procedural Rhetoric resist providing any explicit goals at all to the player, as in the case of Gonzalo Frasca's *September the 12th*.[3]

In the Emergent Dialogue model, the goal is to arrive at some shared narrative or expression of a desired future world. In design this manifests in the form of open-ended tools for creating and sharing content within a community of players, and through establishing this process of dialogue as the central objective of the experience. In *Youtopia*, the players' "goal" is to explore, through game play, how to create a world that reflects their personal values.

4.5 Motivation and Reward

Closely related to the Game Goals marker is the notion of Motivation and Reward. How does the game motivate the player to take action, what types of rewards are provided to the player to remain engaged, and what types of behaviors are reinforced with rewards in the game?

In the Information Deficit model, motivation is almost always extrinsic to the experience of play itself. Rewards take the form of high scores, badges, trophies, and other evaluations of performance. However, these rewards are external to the core game experience; they seldom feed back into the game play experience, or the game system. In the Information Deficit model the player is rewarded for getting the facts right and demonstrating knowledge. In the NASA quiz game, the player is rewarded by being designated as an expert or student based on their performance (extrinsic motivation and external reward).

In the Procedural Rhetoric model, some motivation might be extrinsic but most of the rewards are connected to the actual experience of play. That is, they are internal to the core play experience. An internal reward creates new opportunities for play, or augments existing capabilities: becoming more skilled at the game is thus rewarded by making the game more interesting, complicated and challenging. Procedural Rhetoric oriented games reward the player for taking actions that are in line with the values encoded in the system and punish players in ways that are equally expressive of those values. In Climate Challenge, players are voted out of office for not making decisions to support sustainability.

In an Emergent Dialogue-oriented design, it is harder to imagine how systemic rewards and punishments would be meted out, due to the absence of predefined

[3] http://www.newsgaming.com/games/index12.htm

objectives. Feedback in a system built to facilitate Emergent Dialogue must focus on the *process*, providing rewards and incentives for authentic participation and honest engagement in the experience. In Youtopia, the players are motivated by their personal values (intrinsic motivation) and their reward is being able to work through how to a world that reflects their values.

4.6 Game Path and Outcomes

Our final maker is concerned with the path that players take through the game, and the nature of the game's outcome.

In an Information Deficit oriented design there is often only a single unidirectional route through the game system. Choices are limited to demonstrating knowledge and the outcomes in these systems are predetermined by the designer. Causality is almost always explicit in these systems: the player selects an answer and is told whether or not it was correct, as shown in Fig. 1a.

Procedural Rhetoric provides players with a more complicated simulation, often with multiple paths toward completion and outcomes that arise from the state of the computational model. Paths are still often unidirectional. Players cannot go back and reverse decisions. Cause and effect is often left implicit in a Procedural Rhetoric, however they are still predetermined by the game design. For example, in Climate Challenge there is more than one path (and sets of advice that can be followed) to stay voted in office.

In the Emergent Dialogue model the path to be navigated is less clear and there are no predetermined outcomes for the player to encounter. Causality is secondary to configurability and it is up to the player to determine his or her own stopping point. In Youtopia, players can stop whenever they are satisfied with their world. They can then use a 3D pig object to take a snapshot of their world and final impact (state of human needs and nature). At any time they can start over, change the map, change the population, erase land uses and continue playing. There are multiple bidirectional pathways and the final outcome is determined by the players rather than determined by the game design.

Based on these six design markers, we next discuss general design guidelines that may result in more effective games for behavior change and learning about social and environmental issues.

5 Design Guidelines for Persuasion Through Games for Change

Persuasive technologies mainly fail due to poor design (Fogg 2009). For example, many persuasive projects are too ambitious in that the targeted behavior may be extremely difficult to change (e.g., stop smoking) and/or the design team may be too inexperienced. Persuasive games may also fail because they fail to engage users, either because the core game mechanic is not well developed or the game itself is simply not fun or interesting to play. While concerns about quality and engagement seem reasonable explanations for the failure of many Games for Change to promote desired outcomes, it is possible that failure may, in whole or part, be because the underlying model of behavior change is flawed.

By making explicit what design decisions underlie each persuasion model in each design marker category we aim to support designers to be more intentional in the model they implement. While it is unclear which model is the best in all cases, research suggests that the Information Deficit model is ineffective at supporting behavior change (Robinson 2004). We suggest that using a more value-free approach with the Procedural Rhetoric model may work well for single player games. However, for multiplayer or games with agents, the Emergent Dialogue model may be the more effective. The particulars of each game design context will dictate how, and if, these guidelines can be implemented. In general we suggest the following guidelines based on the six design marker categories:

1. Content: Use value-free content that is about causes and consequences but not right and wrong;
2. Interpretation: Enable players to experience and interpret the consequences of their actions;
3. Mode of Communication: Allow the player to access information and facts on demand and provide mechanisms to support discussion about content;
4. Game Goals: Enable players to set goals and take action in line with their personal values;
5. Motivation and Reward: Reward actions, choices, and outcomes in line with personal values;
6. Game Path and Outcomes: Provide "no cost" opportunities to explore the consequences of a range of choices and finish when satisfied with attainment of personal goals.

Further research is needed to assess the effectiveness of these guidelines. We have one such study underway in which we are investigating the effectiveness of these guidelines in supporting emergent dialogue with 20 pairs of children using Youtopia, the open-ended tabletop sustainability game (He et al. 2010). Initial results from video coding the kind of in-depth discussions associated with emergent dialogue are promising. More work is needed.

6 Conclusion

Our work exploring different models of persuasion for behavior change in Games for Change related to sustainability is early stage and still fairly subjective. However, it contributes by explicitly articulating three models of persuasion that underlie today's Games for Change in terms of observable design strategies associated with each model of persuasion. A second contribution is that we have identified six observable design markers in existing Games for Change that enable us to identify how and when each of these models is being employed. And thirdly, our work contributes by describing our first pass at transforming the Emergent Dialogue model to one that can be used to design digital games. We illustrate our Emergent Dialogue design markers with our first fully functional Game for Change, called Youtopia. We summarize our work with six design guidelines that can be used by designers of Games for Change to support behavior change through Emergent Dialogue in multiplayer or agent-based Games for Change.

Acknowledgments We gratefully acknowledge funding and support from GRAND NCE, PICS, NSERC, and SSHRC and the insightful comments from the reviewers.

References

Antle AN, Wise AF, Hall A, Nowroozi S, Tan P, Warren J, Eckersley R, Fan M (2013) Youtopia: a collaborative, tangible, multi-touch, sustainability learning activity. Paper presented at the interaction design for children conference, New York, 24–27 June 2013

Bizzocchi J, Tanenbaum TJ (2011) Well read: applying close reading techniques to game play experiences. In: Davidson D (ed) Well played 3.0: video games, value, and meaning. ETC-Press, Pittsburgh, pp 262–290

Bogost I (2007) Persuasive games: the expressive power of video games. The MIT Press, Cambridge

Bogost I (2008) The rhetoric of video games. In: Salen K (ed) The ecology of games: connecting youth, games, and learning. The MIT Press, Cambridge, pp 117–140

Carr D (2009) Textual analysis, digital games, Zombies. Paper presented at the DiGRA 2009 conference: breaking new ground: innovation in games, play, practice and theory, London, 1–4 Sep 2009

Carr D, Burn A, Schott G, Buckingham D (2003) Textuality in video games. In: DiGRA 2003 conference: level up. Utrecht, Nov 2003

Fogg BJ (2003) Persuasive technology: using computers to change what we think and do. Morgan Kaufmann, Amsterdam

Fogg BJ (2009) A behavior model for persuasive design. Paper presented at the persuasive technologies, Claremont, 26–29 April 2009

Games for Change Society (G4C) (2014) Games for change conference. http://www.gamesforchange.org. Accessed January 8 2014.

Hayes SC, Strosahl K, Wilson KG (1999) Acceptance and commitment therapy: an experiential approach to behavior change. The Guilford Press, New York

He HA, Greenberg S, Huang EM (2010) One size does not fit all: applying the transtheoretical model to energy feedback technology design. Paper presented at the conference on human factors in computing systems, Atlanta, 10–15 April 2010

Jesper J (2005) Half-real: video games between real rules and fictional worlds. MIT Press, Cambridge

Robinson J (2004) Squaring the circle? Some thoughts on the idea of sustainable development. Ecol Econ 48:369–384

Tanenbaum TJ, Antle AN, Robinson J (2011) Procedural rhetoric meets emergent dialogue: interdisciplinary perspectives on persuasion and behavior change in serious games for sustainability. Presented at 12th annual association of internet researchers conference, Seattle, 10–13 Oct 2011

Tanenbaum TJ, Antle AN, Robinson J (2013) Three perspectives on behavior change for serious games. Paper presented at the conference on human factors in computing systems, Paris, April 27—May 2 2013

Individual and Collaborative Personalization in a Science Museum

Betsy van Dijk, Andreas Lingnau, Geert Vissers and Hub Kockelkorn

Abstract Museums increasingly use interactive technologies to make a museum visit more rewarding. In this chapter, we present opportunities that tabletop environments offer for learning, enjoyment, motivation, collaboration and playful interaction in museums. We discuss experiments with a tabletop interface in a popular science museum. This museum is an open space where visitors walk around and interact with exhibits in various ways. We integrated a tabletop application in the existing museum context that allowed visitors, mostly children, to plan and personalize their visit in a playful way. Personalization was either done individually, in a pilot experiment, or in a small group, in the main experiment. The question to be answered was whether children who follow a personalized route through the museum enjoy the experience more, are more motivated, learn more, and are more collaborative than children who follow a route that was not personalized, individually or collaboratively. We did not find many differences between experimental conditions (personalized versus nonpersonalized groups) on enjoyment and collaboration, possibly due to the fact that our research setting resembled "in the wild" studies more than classical experiments. However, in one experiment we found a learning effect of personalization. Overall, scores on the enjoyment measures were high and the experiments gave rise to engaged behavior and playful interaction. We discuss implications of our work for the study of collaborative learning in tabletop environments.

B. van Dijk (✉)
Human Media Interaction, University of Twente, Enschede, The Netherlands
e-mail: e.m.a.g.vandijk@utwente.nl

A. Lingnau
Computer and Information Science, University of Strathclyde, Glasgow, Scotland

G. Vissers
InnoTeP, Nijmegen, The Netherlands

H. Kockelkorn
Museon, The Hague, The Netherlands

A. Nijholt (ed.), *Playful User Interfaces*, Gaming Media and Social Effects,
DOI: 10.1007/978-981-4560-96-2_9,

Keywords Personalization · Collaboration · Enjoyment · Playfulness · Tabletop · Museum

1 Introduction

Museums are modernizing. They used to conserve, collect, and display objects, and in addition perform tasks in the areas of research and education, but they are increasingly interactive in the attempt to appeal to a wider public. One reason is intensifying competition from cultural events and other activities that combine education and entertainment, another that new insights have emerged about ways to introduce visitors into the world that is represented by a museum's cultural or scientific objects (Hall and Bannon 2005; Roussou 2010; Bieldt 2012; Dahl and Stuedahl 2012). Especially science museums are eager to exploit new technologies (Quistgaard and Kahr-Højland 2010). One of these is personalization, the main subject of the experiments presented in this chapter. Personalization is of considerable interest to science museums. They are visited mainly by groups, either school classes or families with children, which makes it difficult for them to offer personalized tours. Individual personalization is possible, technically, but this may not be satisfactory to all the members of a group.

The experiments that we describe were designed to explore the prospects of "collaborative personalization," a term to indicate that personalization—adapting a technology to the user's behavior and preferences—may refer to collective users even though the word personalization seems to suggest that the systems adapts to individual users. Groups received a recommended route through the museum (by means of a "quest" that consisted of questions to be answered), and this route was based on preferences that, in the main experiment, were the result of a group process. The group process was not enforced, but encouraged by the technology that was used (which includes both the tabletop device that was used and the design of the interface).

A tabletop is a computerized table with a touch screen that can be simultaneously used by several people. In the experiments this table was connected to the museum's own database, and thus to all exhibits in the exhibition room. A main experiment aimed at collaboration processes. Before this, a pilot study was conducted with a first version of the tabletop's personalization application. Part of both experiments was a route through the museum. In the pilot study, this route was personalized, based on children's individual choices. In the main experiment, the tabletop was used for collaborative personalization. Group members worked together, using the tabletop application, to arrive at a route through the museum's exhibition room that reflected the result of their collaborative choice process. The museum context will be described in the third section, the experiment (the pilot study and the main experiment) in the fourth section.

First we will, in the next section, look at other studies that used tabletops for a variety of tasks, such as learning and collecting information, and in a variety of settings. We will focus especially on the issue of user experience, which denotes the perceived qualities of technology use beyond instrumental criteria (Van Dijk et al. 2014). For a museum it is desirable that visitors find their stay in the museum enjoyable, of course, but it is the exhibition that should be the main source of enjoyment. Tabletop designers may have a different perspective. Many studies present tabletop applications that were designed to be very attractive, and user experiences are reported—in terms of enjoyment, fun, engagement, playfulness, and the like—that pertain to the tabletop as an isolated device. This is understandable practice for a number of applications (games, for instance) but the result is that we have limited knowledge of how users experience a plain and sober tabletop application, that is, an application with collaboration-supporting functionality that was not primarily designed to be attractive.

To summarize, this chapter makes two contributions to the study of tabletops. It investigates the possibility of collaborative personalization in the context of a science museum, and it examines user experience as pertaining to the use of a plain and sober tabletop that is supposed to make children's visit experience more enjoyable, through its capacity to personalize the route of the visit and to encourage collaboration among group members. In the final section, we will discuss the results of the pilot and the main experiment.

2 Related Work

As a general overview of this section, we note that most studies that consider the use of a tabletop environment for learning do not fail to mention playfulness as a supportive condition. An indication is the number of studies that refer to the article about tangibles and playful learning by Price et al. (2003). The widely held view is that tabletops, because of their tangible interface and their capacity to accommodate multiple users (Hornecker et al. 2007), will foster enjoyment, engagement, interest, motivation, and playfulness, all of which are supposed to be beneficial for interaction, collaboration and learning. This view seems derived from studies showing that "tangibles" support playful learning. While appealing, this line of reasoning needs clarification. For it is not obvious that enjoyment and playful behavior as observed in learning contexts using tangible objects will also be found in learning contexts using tabletops, which differ in many ways from "ordinary" tangible objects. Nor is it clear whether playfulness-enhancing qualities of tabletop environments should be ascribed to the tabletop device or to the application that is used. Also unclear is the relation between enjoyment, engagement, motivation, playfulness, and various other concepts related to user experience. As a final point, it is unlikely that enjoyment, engagement, and playful interaction are terms that can be used across settings. In particular, they may refer to very different phenomena when groups are small and the tabletop is used for a prolonged period of time as compared to uses in public places. In five subsections, we will briefly review the relevant literature on these issues.

2.1 Tangible Interfaces for Learning

Important parts of educational research support the view that "tangibles" are beneficial for learning (Antle 2007; Van Dijk et al. 2014). An exception is Marshall (2007), who criticizes the application of this view to tangible interfaces. He argues that studies showing the utility of tangible interfaces for learning are scarce, to the effect that designers of learning environments have little guidance and must rely upon their own presumptions about interaction. Perhaps, Marshall's critique must be taken to mean that more robust research is needed to establish the contribution of tabletop environments to learning processes. But the argument ignores evidence in favor of "hands-on activity or manipulation of physical manipulatives" (e.g., Rogers et al. 2002; Price et al. 2003; Zuckerman et al. 2005; Kangas 2010; Manches 2010). It must be noted here that the phrase "tangible interfaces" is not unambiguous. O'Malley and Stanton Fraser (2004), who endorse the use of concrete physical objects for children that fail to solve problems with abstract means, stress the aspect of "disappearing computers" rather than interaction by touch that is typical as well of tangible interfaces.

Very different is the "digital natives-digital immigrants" theory proposed by Prensky (2001). It states that children who have been used to interact with digital technology from a very early age differ profoundly, both in cognitive skills and in preferences, from those who started interacting with digital technology at a later age. The theory is controversial, mainly because of its very general claims (Helsper and Eynon 2010), but it does draw attention to changes in people's direct environment that have taken place in recent years, changes that may affect the applicability of older research on the use of tangibles.

2.2 Tabletops as Appliance and Application

A tabletop has a large shared display that accepts natural and direct interaction from multiple users through touch detection (Dohse et al. 2008), and thus can be considered a form of tangible interface. This study uses a tabletop to create an environment that supports personalization and stimulates playful interaction and collaboration. If such an environment has been created, however, it is difficult to distinguish between the engaging qualities of the tabletop as a tangible, interactive device and the engaging qualities of the application created for it. The difference is significant in view of the fact that many available studies present applications that are meant to be very attractive to prospective users.

For example, Price et al. (2003) describe an adventure game called "The hunting of the snark" that was designed to enable children (aged 6–10) to discover and reflect upon new kinds of experiences. While this study did not use a tabletop environment, it is widely cited as an exemplary case showing the learning benefits of a playful environment. The example has been followed by various studies that do make use of a tabletop environment. Iglesias et al. (2009) present a playful environment for learning intelligent systems, based on programming soccer teams in the Robocup competition. Anstead et al. (2012) developed tabletop "sideshow games" that allow small groups (families) to sort and triage photos taken in already playful circumstances—a day trip to a UK theme park. Marchetti and Petersson Brooks (2012a) present Micro-Culture, a game about urban development that was designed as a mixed reality tangible installation containing playful elements that aimed at keeping children's attention focused on the learning content and the game. Xie et al. (2008) designed an experimental comparison of school-aged children's enjoyment and engagement on three interfaces (traditional, graphical, and tangible) for solving jigsaw puzzles.

In all these studies, an attractive application was offered that would probably have produced enjoyment and playful interaction even if no tabletop device had been used—which leaves us with the question whether a tabletop environment with a plain and sober application is able to provoke playful interaction. Asai et al. (2012) attempt to answer this question. They created an interactive tabletop environment with augmented reality (AR) to explore the properties of a lunar surface browsing system. Design criteria included that the interface was easy to use and intuitive, allowed group interaction, and was attractive to visitors in terms of interaction and visualization. Thus an interactive browsing system was built that enabled users to browse information about the exploration activities of NASA Apollo missions and view the lunar surface. Apart from a few quizzes during browsing tasks no attempt was made to make the system attractive beyond the design criteria. Scores on preference tests (after a preliminary experiment or after the final exhibit) suggest that visitors found the tabletop environment enjoyable, but not much more than a WIMP (windows, icon, mouse, pointer) environment. Thus the study provides some evidence that the attractiveness of a tabletop environment depends on the application, more than on the appliance.

2.3 *Enjoyment, Fun, Engagement, Motivation*

Enjoyment, fun, engagement, involvement, immersion, interest, motivation, flow, presence, awareness, and playfulness are all concepts used to describe aspects of user experience, mostly in the context of digital games. Most of these concepts lack a widely accepted definition. Such definitions are difficult to obtain as the concepts are elusive and multidimensional (IJsselsteijn et al. 2007; Takatalo et al. 2007) and overlaps are frequently observed. We will limit our discussion to the concepts of enjoyment, fun, engagement, motivation, that are often reported in studies of tabletop-based learning.

Many studies emphasize enjoyment and fun as major elements of user experience, often treating these concepts as synonyms. Price et al. (2003) state that fun and enjoyment are effective in children's development, both supporting learning and facilitating engagement and motivation. According to Karimi and Lim (2010) enjoyment and fun enhance children's intrinsic motivation, increase participation in activities, and contribute to engagement and participation. Bell et al. (2010) note that fun creates a desire for the learning experience to reoccur and that it may motivate learners to new experiences.

However, Price and Falcão (2011) stress the need to move beyond "engagement as fun", arguing that measures of fun and enjoyment do not reveal the activities or thoughts of students interacting in new learning environments. Related is Dillenbourg and Evans' (2011) warning against high expectations of tabletop learning. They note that interactive tabletops are a new and an exciting technology, enabling multiple modes of communication and hands-on activities by multiple, co-located users, but 'deep analysis' is needed to disclose the educational merits of this technology. No "intrinsic educational effectiveness" can be assumed. Similar is Yelland's (2011) practice-based observation that playful learning requires more than teachers providing young children with materials—including materials that involve new technology. She insists that children need guidance in choosing how to use such materials.

Price and Falcão (2011) acknowledge that fun and enjoyment are beneficial for learning, and closely related to motivation, but stress that the mechanisms need to be disclosed that are mediating the relation between enjoyment and learning. Price and Falcão propose to study foci of interaction, that is, ways in which children are engaged with a learning environment. Concentrating on interactive as opposed to individual learning, they distinguish three main foci in tabletop-enhanced interactive learning: learning domain concepts, tangential activity ("a general use of the environment, where children are focused on interacting and producing different digital effects with the system, but not engaging with any learning domain-related reflection"), and technology.

By adding a tangential and a technology focus, Price and Falcão extend Kearsley and Shneiderman's (1999) Engagement Theory, a prescriptive model for technology- and project-based collaborative learning that focuses on meaningful learning tasks. Kearsley and Shneiderman claim that students are intrinsically motivated to learn if the learning environment is meaningful, and they insist that students' activities must involve "active cognitive processes" such as creating, problem-solving, reasoning, decision-making, and evaluation. Their model fits in a tradition in educational research that emphasizes interest and motivation (Schiefele 1991). Interest is positively related to the allocation of attention, the use of learning strategies, and the quality of experience (Krapp 1999). Interest and enjoyment have been found to both contribute to intrinsic motivation, but in different ways (Reeve 1989). Interest encourages individuals to explore and investigate, while enjoyment makes them willing to continue and persist in an activity.

Motivation is more narrowly defined than most other concepts pertaining to user experience, especially "intrinsic motivation" as derives from self-determination theory (Ryan and Deci 2000). This theory presupposes a human tendency toward learning and creativity. It states that this tendency may be harmed by external rewards, and will be maintained or strengthened by the feeling of being competent and self-determined (McAuley and Tammen 1989; Eccles and Wigfield 2002). According to Ryan and Deci (2000), intrinsic (or authentic) motivation will contribute to interest, excitement, and confidence, which will result in enhanced performance, persistence, creativity, but also in greater vitality.

Eccles and Wigfield (2002) propose a somewhat different relation between concepts that relate to user experience, stating that intrinsically motivated individuals "engage in an activity because they are interested in and enjoy the activity" (p. 112), and they treat intrinsic motivation as "a reason for engagement". In the interest and motivation tradition, broadly speaking, fun and enjoyment are viewed to be useful for learning but subordinate to engagement (taken as focus of attention) and intrinsic motivation. In contrast, research that emphasizes "flow" suggests a more equal relation. Shernoff et al. (2003), who treat engagement and flow as synonyms, note that "concentration, interest and enjoyment in an activity must be experienced simultaneously in order for flow to occur." O'Brien and Toms (2008) criticize such an equation of engagement and flow. They note that "flow involves intrinsic motivation, yet engaging experiences may come about as well during the nonvoluntary use of a system", and they stress that "flow requires sustained, long-term focus and loss of awareness of the outside world; engagement should still occur in the midst of today's multitasking and dynamic computer environments."

This study does not seek to elucidate the relation between enjoyment, fun, engagement, and motivation. Our discussion may show that the relevance of these concepts to user experience is beyond dispute but theoretical clarification is needed, as well as exploration of the relation between enjoyment/fun/engagement/motivation and performance. This study aims to show that tabletops, used in a way that encourages collaboration, may enhance children's experience of visiting a science museum. As will be explained in the section discussing the experiments that were conducted, we apply some widely used instruments to measure enjoyment/fun and intrinsic motivation.

2.4 Playfulness

The concept of playfulness seems to have a distinct place in tabletop-related studies. Unlike enjoyment, fun, engagement, and motivation, it is used in relation to interactive behavior. (The word "playful" is also used sometimes as an attribute of a tabletop application (see Kidd et al. 2011). We will not discuss this use of the word since our focus is on users' experiences and behaviors. For a discussion of the relation between a system's playfulness and playful interaction, see Barranis 2011).

It is possible to think of playfulness as an individual trait (Martocchio and Webster 1992; Yager et al. 1997), as illustrated by work on gamers' experiences (Takatalo et al. 2010), but examples of playfulness in tabletop-related studies do not sketch such an individual picture. Tuddenham and Robinson (2007) mention participants who "were observed taking words from each other's personal territories. However, this was not unintentional; rather they were enjoying themselves and began deliberately and playfully "stealing" words from each other and moving them into their own personal territories, resulting in exclamations like 'Give that back!'" Hornecker (2010), discussing an exhibit called Tele-Jurascope, observed that "children would playfully pretend to challenge the dinosaurs, telling them 'bite me'." Hinrichs and Carpendale (2011) refer to playfulness as "tossing media items back and forth between each other, gathering as many media items as possible, or trying to delete all media items by frantically flicking them toward the surface boundaries." Walker and Fróes (2011) give examples of "overtly playful" activities: students laughing out loud in order to gauge other visitors' reactions, imagined conversations between the subjects of a painting, secretly photographed other visitors' ankles in front of a Tudor-era bed, wrote messages to other visitors, or set a device in front of a display of plates while it played a video of someone breaking plates, in order to watch other visitors' reactions.

The examples may illustrate the interactive connotations of playfulness in studies of tabletop-related collaboration and learning. They stem from qualitative observations, and we did not find attempts to measure interactive playfulness. Measures for playfulness as an individual trait are available (see Rogers et al. 1998), as well as a teacher rating scale of preschool children's interactive peer play competencies (Fantuzzo et al. 1998), but none of these have been used in tabletop-related research.

As a theoretical underpinning, the work of Roger Caillois is sometimes referred to (Dixon 2009; Koeffel et al. 2010). Caillois distinguished four types of play, representing competition, chance, simulation, or balance/vertigo. Examples from each type can be placed on a continuum between two extremes, one called paidia (diversion, turbulence, free improvisation, and carefree gaiety), the other called ludus (arbitrary, imperative, and purposively tedious conventions to bind this capriciousness with) (Henricks 2010). The paidia-ludus distinction is very relevant in relation to the issue of playful learning in a tabletop environment. It raises the question whether a "script" is needed, or should be avoided, in order to facilitate playful interaction and learning (Streng 2009).

Concerning the relation between playfulness and the concepts discussed above, Xie et al. (2008) argue that "enjoyment and engagement are integral and prerequisite aspects of children's playful learning experiences." Marchetti and Petersson Brooks (2012a) follow Bundy's Model of Playfulness, which relates playfulness to intrinsic motivation, internal control, freedom to suspend reality, and social play cues (Bundy 1993, 2001). In a further paper, Marchetti and

Petersson Brooks (2012b) discuss a tool to enrich children's interaction that targets playful and fun experiences. Children's engagement was elicited by different choices of action, the choices were related to having fun, and a learning tool was obtained that "provided a basis for evolution of playful experiences where the children could find their own ways for interacting" (p. 43). These cases are in line with the remark that "nowadays, play is seen as an integral part of motivating learning and work."

We were able to identify only few tabletop-related studies that seek to specify the relation between playfulness and other concepts pertaining to user experience, which is in agreement with the remark made by Anstead et al. (2012) that studies of tabletop gaming rarely consider how aspects of gameplay and playfulness may be used as interaction mechanisms within other tasks.

2.5 Different Tabletop Settings

Finally, we mention the variety of settings that apply to the study of user experiences and interaction in tabletop environments. Most of the research was conducted in a museum (science, art, or cultural heritage museum), but studies have also been conducted in schools, in other public places, in firms, or even outdoors. These settings are markedly different, in several ways. Concepts to indicate user experience, such as enjoyment, fun, engagement, and motivation may not refer to identical phenomena in different settings. The same holds for interaction. We illustrate our point with two comparisons.

Many tabletop environments are used for collaborative learning. While particularly children very often have difficulties to successfully collaborate in open learning situations, tabletop interfaces do have a high potential to help children engaging in collaborative activities (Fleck et al. 2009). When situated in a school (which is not common presently) tabletops may be used for a range of tasks, many of which being part of the regular curriculum. Children use the tabletop environment as part of their normal routine, in a school setting that is the real, even official site for learning. A tabletop environment in a museum is a different phenomenon. When children use it they are away from school. The museum is probably associated with leisure and discovery, more than the school. The tabletop task to be performed may relate only remotely to the daily lessons. Therefore, working at the museum tabletop will be more playful, more enjoyable, and possibly less silent than working at a tabletop in the school. It is possible that "fun" is hardly relevant as an aspect of user experience at school, but it is relevant in the museum context. The museum visit has to be more fun than school, or it is a failure.

Even much more pronounced is the difference between a tabletop in a school or work setting and a tabletop "in the wild," in a town centre, a high street, or another public space. In the case of an "in the wild" study no arrangements have been made concerning access to and use of the table. Individuals and groups are free to use it.

Marshall et al. (2011) offer a detailed description of users' behaviors at such a "walk-up-and-use" table. These behaviors are very different from behaviors under controlled conditions. In particular, groups approached the table in a dispersed way, and they tended to use it in parallel instead of interactively. The table was designed for retrieving tourist information, and sessions were short (mean length of 2 min 10 s), which is much less than most school or work sessions. As a consequence, several concepts to capture user experience are nearly useless. Enjoyment and fun, if measured in an 'in the wild' study, will refer to short-lived affect.

3 Setting

We conducted experiments in Museon, an educational museum in the Hague, the Netherlands. Thus we join a large branch of research on interactive learning in tabletop environments in science museums, many of which present themselves as sites of exploration and learning. It is not by coincidence that science museums meet HCI research. For HCI researchers, a science museum is an attractive partner. Science museums are permanent institutions in the service of society, open to the public, available for purposes of study, education, enjoyment, and often used to support notions of collective learning as a "mobile" experience (Damala and Kockelkorn 2006). Increasingly, a picture is sketched of museums as an "informal setting," that is, a setting where learning may take place although it is not, unlike a school, a setting designated for learning (Anderson et al. 2003; Martin 2004; Meisner et al. 2007; Aubusson et al. 2012). These are qualities that are consistent with the view of learning as interactive, playful exploration that informs much of the work on tabletops and virtual environments.

The attraction is mutual. Vom Lehn et al. (2007) suggest that museums had already started—independent from developments in HCI—to deploy new technologies that were meant to engage visitors in novel ways with their exhibitions, but encountered difficulties to design interactive technologies that are accessible to a diverse audience, individuals or groups, with very different interests and commitments. This willingness to use new technologies may relate to the fact that museums are shifting from institutions devoted to conserving, collecting, research, exhibiting and educational tasks to institutions that are appealing to a broad public (Kotler and Kotler 2000). They have to compete with other activities and cultural centers that focus both on education and entertainment (Hall and Bannon 2005), but also they are moving from nineteenth century's 'object-based epistemology' toward an "object-based discourse" that seeks to introduce cultural or scientific objects into the "history of the visitor" (Pedretti 2002; Dahl and Stuedahl 2012). The new substance thus given to the museums' unchanged mission of informing the public concurs with the present viewpoint in the field of science education that the narrow focus on practical work and scientific achievements must give way to an emphasis on critical reflection in relation to scientific topics (Dani and Koenig 2008; Quistgaard and Kahr-Højland 2010).

Thus museums—art museums as well as cultural heritage and science museums—have made attempts to make visits more active, dynamic, engaging, and enjoyable, often with the aid of interactive technology. Widespread is the use of electronic guides, handheld devices that provide information and sometimes allow objects to be specified for information retrieval after the museum visit. However, these guides were designed for individual use, and they were general rather than personalized. Hsi (2003) found that they isolated visitors from their surroundings, due to the use of headsets, and that they tended to focus museum visitors on the device, instead of on the rest of the museum. Wakkary and Hatala (2007) note that PDAs (personal digital assistants) is a tool for business and "not a device that lends itself easily to playful interaction". Accordingly, next generation electronic aids and environments have been created—such as virtual reality (VR), AR, haptic displays (including tabletops), and Web3D technologies (Karoulis et al. 2006; Michael et al. 2010)—for the purpose of making museum visits more engaging and active, and more interactive and playful.

As described by Bell (2002) museums differ in their "display culture," the type of visitors they attract, the visitor behaviors they endorse, and the technologies they use. In particular, science museums tend to apply interactive displays and installations (and have been quick to adopt in-museum technology), are a frequent destination of school excursions, and support a wide range of "visitor rituals." Bell's description, based on a few U.S. science museums, also applies to Museon.

Museon has a permanent exhibition "Your World, My World" about humans and their relation with nature, culture, society, science, and technology. One main source of daily visitors are school classes (primary and lower secondary schools) usually visiting the exhibition in the morning, while in the afternoon, at weekends and school holidays the museum is visited by children with their parents or grantparents. Museon has an electronic quest that children can choose to do during their visit. About 120 terminals with touch input and a bar code scanner are available all over the exhibition area. The admission tickets of the museum have a bar code on the back that can be scanned at a terminal. The first time a ticket is used with one of the terminals a quest is generated with random questions from a central database. The visitor is asked to register his or her name to personalize the quest and will then get 12 questions about different topics in the exhibition area. When the admission ticket is scanned at one of the terminals, the next question appears. If the visitor has to move to another area in the exhibition to find the answer, he or she can use an available terminal in that area and the question will re-appear (see Fig. 1). Once the question is answered, the visitor gets immediate feedback whether or not the answer is correct and why this is the case. The question re-appears until the correct answer is given. After a fixed number of attempts the correct answer is given and a new question is displayed.

As part of PuppyIR,[1] a European project with as one of the objectives to develop new paradigms that allow children to interact with information in an

[1] http://www.puppyir.eu/

Fig. 1 Children in exhibition room approaching and using a terminal

intuitive way, an application for a multiuser, multitouch table was created that aims to enrich children's experiences during a museum visit. In addition to the existing infrastructure with terminals near the exhibits, a tabletop device was placed in Museon's entrance area. It has a multitouch surface and can identify fiducial markers, unique identifiers, similar to the concept of 2D bar-codes. Each visitor who wants to use the table gets a ticket with both a fiducial marker and a bar code. The ticket can be used as an identifier on the table and with the terminals in the exhibition space. For this interactive museum environment we developed an application for the multitouch table that can be used by up to four visitors simultaneously. With this application visitors can browse through the different exhibition topics of the permanent exhibition and they choose topics they are interested in. These topics are used to determine the contents of a personal interactive quest. In the main experiment, the groups who used the tabletop were told that their choices would have an impact on the route to be followed. In the pilot study, users did not get this information, which was a requirement of the experimental design.

4 Two Experiments

The tabletop environment as described in the previous section was used in two experiments, of which the first had a pilot status. This first experiment tested the integration of the tabletop in the existing museum context and explored the prospects of personalization by asking children, working individually, to use the tabletop and select the exhibition subjects they preferred. Participants were children from the museum's general public willing to participate in the experiment. The second experiment, which was the main experiment, focused on children's collaborative interaction around the multitouch interface. Small groups of children used the table to choose collaboratively a selection of subjects from the exhibition, thus creating a guided tour that was "personalized" at the group level. Children participating in this experiment were from two classes of a neighboring school.

4.1 Pilot Experiment

This experiment was designed to answer the question if personalization enhances the experience of a child visiting a museum, and if the tabletop-generated quest was suitable to guide children (Perloy 2011). Participants in the experiment were selected from the museum's normal visitors, often families. Participants were recruited by asking visitors if they wanted to help in a test for a new exhibit. Only groups with at least two children and at least three people in total were asked to participate. If they agreed they received a new entrance ticket in return. In total 58 people participated, 36 of whom were children. As the research focus was on children, no data of the adults' behaviors were collected.

Participants were randomly assigned to one of two conditions. Each condition had a total of eight groups. Groups consisted of three or four visitors that started their museum visit using the multitouch table to select interesting images from the permanent exhibition ("initial game"). Each group member had to indicate its preferences by selecting six images out of twelve categories. Categories were subjects such as "stones," "in the air," or "water" (for a complete list, see Perloy 2011, p. 82). With the selected images as input, personalized routes through the museum's permanent exhibition were generated. In the first condition, the selected items were used to generate a route through the exhibition. In the second condition, a route was not based on the selected categories but on the remaining six (that is, the categories the participant was least interested in). The six selected categories represented twelve exhibits. As part of the route, twelve questions had to be answered that related to these exhibits. After each good answer, the children could choose a virtual object of their liking.

After completing their route through the exhibition room, the children returned to the multitouch table to play an "end game," which in both experimental conditions was based on the route that was actually followed. The end game was played collectively. From the virtual objects collected during the quest the group members chose twelve different objects. In the end game, these objects were in the middle of the table, while 12 boxes with words were at the edges of the table. The task was to draw lines between these words and the corresponding virtual objects. Time was limited; after 2 min the connections were checked and an animation was shown. One by one the virtual objects were highlighted. The lines became green when a connection was correct, red when it was incorrect, and finally the group's end score was displayed.

The experiment was designed to answer the question whether children who followed a personalized route through the museum (first condition) enjoyed the experience more and learned more than children who followed a route that did not contain the exhibits they had chosen (second condition). Noticeably, the children in the second condition did not seem to notice that the route they followed did not match their preferences (although their parents did). Children's experiences in the two conditions were quite comparable. No difference was found in enjoyment, measured using Read and MacFarlane's (2002) dimensions of "expectations" (the Smileyometer) and "endurability," which comprises 'remembrance' (questions to see how well the children remembered things they had enjoyed) and "returnance" (the Again–Again table to see if the children would like to go through the experience again, as an indicator of fun), and by the subscale interest/enjoyment of the Intrinsic Motivation Inventory (IMI) (McAuley et al. 1989; University of Rochester, (n.d.)). On the whole, scores on these enjoyment measures were very high, which suggests a ceiling effect that prevents differences in enjoyment between the two conditions. These high scores were taken to indicate that children liked the touch table interface and the quest, which was also a question that the experiment sought to answer.

A significant difference in learning ('understanding') was found between the conditions in the end game, measured by the categories "remembering" and "understanding" of a revised version of Bloom's Taxonomy (Krathwohl 2002; Forehand 2005). No significant difference was found for "remembering," which concerned information not directly related to the questions that had been answered during the quest. Understanding, in contrast, measured as a group's score in the end game, differed considerably between the conditions. The number of good connections made by groups in the first (preferred categories) condition was significantly higher than the number of good connections made by groups in the second (nonpreferred categories) condition. A likely explanation is that participants in the first condition, whose preferences were used, had greater interest in the topics encountered during the tour (and in the end game), and had acquired more knowledge than participants in the other condition. An alternative explanation, that participants in the first condition already knew more about the topics they indicated as most interesting, is to be rejected. If this were the case, participants in the first condition would also have had higher scores on the questions that were asked during the tour, which tested factual knowledge, and they would have answered these questions faster. Both were not the case.

4.2 Main Experiment

This experiment examined the effects of "collaborative personalization" on children's museum experience. It was designed to find out if a collaborative 'initial game' at the multitouch table and an ensuing personalized quest would encourage interaction and collaboration between participants, and if it would have effects in terms of enjoyment of and engagement in the tabletop-supported museum visit. Collaboration was not enforced in any way, except that the design of the tabletop application encouraged group members to discuss and align their choices—through a shared area in the middle of the tabletop, used to store subjects of joint interest. This shared area was then used for personalization of the group's route through the museum. Discussions at the table were expected, as well as joint answering of questions during the quest, but groups were not asked to collaborate because such a

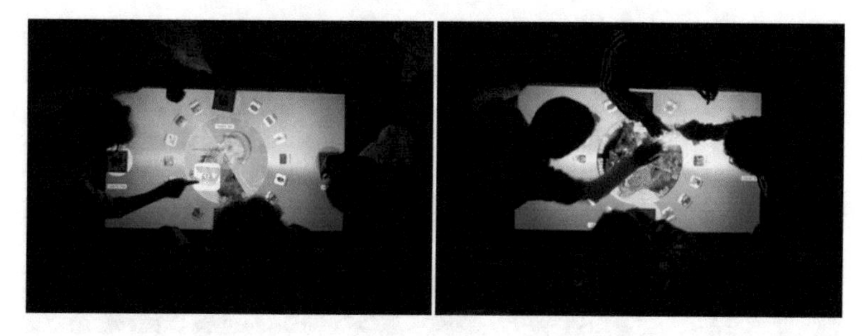

Fig. 2 Groups at the tabletop, choosing objects sometimes sequentially and sometimes simultaneously

request would have made it impossible to find out if it was the tabletop application or the collaboration request that encouraged collaboration.

Participants were 48 children from two classes of a Dutch primary school located close to Museon. 27 children, aged 10–11, were from the pre-final year of primary school, 21 children, aged 11–12, from the final year. The children were accompanied by their teachers. The teachers formed groups of four children (three if necessary) before the museum visit. The pre-final year class was divided in seven, the final year class in six groups.

The experimental setup involved two conditions. In the "table condition" the children started the museum visit at the multitouch table, where they selected topics of joint interest by choosing images that represented these topics (see Fig. 2). All topics were part of Museon's permanent exhibition. The selected topics were used to automatically generate a personalized route through the museum's exhibition room. Seven teams (four from the pre-final year, three from the final year, in total 26 children) were assigned to the condition that started at the multitouch table. Six teams (three from both pre-final; and final year, in total 22 children) were assigned to the no-table condition that started with an electronic quest generated with random questions from a central database. In this condition the children did not start at the tabletop and their quest was not personalized. In both conditions, the quest included 12 questions to be answered at 12 different exhibits.

The teams received instructions about the procedure of the experiment. After all members of a team had finished the quest, they went to the multitouch table to get further information about the visited exhibits and to choose again topics/exhibits they were most interested in (see Fig. 3). All team members performed this task individually at the table, also the teams from the no-table condition. The results could be used later, in school, as a personal catalogue (Van Dijk et al. 2012). After completing this task, the children handed in their tickets and filled in a questionnaire.

A questionnaire was used to measure enjoyment/fun and collaboration, and contained four open questions relating to learning outcomes.[2] It also contained

[2] The complete questionnaire can be found in the user evaluation toolkit on the PuppyIR site (see http://www.puppyir.eu/results/user-evaluation-toolkit).

Fig. 3 Children collaborating at the tabletop

three questions to measure collaboration: (1) I tried very hard to support others on doing the quest, (2) I collaborated much with my classmates, (3) I liked supporting others during the quest. Moreover, datalogs were used to derive a quest score, and the attempt was made to observe children during the quest. Observation as a data collection method is needed when children's collaborative behaviors are to be studied, as in the main experiment, or changes in students' focus of interaction during tabletop-centered learning (Price and Falcão 2011). Subjective measures (questionnaire, interview) are less suitable for this, which includes concurrent task rating as proposed by Teague et al. (2001) because its requirement to give concurrent ratings will interrupt the interaction process that is being studied. However, in the experiment no systematic observation results could be obtained as it turned out that several teams did the quest almost simultaneously.

To measure the aspects of enjoyment/fun, Read and MacFarlane's (2002) Smileyometer and Again–Again table were used, and an adapted version of the subscale interest/enjoyment of the IMI. Results of the Again–Again table are often highly correlated with results of the Smileyometer, but both measures were used because of their difference in emphasis. Conceptually, judging parts of the museum tour is different from giving an opinion on what you like to do again.

The IMI interest/enjoyment subscale is considered the self-report measure of intrinsic motivation and is developed for use by adults. It also measures fun and enjoyment, like the Smileyometer and the Again–Again table, but it measures interest as well. The IMI interest/enjoyment scale was adapted to make it suitable for children (the negatively formulated statements were reversed and smileys were used instead of numbers on a 5-point scale). Thus, an instrument was obtained that couples the enjoyment and interest constructs to a reliable scale for measuring children's intrinsic motivation, a very important factor for learning (Van Dijk et al. 2012).

The questionnaire's open questions, focusing at learning outcomes, concerned a dinosaur in the museum. The children in the table condition were asked to include the dinosaur in their collection. The children in the no-table condition all had the dinosaur included in their route.

The table condition was expected to stimulate interaction and collaboration between the children in a team, to enhance learning outcomes, and to increase enjoyment and intrinsic motivation. No significant differences were found between the two experimental groups on enjoyment as measured by the Smileyometer and the Children IMI interest/enjoyment scale. Moreover we did not find any significant differences between the experimental groups on collaboration and on how much they learned. A significant difference between the conditions was found on the Again–Again table measure for the end game: children who already used the multitouch table at the start of the visit were more positive to use the table again than children who did not start at the table. We presume, based on observations, that the children found the purpose of the final game less clear than the purpose of the initial game, and that the children that played the initial game still remembered the positive experience of playing the initial game. However, the results of the other enjoyment measures on the final game were similar for the two experimental conditions hence no conclusive explanation can be given.

While we found no significant differences between the experimental conditions on enjoyment of quest and end game as measured with the Smileyometer, we did find a significant difference between age groups: children from the pre-final year scored higher than children from the final year on enjoyment of the quest and initial game. Moreover, the overall mean score on the Children IMI interest/enjoyment scale was significantly higher for children from the pre-final than for children from the final year.

Getting back to our main research question: we did not find clear evidence that the collaborative initial game at the multitouch table supported and invoked interaction and collaboration between the participants of the experiment and enriched the children's experience during the museum visit. The results on the enjoyment scales were quite high and where we found small differences these were in favour of the condition with the multitouch table at the start. However, there is not enough evidence to substantiate the claim that team-based personalization using a multitouch table enhances childrens' visiting experience. This conclusion does not, of course, discredit the use of multitouch tables as an extra attraction or an element of science museums' exhibitions.

5 Discussion

The experiments described in the previous section were not true experiments. Many factors could not be controlled for, especially not in the main experiment that was conducted in a very condensed time period. As a result the museum visit was poorly integrated in the school curriculum, and many children were simultaneously present at the site, which precluded systematic observation. Nonetheless, the experiments were fairly successful. Children enjoyed using the multitouch application and the subsequent quest and end game, they did work in groups (and sometimes even complained that individual questions during the quest thwarted collaboration), and some of the differences between experimental conditions were statistically significant. Here we note that most differences between the experimental conditions were in the expected direction. That many of these differences were not significant may be due to the fact that the number of participants was rather small. A further point is that our research setting resembled "in the wild" studies more than classical experiments, which is supposed to increase ecological validity (Hornecker and Nicol 2011) but goes at the expense of clear, uncontaminated effects.

We emphasize two contributions to the study of collaborative learning in tabletop environments. The first is our use of a tabletop application to shape collaborative personalization and learning, even though few significant effects can be reported. This lack of results is consistent with the remark that recent studies on collaborative learning around tabletops suggest effects to be small (Fleck et al. 2009) and the observation that collaboration outcomes of tabletops in the context of interactive learning are hard to show (Dillenbourg and Evans 2011). Still, we want to think of reasons that could explain the small number of significant effects in our study.

(i) The discriminating power of collaboration questions may be insufficient as the questions were asked after the museum visit had been completed. Systematic observation might have produced more evidence, but the circumstances of the experiment did not allow. Therefore it is likely that examples of collaboration went unnoticed.
(ii) In both conditions all participants were members of a group. In the pilot experiment, these were mostly natural groups (often families), in the main experiment children were assigned to a group. None of these groups received instructions on how to behave as a group, but it is possible that "being in a group" was not the weak cue it was intended to be, but something that overshadowed the activities at the multitouch table in the table condition. An indication of the importance of "groupness" (Meneses et al. 2008) appears from a comparison of the natural groups in the pilot experiment and the created groups in the main experiment. Some measures were used in both experiments, one of these being the Smileyometer. It was used to measure enjoyment/fun of the three parts of the visit (tabletop, quest, end game). Of these three, only the quest can be used for comparison (the tabletop was not used in the main experiment's no-table condition, and the end game differed between pilot and main experiment). The quest was more or less similar in all cases, though there was a difference in the specific exhibits to be visited. On the Smileyometer's 5-point scale, mean scores were 4.22 in the first condition and 4.62 in the second condition of the pilot experiment. In the main experiment, mean scores were 3.50 in the table condition and 3.00 in the no-table condition. The difference between the experiments is striking, but equally striking is that in the pilot experiment the route of nonchosen exhibits (second

condition) was enjoyed even more by children than the route of chosen exhibits. It is possible that familiar exhibits had been chosen at the table and that unfamiliar exhibits turned out to generate more excitement and enjoyment. However, no such effect was found in the main experiment. More likely is that the parents in the pilot study's second group did their very best to make the unfamiliar exhibits appealing to the children.

(iii) That no instructions were given on how to behave as a group stemmed from the desire to keep groups in the different conditions comparable and "let the context speak for itself". We were aware of the fact that the experiment involved a certain degree of scripting (assigned groups, use of the tabletop, a tour with questions that directed attention to selected exhibits, an end game) and we did not want to add further questions or guidelines that might prevent playful interaction between the children. In the end we may have underestimated the strength of the museum context (or overestimated the strength of the tabletop context).

The second contribution is that the experiment aimed at enjoyable, engaging, playful, interactive behavior, but not as a result of adding a device that was designed to induce such behavior. The tabletop application used was attractive but not exhilarating.[3] As reported we found a few enjoyment effects. Playfulness, engagement, and interactive behavior are phenomena that require systematic observation (Read and McFarlane 2002; Price and Falcão 2011). What can be reported here, based on free (as opposed to systematic) observations, is that the experiments did encourage engaged behavior and gave rise to playful interaction despite considerable scripting. In this respect the experiments are comparable with many other studies that consider tabletop-related interaction. Questions that remain to be answered are how much scripting is possible before enjoyable, engaged, playful interaction is hampered, and whether tabletop scripts have similar effects as behavioral scripts.

In addition to these intended contributions we mention the importance of age. In the main experiment we found clear differences between the pre-final and final year of primary school. These differences were not anticipated—they could only appear because the participating school sent children from these two classes—but they suggest a large difference between children (pre-final year) and early adolescents (final year). At least the difference shows the risks of drawing conclusions about tabletop-related enjoyment, engagement, and playful interaction if differences between (groups of) participants are not taken into account.

Finally we comment on our use of a tabletop in a science museum. The experiments we described were a first attempt to use a tabletop as a device to support personalization, directing visitors of the science museum to exhibits of their own choice. Perhaps it was alerting in general, rather than personalized alerting, that gave rise to a high level of enjoyment, and perhaps this was only

[3] http://hmi.ewi.utwente.nl/puppyir/results/demos/expedition-museon/

possible in an exciting environment such as a science museum. Not a very stimulating conclusion would be that the tabletop did not distract from what the museum has to offer. More tempting, we think, is the conclusion that the very presence of a tabletop works as an incentive to think of new ways to intensify visitors' experiences.

Acknowledgments The research leading to these results has received funding from the European Community's Seventh Framework Programme FP7/2007–2013 under grant agreement n° 231507 and from the Pieken in de Delta project (number PID092064) SEA (Smart Experience Actuator).

References

Anderson D, Lucas KB, Ginns IS (2003) Theoretical perspectives on learning in an informal setting. J Res Sci Teach 40(2):177–199
Anstead E, Durrant A, Benford S, Kirk D (2012) Tabletop games for photo consumption at theme parks. In: Proceedings of the 2012 ACM international conference on Interactive tabletops and surfaces, pp 61–70
Antle AN (2007) The CTI framework: informing the design of tangible systems for children. In: Proceedings of the 2006 ACM international conference on tangible and embedded interaction, pp 195–202
Asai K, Kondo T, Kobayashi H, Sugimoto YY (2012) Browsing lunar surface using tabletop augmented reality at exhibit in science museum. J Inf Syst Educ 10(1):11–31
Aubusson P, Griffin J, Kearney M (2012) Learning beyond the classroom: implications for school science. In: Fraser BJ, Tobin KG, McRobbie CJ (eds) Second international handbook of science education. Springer, Netherlands, pp 1123–1134
Barranis NJ (2011) Altering user perceptions of applications: how system design can impact playfulness and anxiety. University of Illinois, Urbana
Bell G (2002) Making sense of museums: the museum as 'cultural ecology'. Intel White Paper, pp 1–17. http://echo.iat.sfu.ca/library/. Accessed 02 June 2013
Bell MW, Smith-Robbins S, Withnail G (2010). This is not a game—social virtual worlds, fun, and learning. In: Peachey A et al (eds) Researching learning in virtual worlds. Human-computer interaction series. Springer, London, pp 177–191
Bieldt N (2012) Building a transformative museum? Getting to 'Our Place' through the creative industries lens: a case study from New Zealand. In: Kristiansen E (ed) The transformative museum—proceedings of the DREAM conference, May 2012, Roskilde University, Denmark, pp 1–14
Bundy AC (1993) Assessment of play and leisure: delineation of the problem. Am J Occup Ther 47(3):217–222
Dahl MI, Stuedahl D (2012) Transforming children's participation and learning in museums: from singular dialogues to a multilayered explorative experience. In: Kristiansen E (ed) The transformative museum—proceedings of the DREAM conference, May 2012, Roskilde University, Denmark, pp 96–109
Damala A, Kockelkorn H (2006) A taxonomy for the evaluation of mobile museum guides. In: Proceedings of the 8th conference on human-computer interaction with mobile devices and services. ACM, pp 273–274
Dani DE, Koenig KM (2008) Technology and reform-based science education. Theor Pract 47(3):204–211

Dillenbourg P, Evans M (2011) Interactive tabletops in education. Int J Computer-Support Collab Learn 6(4):491–514

Dixon D (2009) Nietzsche contra Caillois: beyond play and games. In: The philosophy of computer games conference, University of Oslo

Dohse KC, Dohse T, Still JD, Parkhurst DJ (2008) Enhancing multi-user interaction with multi-touch tabletop displays using hand tracking. In: IEEE proceedings of first international conference on advances in computer-human interaction, pp 297–302

Eccles JS, Wigfield A (2002) Motivational beliefs, values, and goals. Annu Rev Psychol 53(1):109–132

Fantuzzo J, Coolahan K, Mendez J, McDermott P, Sutton-Smith B (1998) Contextually-relevant validation of peer play constructs with African American Head Start children: Penn interactive peer play scale. Early Childhood Res Q 13(3):411–431

Fleck R, Rogers Y, Yuill N, Marshall P, Carr A, Rick J, Bonnett V (2009) Actions speak loudly with words: unpacking collaboration around the table. In: Proceedings of the ACM international conference on interactive tabletops and surfaces, pp 189–196

Forehand M (2005) Bloom's taxonomy: original and revised. In: Orey M (ed) Emerging perspectives on learning, teaching, and technology. http://epltt.coe.uga.edu/index.php?title=Bloom%27s_Taxonomy. Accessed 31 May 2013

Hall T, Bannon L (2005) Designing ubiquitous computing to enhance children's interaction in museums. In: Proceedings of the 2005 ACM conference on interaction design and children, pp 62–69

Helsper EJ, Eynon R (2010) Digital natives: where is the evidence? Br Educ Res J 36(3):503–520

Henricks TS (2010) Caillois's man, play, and games—an appreciation and evaluation. Am J Play 3(2):157–185

Hinrichs U, Carpendale S (2011) Gestures in the wild: studying multi-touch gesture sequences on interactive tabletop exhibits. In: Proceedings of CHI 2011, Vancouver, BC, Canada, 7–12 May 2011, pp 3023–3032

Hornecker E (2010) Interactions around a contextually embedded system. In: TEI 2010, Cambridge, Massachusetts, USA, 25–27 Jan 2010, pp 169–176

Hornecker E, Nicol E (2011) Towards the wild: evaluating museum installations in semi-realistic situations. In: Proceedings of re-thinking technology in museums conference. http://www.ehornecker.de/Papers/HorneckerNicolFinal.pdf. Accessed 31 May 2013

Hornecker E, Marshall P, Rogers Y (2007) From entry to access: how shareability comes about. In: Proceedings of the 2007 conference on designing pleasurable products and interfaces, pp 328–342

Hsi S (2003) A study of user experiences mediated by nomadic web content in a museum. J Comput Assist Learn 19(3):308–319

Iglesias CA, Garijo M, Santiago F (2009) A playful approach for learning intelligent systems in engineering. In: EAEEIE annual conference, 2009, pp 1–4

IJsselsteijn W, de Kort Y, Poels K, Jurgelionis A, Bellotti F (2007) Characterising and measuring user experiences in digital games. In: Proceedings of ACE 2007, Salzburg, Austria. http://www.yvonnedekort.nl/pdfs/ACE%202007%20workshop%20submission%20TUe%20final.pdf. Accessed 31 May 2013

Kangas M (2010) Creative and playful learning: learning through game co-creation and games in a playful learning environment. Thinking Skills Creativity 5(1):1–15

Karimi A, Lim YP (2010) Children, engagement and enjoyment in digital narrative. Curriculum, technology & transformation for an unknown future. In: Steel CH, Keppell MJ, Gerbic P, Housego S (eds) Proceedings of the Australasian society for computers in learning in tertiary education (ASCILITE), Sydney, pp 475–483

Karoulis A, Sylaiou S, White M (2006) Usability evaluation of a virtual museum interface. Informatica 17(3):363–380

Kearsley G, Shneiderman B (1999) Engagement theory: a framework for technology-based learning and teaching. http://home.sprynet.com/~gkearsley/engage.htm. Accessed 28 May 2012

Kidd J, Ntala I, Lyons W (2011) Multi-touch interfaces in museum spaces: reporting preliminary findings on the nature of interaction. In: Rethinking technology in museums: emerging experiences, University of Limerick

Koeffel C, Hochleitner W, Leitner J, Haller M, Geven A, Tscheligi M (2010) Using heuristics to evaluate the overall user experience of video games and advanced interaction games. In: Evaluating user experience in games. Springer, London, pp 233–256

Kotler N, Kotler P (2000) Can museums be all things to all people?: missions, goals, and marketing's role. Mus Manage Curatorship 18(3):271–287

Krapp A (1999) Interest, motivation and learning: an educational-psychological perspective. Eur J Psychol Educ 14(1):23–40

Krathwohl DR (2002) A revision of Bloom's taxonomy: an overview. Theor Pract 41(4):212–218

Manches AD (2010) The effect of physical manipulation on children's numerical strategies: evaluating the potential for tangible technology. Ph.D. thesis, University of Nottingham

Marchetti E, Petersson Brooks E (2012a) From lecturing to apprenticeship; introducing play in museum learning practice. In: eLmL 2012, fourth international conference on mobile, hybrid, and on-line learning, pp 94–99

Marchetti E, Petersson Brooks E (2012b) Playfulness and openness: reflections on the design of learning technologies. In: ArtsIT 2011, LNICST 101. Springer, Heidelberg, pp 38–45

Marshall P (2007) Do tangible interfaces enhance learning? In: Proceedings of TEI'07. ACM, New York, pp 163–170

Marshall P, Morris R, Rogers Y, Kreitmayer S, Davies M (2011). Rethinking 'multi-user': an in-the-wild study of how groups approach a walk-up-and-use tabletop interface. In: Proceedings of the 2011 annual conference on human factors in computing systems, May 2011, pp 3033–3042

Martin LMW (2004) An emerging research framework for studying informal learning and schools. Sci Educ 88 (Special Issue):71–82

Martocchio JJ, Webster J (1992) Effects of feedback and cognitive playfulness on performance in microcomputer software training. Pers Psychol 45(3):553–578

McAuley E, Tammen VV (1989) The effects of subjective and objective competitive outcomes on intrinsic motivation. J Sport Exerc Psychol 11(1):84–93

McAuley E, Duncan T, Tammen VV (1989) Psychometric properties of the intrinsic motivation inventory in a competitive sport setting: a confirmatory factor analysis. Res Q Exerc Sport 60(1):48–58

Meisner R, vom Lehn D, Heath C, Burch A, Gammon B, Reisman M (2007) Exhibiting performance: co-participation in science centres and museums. Int J Sci Educ 29(12):1531–1555

Meneses R, Ortega R, Navarro J, de Quijano SD (2008) Criteria for assessing the level of group development (LGD) of work groups groupness, entitativity, and groupality as theoretical perspectives. Small Group Res 39(4):492–514

Michael D, Pelekanos N, Chrysanthou I, Zaharias P, Hadjigavriel LL, Chrysanthou Y (2010) Comparative study of interactive systems in a museum. In: Digital heritage. Springer, Heidelberg, pp 250–261

O'Brien HL, Toms EG (2008) What is user engagement? a conceptual framework for defining user engagement with technology. J Am Soc Inform Sci Technol 59(6):938–955

O'Malley C, Stanton Fraser D (2004) Literature review in learning with tangible technologies. Futurelab, Bristol, Report 12

Pedretti E (2002) T. Kuhn meets T. Rex: critical conversations and new directions in science centres and science museums. Stud Sci Educ 37(1):1–41

Perloy LM (2011) The influence of personalization on education and enjoyment in a museum. Master thesis, Human Media Interaction, Twente University, Enschede

Prensky M (2001) Digital natives, digital immigrants: part 1. On the Horiz 9(5):1–6

Price S, Pontual Falcão T (2011) Where the attention is: discovery learning in novel tangible environments. Interact Comput 23:499–512

Price S, Rogers Y, Scaife M, Stanton D, Neale H (2003) Using 'tangibles' to promote novel forms of playful learning. Interact Comput 15(2):169–185
Quistgaard N, Kahr-Højland A (2010) New and innovative exhibition concepts at science centres using communication technologies. Mus Manage Curatorship 25(4):423–436
Read JC, MacFarlane S (2002) Endurability, engagement and expectations: measuring children's fun. In: Proceedings of the 2002 conference on interaction design and children, IDC'02. ACM, New York, NY, pp 1–23
Reeve J (1989) The interest-enjoyment distinction in intrinsic motivation. Motiv Emot 13(2):83–103
Rogers CS, Impara JC, Frary RB, Harris T, Meeks A, Semanic-Lauth S, Reynolds M (1998) Measuring playfulness: development of the child behaviors inventory of playfulness. Play Cult Stud 1:121–135
Rogers Y, Scaife M, Gabrielli S, Smith H, Harris E (2002) A conceptual framework for mixed reality environments: designing novel learning activities for young children. Presence: Teleop Virtual Environ 11(6):677–686
Roussou M (2010) Learning by doing and learning through play: an exploration of interactivity in virtual environments for children. In: Parry R (ed) Museums in a digital age. Routledge, London, pp 247–265
Ryan RM, Deci EL (2000) Self-determination theory and the facilitation of intrinsic motivation, social development, and well-being. Am Psychol 55(1):68–78
Schiefele U (1991) Interest, learning, and motivation. Educ Psychol 26(3/4):299–323
Shernoff DJ, Csikszentmihalyi M, Schneider B, Shernoff ES (2003) Student engagement in high school classrooms from the perspective of flow theory. Sch Psychol Q 18(2):158–176
Streng S (2009) The role of personal and shared displays in scripted collaborative learning. In: Gross T et al (eds) INTERACT 2009, Part II, LNCS 5727, pp 876–879
Takatalo J, Häkkinen J, Kaistinen J, Nyman G (2007) Measuring user experience in digital gaming: theoretical and methodological issues. In: Electronic imaging 2007, international society for optics and photonics, 649402, pp 1–13
Takatalo J, Häkkinen J, Kaistinen J, Nyman G (2010) Presence, involvement, and flow in digital games. In: Evaluating user experience in games. Springer, London, pp 23–46
Teague R, De Jesus K, Ueno MN (2001) Concurrent vs. post-task usability test ratings. In: CHI'01 extended abstracts on human factors in computing systems, pp 289–290
Tuddenham P, Robinson P (2007) Distributed tabletops: supporting remote and mixed-presence tabletop collaboration. In: TABLETOP'07, second annual IEEE international workshop on horizontal interactive human-computer systems, pp 19–26
University of Rochester (n.d.). Intrinsic motivation inventory. http://www.selfdeterminationtheory.org/questionnaires/10-questionnaires/50. Accessed 2 Dec 2011
Van Dijk EM, Lingnau A, Kockelkorn H (2012) Measuring enjoyment of an interactive museum experience. In: Proceedings of the 14th ACM international conference on multimodal interaction, pp 249–256
Van Dijk EMAG, van der Sluis F, Perloy LM, Nijholt A (2014) A user experience model for tangible interfaces for children. International Journal of Arts and Technology, Special Issue on Expressive Interactive Systems for Arts and Entertainment (selected contributions from Intetain 2011), to appear
Vom Lehn D, Hindmarsh J, Luff P, Heath C (2007) Engaging constable: revealing art with new technology. In: Proceedings of the SIGCHI conference on human factors in computing systems, pp 1485–1494
Wakkary R, Hatala M (2007) Situated play in a tangible interface and adaptive audio museum guide. Pers Ubiquit Comput 11(3):171–191
Walker K, Fróes I (2011) The art of play: exploring the roles of technology and social play in museums. In: Beale K (ed) Museums at play: games, interaction and learning. Museums Etc, Edinburgh. http://www.lkl.ac.uk/kevin/play.pdf. Accessed 30 May 2013

Xie L, Antle AN, Motamedi N (2008) Are tangibles more fun?: comparing children's enjoyment and engagement using physical, graphical and tangible user interfaces. In: Proceedings of the 2nd international conference on tangible and embedded interaction, pp 191–198

Yager SE, Kappelman LA, Maples GA, Prybutok VR (1997) Microcomputer playfulness: stable or dynamic trait? ACM SIGMIS Database 28(2):43–52

Yelland N (2011) Reconceptualising play and learning in the lives of young children. Australas J Early Childhood 36(2):4–12

Zuckerman O, Arida S, Resnick M (2005) Extending tangible interfaces for education: digital montessori-inspired manipulatives. In: Proceedings of the SIGCHI conference on human factors in computing systems, CHI'05. New York, NY, pp 859–868

NoProblem! A Collaborative Interface for Teaching Conversation Skills to Children with High Functioning Autism Spectrum Disorder

Massimo Zancanaro, Leonardo Giusti, Nirit Bauminger-Zviely, Sigal Eden, Eynat Gal and Patrice L. Weiss

Abstract This chapter presents *NoProblem!*, a multi-user interface designed to teach social conversation and social interaction skills to children with High Functioning Autism Spectrum Disorder (HFASD). HFASD is defined as a pervasive developmental disorder which involves deficits in socio-communication skills as well as with repetitive behaviors and restricted interests but with a relative high IQ (usually higher than 75). *NoProblem!* implements the principles of Cognitive-Behavioral Therapy (CBT) by interleaving learning and experience techniques. The process entails the display of short vignettes to present social tasks for the children to learn about and problem solve. In the learning part, a facilitator (teacher or therapist) uses the system to teach pairs of children about the phases of social conversation in different settings (e.g., at school, in the playground). The children consider and discuss prepared solutions, some of which are more appropriate than others; they may also propose and audio-record their own solutions. In the experience part, role-play of various conversational solutions is used. *NoProblem*! aims to provide opportunities to act out various conversational responses in a social setting that was selected as the most appropriate one and to practice it in a safe environment with partners who will cooperate. The results of formative and evaluation studies that demonstrate system usage and effectiveness in the teaching of conversational skills for HFASD children are presented. The children involved were interested, felt very competent doing this task, perceived that they could make choices, and felt minimal pressure and tension.

M. Zancanaro (✉) · L. Giusti
FBK, Trento, Italy
e-mail: zancana@fbk.eu

N. Bauminger-Zviely · S. Eden
Bar-Ilan University, Tel Aviv, Israel

E. Gal · P. L. Weiss
University of Haifa, Haifa, Israel

A. Nijholt (ed.), *Playful User Interfaces*, Gaming Media and Social Effects,
DOI: 10.1007/978-981-4560-96-2_10, © Springer Science+Business Media Singapore 2014

1 Introduction

Autism spectrum disorder (ASD) is defined as a pervasive developmental disorder which involves deficits in social relationships, communication impairments, repetitive behaviors, and restricted interests (Bailey et al. 1996). The severity and range of disordered thought processes, communication interactions and behaviors vary from one child to another, ranging from very low to very high functioning. High Functioning Autism Spectrum Disorder (HFASD) is defined as a pervasive developmental disorder which involves deficits in socio-communication skills as well as with repetitive behaviors and restricted interests but with a relative high IQ (usually higher than 75).

Children with HFASD make or accept fewer social initiations and spend more time playing alone compared to their peers (Koegel et al. 2001). These children typically exhibit difficulties in tasks requiring social (Pierce and Schreibman 1995), affective (Hobson et al. 1988) and motivational (Schreibman 1988) competencies.

Recently, the principles and techniques of cognitive behavioral therapy (CBT; Dobson and Dobson 2009) have been adapted to help children with HFASD engage in more effective interactions with peers (Solomon et al. 2004; Bauminger 2007). CBT describes social competence as a multidimensional concept and assumes reciprocity between the ways an individual thinks, feels, and behaves in social situations (e.g., Dobson and Dobson 2009). The approach presumes that a more efficient cognitive understanding of the social world will lead to successful social adjustment in future situations.

CBT uses both cognitive and behavioral techniques to enhance social competence. In a typical CBT session, a facilitator (either a therapist or a teacher with specific training) involves children in one or more learning techniques such as concept clarifications, self-instruction, and reflection, followed by or intermingled with behavioral techniques such as practicing via role play, rehearsal, and reinforcement. These allow children to experience some social constructs which will then be reflected upon. The two parts should therefore be interleaved.

Among the cognitive techniques that appear to be effective for children with HFASD is "problem solving" which consists of suggesting a social schema to perceive and learn about various social situations (Bauminger 2002). Among the behavioral techniques that have been demonstrated to be effective in enhancing interactive/collaborative skills in HFASD is "rehearsal through role-play" where children practice these skills in a safe, controllable environment (Mackay et al. 2007).

Another technique often used in CBT is called modeling. The concept of modeling or observational learning is based on the idea that children may acquire new skills by observing others rather than only by personal experience. Modeling as an intervention technique was first introduced by Bandura and Menlove (1968). A video modeling intervention typically involves an individual watching a video demonstration and then imitating the behavior just viewed. The specific approach used in *NoProblem!* is called "video self-modeling" (VSM) and it presents the "self" in the video, allowing individuals to imitate targeted behaviors by observing themselves successfully performing an appropriate behavior (Dowrick 1999). Video modelling and video self-modelling have been used across multiple disciplines and populations to teach a wide variety of skills including motor behaviors, social skills, communication, self-monitoring, functional skills, vocational skills, athletic performance, and emotional regulation (Bellini and Akullian 2007).

From the technological point of view, the approach used in the current study is that of multi-user interfaces (Yuill and Rogers 2012) using either multiple mice on a desktop computer or a multi-touch tabletop device that is specifically designed to allow simultaneous interactions by multiple users (Dietz and Leigh 2001). These approaches exploit the concept of "working together" in the design of computer programs aimed at supporting collaboration (Morris et al. 2006). *NoProblem!* is a multi-user application designed to support a facilitator to conduct a CBT-based session where the learning part makes use of the "problem solving" technique and the experience part makes use of video-modeling and role playing, presently focusing on practicing social conversations. An authoring tool is provided for the facilitator to adapt the multimedia material such that different pictures may be displayed and other social situations may be defined.

2 Related Work

Researchers and clinicians have noted the value of technology, in general, and computer-based activities, specifically, as therapeutic and educational tools for people with ASD (Grynszpan et al. 2005). Children with ASD are usually highly motivated by computer-based tasks (Hart 2005). Focusing on a computer screen, where only necessary and relevant information are provided, may help people with ASD reduce distractions from extraneous sensory stimuli. Furthermore, computer programs are generally free from social demands and can provide consistent and predictable responses (Moore 1998). A clearly defined task appears to help people with ASD concentrate on a computer-based activity (Murray 1997). Nevertheless, the responses from some professionals and parents to technology have been mixed due to the fear of increased social withdrawal (Panyan 1984; Bernard-Opitz et al. 1990).

At present, there are few examples of technologies designed to support CBT for children with HFASD, and they have been explored primarily via the use of computer games to teach emotion recognition and regulation skills (e.g., Beaumont and Sofronoff 2008; Golan and Baron-Cohen 2006; Golan et al. 2008). In our own work, we have experimented with a tabletop computer game called *Join-In*, inspired by CBT, to teach social competencies to children with ASD (Giusti et al. 2011; Zancanaro et al. 2011; Bauminger-Zvieli et al. 2013). Piper et al. (2006) investigated how a four-player cooperative computer game that runs on tabletop technology was used to teach effective group work skills in a middle school social group therapy class of children with Asperger's Syndrome. Gal et al. (2009)

evaluated the effectiveness of a three-week intervention in which a co-located tabletop interface was used to facilitate collaboration and positive social interaction for children with ASD. Significant improvements in key positive social skills were achieved. Similarly, Battocchi et al. (2010) studied the ability of a digital puzzle game to foster collaboration among children with ASD; in order to be moved, puzzle pieces had to be touched and dragged simultaneously by the two players. Hourcade et al. (2012) explored the potential of multi-touch tablets to engage children with ASD in face-to-face social activities. Results showed how these applications led to pro-social behavior and fostered the development of appropriate social skills.

3 Design Objectives

Social conversation, the ability to initiate and maintain an effective conversation with another participant, is a core skill impaired in children with HFASD.

Focus of multi-usage: Since these children face great challenges in coping with social interactions, they often prefer to engage in solitary, computer-based activities. The first design objective of *NoProblem!* was to develop a technology paradigm able to foster collaboration with another child. A key design objective was thus to employ a multi-user interface where the task is inherently social: the interface requires two children to use it.

Make explicit the role of facilitator: Although there is increasing evidence about the beneficial effect of the presence of a facilitator in educational interventions for children with ASD (Kroeger et al. 2007) for most educational games, the facilitator acts as a secondary user by moderating use of the game but without actually using the interface. In our previous work on Join-In, we explicitly addressed this issue (Zancanaro et al. 2011); *NoProblem!* was also designed as an interface for the facilitator to smoothly run the session and manage children's interactions, yet enable their active control.

Allows for personalization and open use: Children with ASD have widely varying abilities and limitations and require personal adaptation of tools. They also require opportunities to learn about and practice a variety of tasks. Thus, *NoProblem* was designed with a focus on social conversation as a prototypical example of social interaction but also enabled a facilitator to create content for other situations. The software's open structure and authoring tool capabilities (via the creation of new multimedia content) meets this design objective.

4 The Prototype

NoProblem!, shown schematically in Fig. 1, is a multi-user application designed to be run on a multi-user touch table (i.e., DiamondTouch Dietz 2001) or on a standard desktop computer using a special component enabled to recognize three

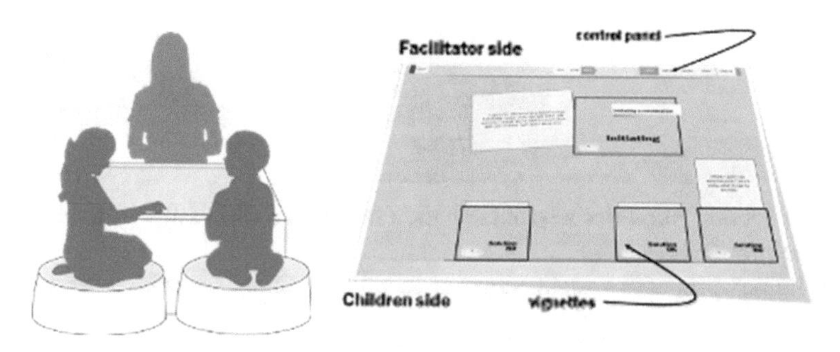

Fig. 1 *NoProblem!* setup showing location of children and facilitator (*left*) and appearance of surface during the discussion of the alternatives solutions for a social conversation (*right*)

individual mice. It is divided into two tightly integrated parts: a learning part which realizes a structured version of the CBT social problem solving technique and an experience part based on the role play technique.

The social problem solving technique presents a series of social vignettes that display the four phases of a social conversion (initiating, maintaining, switching and ending) in three meaningful settings (at school, after school, and at home). The children, guided by the facilitator, explore alternative solutions or suggest new ways of managing each conversation phase. The experience part consists of a recording tool for the facilitator to involve the children in a role-playing session, video-recording their performance, and reviewing it together with them; thus children obtain direct feedback from the facilitators or their peers on their actual performance of social conversation in an activity that is engaging.

The control panel (Fig. 2) is used by the facilitator to switch between the two parts (social problem solving and role-play) and to move along the different steps of the learning part.

NoProblem! takes advantage of the ability to recognize multiple touches by different users of DiamondTouch (replicated by the multi-mice component when the interface works on a standard desktop computer) in order to constrain interactions in a variety of ways. For example, in some cases, to operate the system the children and the facilitator need to tap on the surface together (if the facilitator does not touch the surface, the system is not activated) or in other cases, only the facilitator can activate certain functions (such as the control panel) and a child will not be able to operate them until they are activated.

When run on the tabletop device, the interface is oriented in such a way that both children sit on one side and the facilitator sits on the other side facing them (Fig. 1, left panel) while when it is used on a desktop computer, the three users sit on the same side.

At present, three social settings are included with each setting exploring all four phases of a social conversation (initiating, maintaining, switching, and ending).

- "At school" provides a social situation in which a child talks with friends during the school recess.

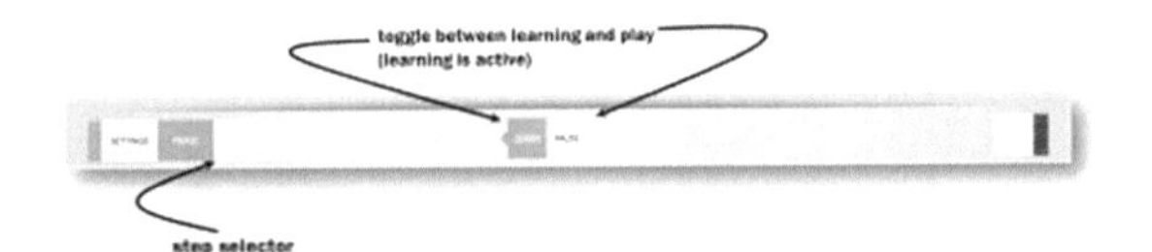

Fig. 2 Facilitator's control panel

- "After school" provides a social situation in which a child talks with friends at a soccer practice session.
- "At home" proposes a social situation in which a child talks with a friend whom he is visiting.

4.1 Learning Part

The learning part consists of five steps.

1. *Setting* step is where the setting for the social conversation is chosen among the several available (currently three, see above);
2. *Phase* step is where the specific phase of the conversation is chosen (see Fig. 3);
3. *Present* step is where the relevant aspects of the chosen phase are presented within the specific setting; these aspects may then be discussed by the children and the facilitator;
4. *Record* step is where the children can record their own solution to the social problem; this step is optional and the facilitator can skip or postpone it until after the Select step;
5. *Select* step is where the children are asked to consider three different alternatives provided by the system related to the specific phase in the conversation and to choose the most appropriate one.

In the *Setting* step, the children and the facilitator decide which setting to choose for the social conversation. Each setting has an introductory story describing it. In order to choose a social setting to continue the session, the three have to tap the corresponding card together (triple tap).

In the *Phase* step, four cards that represent the four phases are displayed (Fig. 3). As before, audio and textual description supports the facilitator and the children to choose.

After the phase is chosen by the children and the facilitator (by tapping together on a card), the system progresses to the *Present* phase. The facilitator can return to the *Phase* step at any time in order to show and discuss its rules.

In the *Present* step, a presentation of the social conversation for the chosen phase in the chosen setting is provided. Again, the children can listen to and read a detailed description of the social task by clicking on the corresponding card.

The facilitator may stay at this step for as long as needed in order to elaborate on the social task. The facilitator then operates the control panel to move to either the *Record* step or the *Phase* step (to choose another phase for the same setting) or the *Setting* step (to choose another setting).

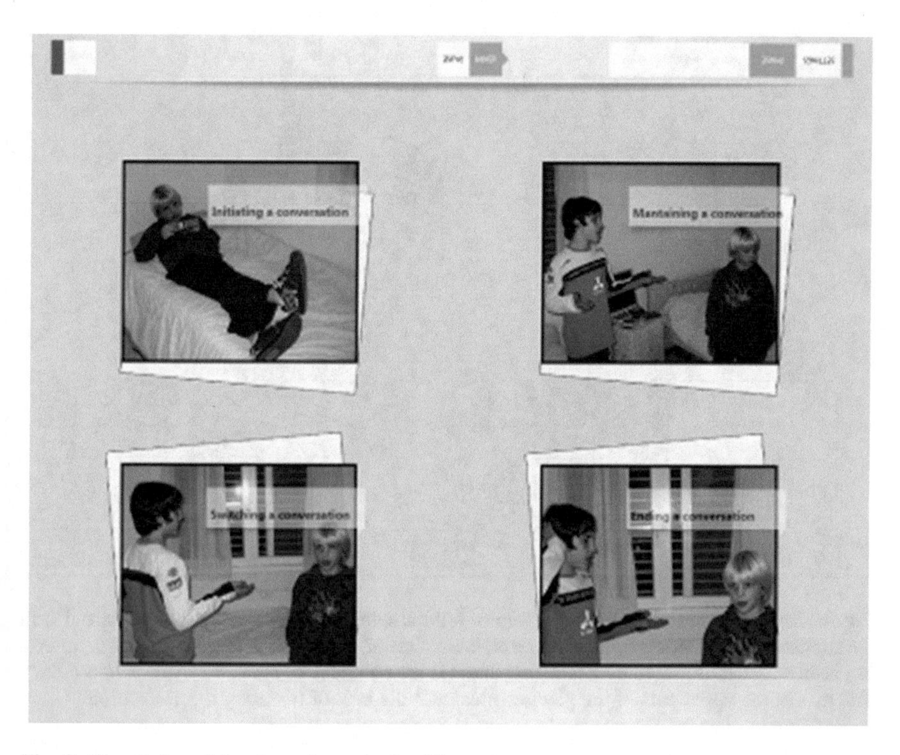

Fig. 3 The choice of the four phases in the *Phase* step

In the *Record* phase, a small control panel for audio-recording appears below the Record button in the control panel step selector. Only the facilitator can use this button to enable the children to record their own narrated solution. The system displays a card that will then be shown together with the system generated alternatives in the *Select* step. This card can be re-recorded as many times as needed. A new recording will supersede any previous ones (though all the recordings are stored by the system for the purposes of logging and analysis).

In the *Select* step (Fig. 4), the system presents the children with three alternate solutions to the social problem and, if available, the card recorded by the children. The children can listen to each alternative by tapping the card's play button or by reading the textual description which appears when they tap on the panes located behind the cards; the system also allows to play videos in the cards but this feature was not used in this implementation.

They can then together select the card that they think contains the most appropriate solution. When they tap together on a card, it flips over to show an orange background (if it is a not appropriate solution) or a yellow background (if it is an appropriate one) together with a short textual feedback message (Fig. 4). The card recorded by the children can either have an empty textual content or contain a neutral message.

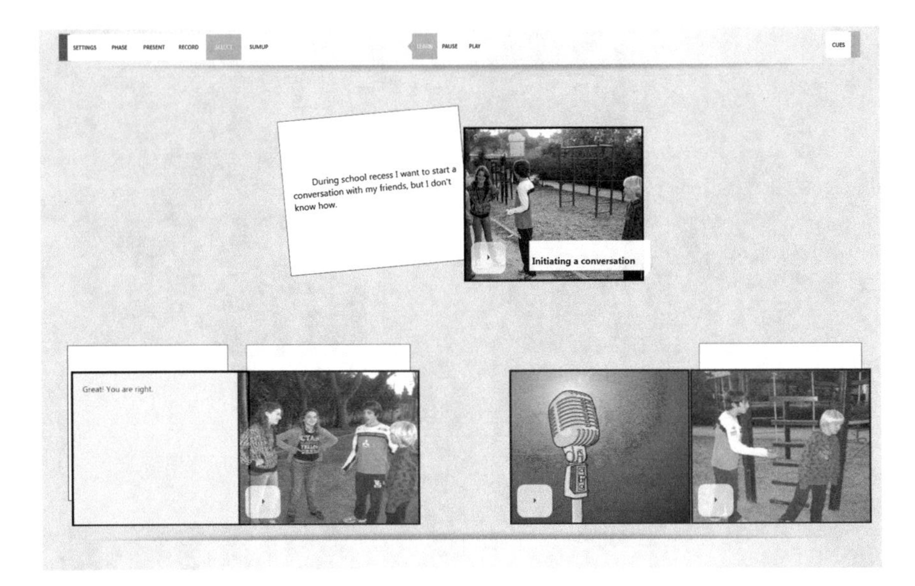

Fig. 4 Screenshot showing the *Select* step with three solutions presented by the system and with the addition of a solution recorded by the children. One of the solutions (second from *left*), after it has been selected, is turned over to reveal that it is the correct one (indicated by its *yellow color*). The text in the upper part of the pictures illustrates the task of initiating a conversation

The facilitator may stay at this step for as long as needed to explore the alternatives and discuss them. The facilitator can then move back to either the *Phase* or the *Setting* steps to change the phase or setting, respectively, or go to the *Record* step to have the children record (or re-record) their own alternate solutions.

4.2 Experience Part

Once the setting and the phase of the social conversation have been chosen, the facilitator can access the experience part (role-play) at any time and then return freely to the learning part.

When entering the experience part, the control panel changes: the step selector is hidden and a few new buttons appear on a video console to record and play the role-played social conversations. Note that only the facilitator can operate the video camera and the record button; there is no need for the children to stay near the table during the role play.

By tapping the *Cue* button at the rightmost position, the facilitator can display a window on the display surface (table or computer screen) that shows hints for appropriate social conversation in the given phase and setting.

5 Evaluation

5.1 Formative Evaluation

The goals of the formative evaluation were to provide end user feedback on the prototype. We were also interested in observing how the children with HFASD understood and felt about the problem and solution parts of the task, and the children's interest in and enjoyment of the task, perceived competence, choice, and feelings of tension.

Nine boys and one girl with HFASD, aged 9–13 years, enrolled in special education classes (Grades 2–5) within a mainstream elementary school, participated in a single session. The gender unbalance in the sample reflects the proportion of HFASD in the population. Three questionnaires were used in the evaluation study. The Scenario Experience Feedback Questionnaire (SEFQ; see Weiss et al. 2011 and appendix) consists of 14 items, rated on a 5-point scale, to query the children's enjoyment, understanding, ease of use, and other usability issues while playing the games; the Scenario Learning Feedback Questionnaire (SLFQ; see Weiss et al. 2011 and appendix) consists of 5 items to query how well the children understood and felt about the problem and solution part that precedes each game and the Intrinsic Motivation Inventory (IMI) task evaluation questionnaire (Plant and Ryan 1985) consists of 22 items, rated on a 7-point scale, designed to assess a user's interest in and enjoyment of the task, perceived competence, perceived choice, and feelings of pressure or tension while doing the task.

The usability of the prototype was examined in two different platforms: a tabletop touch-based device and a multi-mice desktop version. The questionnaires were administered to the children after each session. At the end of the experience, the children were interviewed about what they learned from the session and about their preferences related to using the tabletop touch-based device versus the multi-mice desktop computer.

The results demonstrated (see Fig. 5) that the children were motivated by the task, both when presented on the tabletop and the multi-mice platform. They felt competent doing it, perceived that they could make choices during the task, and felt minimal tension. They understood the main aims of the conversation tasks, and the various phases. Most of the children preferred using the tabletop rather than the computer, but enjoyed both of them. The perceived competence was somewhat lower with the multi-mice configuration presumably because of the reduced immediacy of the mouse compared with the touch modality (Weiss et al. 2011).

In conclusion, the results of the formative evaluation study helped to ensure that *NoProblem!* is a usable and enjoyable application and suitable to achieve its therapeutic goals. While it seems to be more appealing when used on the tabeltop device, it appears to be also feasible when implemented via a less costly and technically complex platform.

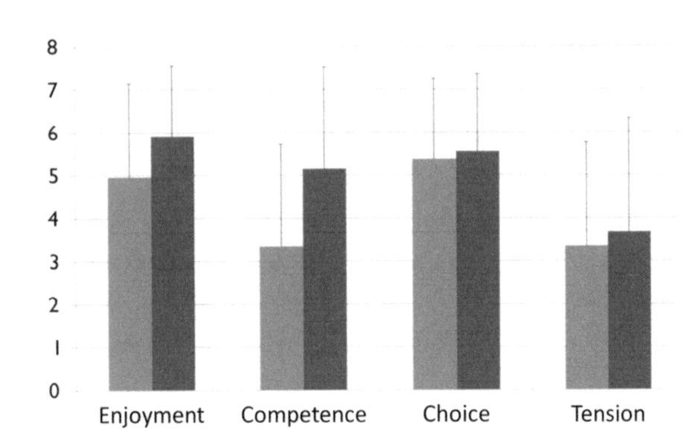

Fig. 5 Mean plus 1 standard deviation of the four components of the intrinsic motivation questionnaire when using the multi-touch table (*dark histogram*) or multi-mice computer (*light histogram*)

5.2 Intervention Study

The goal of the study was to examine the effectiveness of collaborative technology intervention combined with CBT to teach the concepts of social collaboration and social conversation to children with HFASD. The *NoProblem!* prototype was used together with another prototype (Join-In, Giusti et al. 2011), developed during the same project) for a 12-week intervention. The study is briefly summarized here, for further information see Bauminger-Zvieli et al. (2013).

Twenty-two children with HFASD (11 pairs), 18 male and 4 female (mean age 9.83 years), participated in the study; all had an IQ of 70 or above and no serious behavioral problems. The gender unbalance in the sample reflects the proportion of HFASD in the population. The intervention spanned twelve 45-min lessons, with six lessons for the social task-collaboration (Join-In) and six lessons for social conversation (*NoProblem!*). The participants were divided to 2 groups, each received a different intervention order. Two special education teachers and one occupational therapist, all experts in working with children with HFASD, were trained to implement the intervention.

In order to assess changes in social cognition, problem solving and concept clarification, several measures were used including *concept clarification* for cooperation and social conversation (Bauminger et al. 2003, 2004) and a drawing task to assess the children's ability to engage socially with their friends (Bauminger 2007b). In addition, to examine generalization into another domain of social cognition (that was not directly taught in the intervention), a measure of Theory of Mind (TOM), reflecting children's ability to understand mental states such as their belief in others, was utilized via the "Strange Story" (Happé 1994).

With regard to the social cognitive results, mean values of social cognitive measures improved significantly. The children demonstrated a better understanding of the concepts of collaboration and social conversation and could make a more relevant definition of this concept and provide examples of times they experienced collaboration with peers. Thus, children were able to recognize the social situation of a social conversation in a picture; they were able to provide a definition and examples of social conversations and to suggest relevant conversation topics based on the topics that they learned during the intervention. However, they were less able to suggest conversation topics different from those learned during the intervention. Interestingly, they also showed improvements in higher order TOM following the intervention, indicating an indirect effect of this intervention as well as a direct one. With regard to the social engagement, improvement was achieved in the summary social engagement behavior variables, specifically for cooperative behaviors.

In conclusion, the results of this study provided important preliminary insight into the implementation of CBT intervention supported by collaborative technology and computer games. The findings that children improved significantly from baseline to post intervention in socio-cognitive awareness and understanding has important implications for future research and educational applications.

6 Discussion

The *NoProblem!* studies provided valuable insight regarding the possibility of using collaborative technologies as a basis for teaching social competence skills. An important aspect that emerged from the formative study is the functional equivalence of the multi-mice version with respect to the previously thought to be more engaging tabletop version. Indeed, several studies had suggested that for collaborative tasks a direct touch interface is more effective than the use of multi-mice, in particular for what concerns higher levels of awareness, fluidity of interactions and spatial memory (Hornecker et al. 2008; Antle and Droumeva 2009). In our case, the interaction was considered just slightly less engaging when performed with the multi-mice version, an occurrence that may have been related to the children's considerable involvement with the task itself (role-playing) rather than with the use of the computer.

It is worth noting that the strengths of *NoProblem!* derive not only from the motivational and engagement value of computer-based tasks (see, for example, Pennington 2010) but also from the provision of new tools that intrinsically support a facilitator while conducting a session. The need to support the facilitators in shaping the children's experience by giving them a fundamental and dynamic role has already been discussed in our previous work on Join-In (Zancanaro et al. 2011). With *NoProblem!*, we provided an even larger role for the facilitators, that of "director" of a movie: the facilitator was encouraged to instruct the children (by involving them in a discussion about the effectiveness of different behaviors) and coordinating their role play. Our studies provide some evidence that this approach was both helpful to the children in supporting them to learn new skills and made the task of the facilitator easer by providing additional tools for rehearsal and feedback.

One of the key aspects for *NoProblem!* was the possibility for the facilitators to customize and add different stories to teach different types of social competence skills beyond the current one (social conversation). To meet this requirement, we implemented a desktop application (for MS Windows) that allows the construction

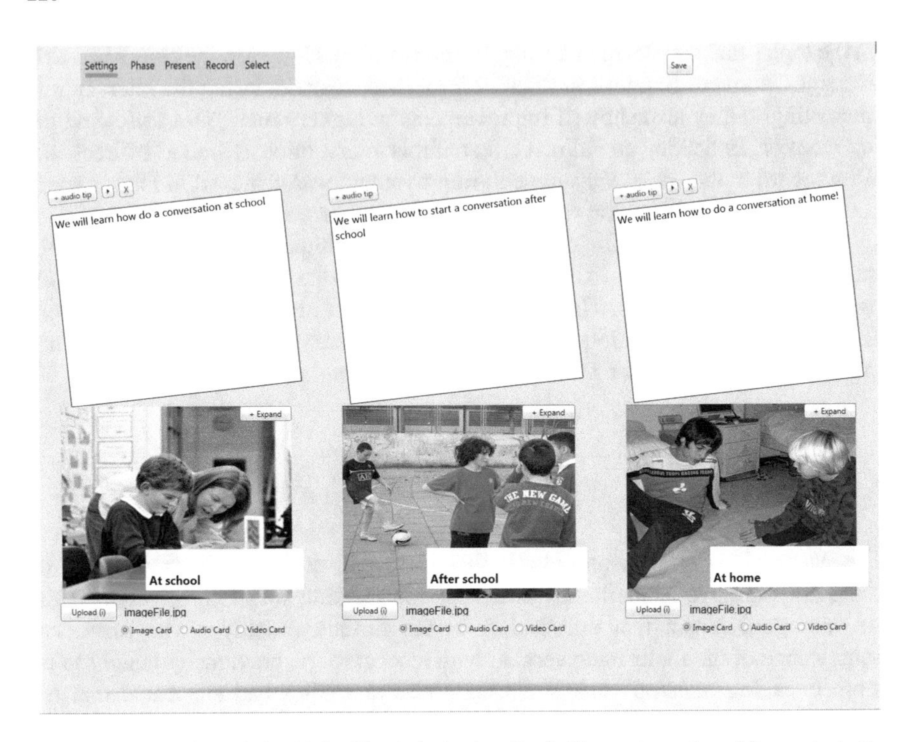

Fig. 6 A screenshot of the *NoProblem!* Authoring Tool. The text on the white cards is the descriptions of the problems presented in the multimedia cards below

of social stories by uploading pictures, videos, texts and audio content into multimedia vignettes. In order to simplify the authoring tool, the content can be fully customized but the structure is fixed. That is, three "stories" are offered in the setting phase (as a multimedia vignette), for each story four possible variations may be defined (as a multimedia vignette), and three outcomes or stories endings for each variation are created (as a multimedia vignette). Therefore, the construction of any new content application entails the preparation of 36 multimedia vignettes. Each multimedia vignette can be as simple as a picture with a text. Audio content and video clips may also be added for some specific vignette if needed. In order to improve usability and learnability, the authoring tool closely resembles the *No-Problem!* interface (see Fig. 6). The vignettes for the social stories, the variations and the outcomes are in the same position as in the *No-Problem!* interface (even if the authoring tool is a standard single-mouse desktop application) and they can be authored independently from each other by clicking on the corresponding buttons and browse the file system to locate the appropriate multimedia file.

Although the studies presented here did not focus on this aspect, this tool was used to support the acquisition of conflict resolution skills by adolescents with HFASD (Hochhauser et al. 2013). In this context, the potential of *No-Problem!* was better exploited by using videos in the multimedia cards too.

7 Conclusions

In this chapter, we presented *NoProblem!,* a collaborative tool inspired by the principles of Cognitive-Behavioral Therapy to provide opportunities to act out various conversational responses in a social setting. The results of a formative evaluation and of an intervention study have been briefly summarized. The studies demonstrated that the children are interested and feel very competent doing this task and that the system usage is effective in the teaching of conversational skills for HFASD children.

In our future work, we aim to improve the system, in particular to provide better support for revealing cues to the children and to enhance discussion. We also aim to investigate more fully how the facilitators use the system in order to achieve their educational objectives.

Acknowledgments This work was partially funded by the European project COSPATIAL (FP VII - 231266). We want to thanks the children and the teachers that participated in the study.

Appendix

Scenario experience feedback questionnaire (SEFQ)

Please circle the number that best reflects your response:

	Not at all ☹		😐	Very much ☺	
Did you enjoy the game?	1	2	3	4	5
Did you succeed in the game?	1	2	3	4	5
Was the game was easy for you?	1	2	3	4	5
Would you like to play the game again?	1	2	3	4	5
Did you feel you could control the game?	1	2	3	4	5
Did the game respond to you as you expected?	1	2	3	4	5
Did you have to wait too much time for the game to respond?	1	2	3	4	5
Did the game seem realistic to you?	1	2	3	4	5
How clear was the computer's response during the game?	1	2	3	4	5
Did you feel that you were an active player in the game?	1	2	3	4	5
How quickly did you get used to playing the game?	1	2	3	4	5
Did you feel comfortable during the game?	1	2	3	4	5
Did you like being with your partner during the game?	1	2	3	4	5
How much did the moving objects and sounds distract you during the game?	1	2	3	4	5

Scenario learning feedback questionnaire (SLFQ)

Please circle the number that best reflects your response:

	Not at all			Very much	
	☹		😐	☺	
Did you succeed in solving the problem?	1	2	3	4	5
Was solving the problem easy for you?	1	2	3	4	5
Did the problem and answers seem realistic to you?	1	2	3	4	5
How quickly did you get used to hearing about and suggesting answers to the problem?	1	2	3	4	5
Did you like solving the problem with your partner?	1	2	3	4	5

References

Antle AN, Droumeva M, Ha D (2009) Hands on what? Comparing children's mouse-based and tangible-based interaction. Proceedings of IDC 2009, ACM Press, pp 80–88

Bailey A, Phillips W, Rutter M (1996) Autism: towards an integration of clinical genetic, neuropsychological, and neurobiological perspectives. J Child Psychol Psychiatry 37:89–126

Bandura A, Menlove FL (1968) Factors determining vicarious extinction of avoidance behavior through symbolic modeling. J Pers Soc Psychol 8:99–108

Battocchi A, Ben-Sasson A, Esposito G, Gal E, Pianesi F, Tomasini D, Venuti P, Weiss PL, Zancanaro M (2010) Collaborative Puzzle game: a tabletop interface for fostering collaborative skills in children with autism spectrum disorders. J Assist Technol 4(1):4:14

Bauminger N (2002) The facilitation of social-emotional understanding and social interaction in high functioning children with autism: intervention outcomes. J Autism Dev Disord 32:283–298

Bauminger N (2007) Group social-multimodal intervention for HFASD. J Autism Dev Disord 37(8):1605–1615

Bauminger N (2007b) Individual social-multi-modal intervention for HFASD. J Autism Dev Disord 37:1593–1604

Bauminger N, Shulman C, Agam G (2003) Peer interaction and loneliness in high functioning children with autism. J Autism Dev Disord 33:489–507

Bauminger-Zvieli N, Eden S, Zancanaro M, Weiss PL, Gal E (2013) Increasing social engagement in children with high-functioning autism spectrum disorder using collaborative technologies in the school environment. Autism (in press, published online 24 April 2013)

Beaumont R, Sofronoff K (2008) A multi-component social skills intervention for children with Asperger syndrome: the junior detective training program. J Child Psychol Psychiatry 49:743–753

Bellini S, Akullian J (2007) A meta analysis of video modeling and video self modeling interventions for children and adolescent with ASD. Exceptional Children 73:264–287

Bernard-Opitz V, Ross K, Tuttas ML (1990) Computer assisted instruction for children with autism. Ann Acad Med 19:611–616

Dietz PH, Leigh DL (2001) DiamondTouch: a multi-user touch technology. In: Proceedings of UIST 2001, Orlando, pp 219–226, 11–14 Nov 2001

Dobson D, Dobson K (2009) Evidence-based practice of cognitive behavioral therapy. Guilford Press, New York

Dowrick PW (1999) A review of self-modeling and related interventions. Appl Prev Psychol 8:23–39

Gal E, Bauminger N, Goren-Bar D, Pianesi F, Stock O, Zancanaro M, Weiss PL (2009) Enhancing social communication of children with high functioning autism through a co-located interface. Artif Intell Soc 24:75–84
Giusti L, Zancanaro M, Gal E, Weiss PL (2011) Dimensions of collaboration on a tabletop interface for children with autism spectrum disorder. In: Proceedings of CHI 2011, Vancouver, Canada, ACM Press, New York, 7–12 May 2011
Golan O, Baron-Cohen S (2006) Systemizing empathy: teaching adults with Asperger syndrome or high-functioning autism to recognize complex emotions using interactive multimedia. Dev Psychopathol 18:591–617
Golan O, Baron-Cohen S, Golan Y (2008) The 'reading the mind in films' task (child version): complex emotion and mental state recognition in children with and without autism spectrum conditions. J Autism Dev Disord 38:1534–1541
Grynszpan O, Martin JC, Nadel J (2005) Designing educational software dedicated to people with autism. In: Pruski A, Knops H (eds) Assistive technology: from virtuality to reality, proceedings of AAATE 2005, IOS Press, Lille, France, pp 456–460
Happé F (1994) An advanced test of theory of mind: understanding of story characters' thoughts and feelings by able autistic, mentally handicapped, and normal children and adults. J Autism Dev Disord 24:129–154
Hart M (2005) Autism/excel study. In: ASSETS 2005: the seventh international ACM SIGACCESS conference on computers and accessibility
Hochhauser M, Gal E, Zancanaro M, Weiss PL (2013) Computer supported collaborative conflict negotiation strategy application for adolescents with high functioning autism spectrum disorders (HFASD). International Meeting of the Federation of Autism Research, Spain
Hornecker E, Marshall P, Sheep Dalton N, Rogers Y (2008) Collaboration and interference: awareness with mice or touch input. Proceedings of CSCW'08, ACM Press, pp 167–176
Hourcade JP, Bullock-Rest NE, Hansen TE (2012) Multitouch tablet applications and activities to enhance the social skills of children with autism spectrum disorders. Pers Ubiquit Comput 16(2):157–168
Koegel LK, Koegel RL, Frea WD, Fredeen RM (2001) Identifying early intervention target for children with autism in inclusive school settings. Behav Modif 25:754–761
Kroeger KA, Schultz JR, Newsom C (2007) A comparison of two group-delivered social skills programs for young children with autism. J Autism Dev Disord 37:808–817
Mackay T, Knott F, Dunlop A (2007) Developing social interaction and understanding in individuals with autism spectrum disorder: a group work intervention. J Intell Dev Disabil 32:279–290
Morris RM, Huang A, Paepcke A, Winograd T (2006) Cooperative gestures: multi-user gestural interactions for co-located groupware. Proceedings of CHI'06, ACM Press
Moore D (1998) Computers and people with autism/Asperger syndrome. Communication(magazine of The National Autistic Society) Summer, pp 20–21
Murray DKC (1997) Autism and information technology: therapy with computers. In: Powell S, Jordan R (eds) Autism and learning: a guide to good practice. David Fulton, London, pp 100–117
Plant RW, Ryan RM (1985) Intrinsic motivation and the effects of selfconsciousness, self-awareness, and egoinvolvement: an investigation of internally controlling styles. J Pers 53:435–449
Panyan M (1984) Computer technology for autistic students. J Autism Dev Disord 14:357–382
Pennington R (2010) Computer-assisted instruction for teaching academic skills to students with autism spectrum. Focus Autism Other Dev Disabl 25(4): 239–248
Pierce K, Schreibman L (1995) Increasing complex social behaviors in children with autism: effects of peer-implemented pivotal response training. J Appl Behav Anal 28(3):285–295
Piper AM, O'Brien E, Morris MR, Winograd T (2006) SIDES: a cooperative tabletop computer game for social skills development. In: Proceedings of CSCW'06, ACM Press, pp 1–10
Schreibman L (1988) Autism. Sage Publications, Newbury Park

Solomon M, Goodlin-Jones BL, Anders TF (2004) A social adjustment enhancement intervention for high functioning autism, Asperger's syndrome, and pervasive developmental disorder NOS. J Autism Dev Disord 34:649–668

Weiss PL, Cobb S, Gal E, Millen L, Hawkins T, Glover T, Sanassy D, Eden S, Giusti L, Zancanaro M (2011) Usability of technology supported social competence training for children on the autism spectrum. International conference on virtual rehabilitation, Zurich, Switzerland, June 2011

Yuill N, Rogers Y (2012) Mechanisms for collaboration: a design and evaluation framework for multi-user interfaces. ACM Trans Computer-Hum Interact 19(1):Article 1

Zancanaro M, Giusti L, Gal E, Weiss PL (2011) Three around a table: the facilitator role in a co-located interface for social competence training of children with autism spectrum disorder. In: Proceedings of INTERACT 2011

Part IV
Health and Sports

Designing for Social and Physical Interaction in Exertion Games

Florian 'Floyd' Mueller, Martin R. Gibbs and Frank Vetere

Abstract Exertion games lend themselves to facilitating social and physical interactions, in particular when compared to button-press games. However, there is little understanding of how specific aspects of an exertion game's design can facilitate these social and physical interactions. In response, we present a set of design themes based on our analysis of players' experiences with a distributed table tennis game. The themes are: Shared Object, Anticipation, Secondary Performance, Movement Variety, and Uncertainty. We hope that these themes can guide other designers who aim to support social and physical interaction in order to support players profiting from the many associated benefits.

Keywords Exertion games · Social interaction · Physical interaction · Design · Shared objects

1 Introduction

Over the last decades, user interface research on social and physical interaction has gained an increased focus in the field of human–computer interaction (Dourish 2001). In particular, designers and researchers are interested in developing playful experiences that support social and physical interactions and promise to be "fun" (Bekker et al. 2010). In this chapter, we focus on playful interactions that involve more than one person interacting and that support physical engagement with the

F. F. Mueller (✉)
Exertion Games Lab, RMIT University, Melbourne, VIC, Australia
e-mail: floyd@exertiongameslab.org

M. R. Gibbs · F. Vetere
Interaction Design Lab, The University of Melbourne, Melbourne, VIC, Australia
e-mail: martin.gibbs@unimelb.edu.au

F. Vetere
e-mail: f.vetere@unimelb.edu.au

A. Nijholt (ed.), *Playful User Interfaces*, Gaming Media and Social Effects,
DOI: 10.1007/978-981-4560-96-2_11,

environment. The physical interaction borrows from tangible computing (Hornecker and Buur 2006) as well as physical exertion. Exertion interactions are interactions that require intense physical effort from players (Mueller et al. 2003).

In the past, several attempts have been made to understand the relationship between the interface and social and physical experiences (Lindley et al. 2008; Mueller et al. 2011; Bekker et al. 2010), however, there still exists only a limited knowledge on how to design for social and physical exertion play.

In order to contribute this knowledge, we have created and studied a novel exertion game based on table tennis, which resulted in several contributions around our analysis of the design (Mueller et al. 2008b, 2009, 2010). We now put forward a set of themes that we have derived from the analysis of player interactions with this game, which we believe can be useful for designers of user interfaces that aim to support social and physical interaction.

The presented case study is of a physical game for three distributed players called "Table Tennis for Three." The game is loosely based on table tennis. However, it offers a gameplay experience through virtual augmentation, which is quite different to traditional table tennis. The virtual augmentation utilizes a videoconferencing component to allow for distributed play while the affordances of the bat and ball support movements that aim to maintain the benefits of physical play. This mixed approach can offer novel user experiences such as supporting geographically distributed participants. The game also supports three players at the same time equally, inspired by the opportunity of networked games to support large user numbers. We note that, first, supporting three players is not easily achieved in traditional table tennis (where it is often two against one). Second, although we acknowledge that three players is not comparable to the large number of players supported by many networked games, we see this as a first step in exploring the scaling of numbers in exertion games.

2 Related Work

A heightened consideration of social and physical aspects in digital play has a history in human–computer interaction with a focus on social and bodily interactions (Dourish 2001). Going back even further, we can see how Merleau-Ponty proposes a perspective highlighting how the human body is mutually engaged with the mind and that an intertwined connection exists helping us interact with our environment and other embodied beings. He argues that we need to consider our bodies and how we interact with other bodies if we want to understand what it means to be human (Merleau-Ponty 1945).

Researchers in human–computer interaction similarly stress the importance of the human body when it comes to the digital world, such as Winograd et al. who argue for a more nuanced view on the embodied user when interacting with computers (Winograd and Flores 1986). Dourish identified an embodied focus in the tangible and social computing systems developed around him (Dourish 2001), arguing that embodied features of interactive systems are related to the features of social settings. These high-level investigations are often based on earlier philosophical arguments made by Merleau-Ponty (Merleau-Ponty 1945), which are then applied to interactions with computers. Many of these arguments highlight the role of play in social and physical interactions. This role of play we will investigate in more detail next.

Lindley et al. suggest that a game's nature can change when players are involved in full bodily actions, rather than button presses, from a traditional virtual game experience of "hard fun" to more social play (2008). Similarly, de Kort et al. promote the consideration of full body actions because they believe players have an intrinsic need to experience their physical environment kinesthetically. The authors argue for a relationship between physical environment and virtual gameplay and propose "sociality characteristics" for games, which include a consideration of exertion actions, as they can "radically" impact social play (de Kort and IJsselsteijn 2008). Moen (2006) highlights the role of free-form movements and presents a framework for kinesthetic movement interaction, arguing that technological augmentation can support novel experiences. However, the author concentrates on single-user interactions, only recently looking at social play (Segura et al. 2013).

There have been several design-oriented explorations on the topic of social and physical interaction in exertion games from which we learn. We articulate these next.

Fish'N'Steps is a social approach to combining physical bodily actions with virtual content to enhance healthy activity. It is a mobile application that is aimed at encouraging participation in walking activities via social comparison (Lin et al. 2006). The proposed system works in combination with a pedometer to motivate an increase in a participant's daily energy expenditure. It separates the physical activity from the social activity; only after the participants walked all day can they share their progress through a fish tank metaphor displayed on a screen in an office. Their bodies rest while they assess their relative progress.

Consolvo et al. present another distributed pedometer-based system implemented on a mobile phone (Consolvo et al. 2006, 2008). The authors identified design requirements that include aspects regarding the body, the users' environment, and virtual augmentation. Systems that combine real-world physical exertion with virtual aspects of a videoconference have suggested that there might be the possibility for social facilitation. For example, exercise bikes have been networked to allow for distributed races in a competitive environment. The bodily component suggests the use of physiological data to enhance the distributed experience, as participants reported that visualized heart rate from a remote rider could motivate them to cycle faster (Bikeboard). The presence of a remote participant appears to affect the exertion performance; however, in exercise bike cycling the participants cannot interfere with one another physically. In many team sports, players can actively prevent their opponents from achieving the game's goals (Vossen 2004). This aspect of the shared experience is missing in networked bike riding; however, it has been explored in an early game of ours, Breakout for

Two (Mueller et al. 2003). Breakout for Two is a synchronous exertion game for two players with an integrated video communication channel. An evaluation showed that players were able to form a social bond with each other despite the geographical distance between them. However, what design aspects contributed to the social play and what role the physical actions as well as the physical environment, such as the two-location setup, had on the resulting experience remain unexplored. Besides Breakout for Two, there is a shortage of projects that look into exertion experiences where participants can interfere with each other. We are interested in table tennis, which is such a game where participants interfere with each other, hence we are looking into table tennis projects next.

The game of table tennis has inspired other research projects (Ishii et al. 1999; Knoerlein et al. 2007; Lawn and Takeda 1998; Woodward et al. 2004). Most of these implementations focus on the demonstration of the technologies' capabilities, showcasing how mechanical and computational advances can simulate certain aspects of table tennis, for example force-feedback (Knoerlein et al. 2007). The outcomes of these projects suggest that simulating and recreating a traditional bodily game such as table tennis over a computer network is probably still not technically feasible and costly.

In summary, previous research suggests that the consideration of the human body and its physically effortful interactions in interactive systems can facilitate beneficial experiences. Most utilized approaches are based on an embodied perspective that highlights the importance of the users' body being situated in a physical environment, interacting with other beings. However, there is a limited understanding of what role design plays when bodily actions are augmented with technology. The approach taken in this work is to suggest an interrelationship between social and physical exertion play and this article examines how the game's design can facilitate this interrelationship to provide a benefit to the experience. This approach is explored through a case study of a prototypal system called "Table Tennis for Three." This case study highlights the potential for novel experiences such as unique social support in terms of number and geographical location of participants, difficult to achieve without technological augmentation. Investigating this can contribute toward an understanding of the role of design in supporting social and physical exertion play, which in turn can support the leveraging of the many benefits physical activity affords, guiding future advances in this field.

3 Table Tennis for Three

We present the analysis of a study of Table Tennis for Three. It shows how social and physical exertion play can be supported and facilitated by certain aspects of the design and articulates what these aspects are, as expressed through a set of design themes. The intention is that these themes are to be used by designers who aim to support social and physical exertion play in future interfaces.

Fig. 1 Table tennis for three

3.1 Playing Table Tennis for Three

We now describe Table Tennis for Three, an exertion game that uses a physical ball, bat, and table for play and supports players from three geographically distant locations. In particular, this prototypal system highlights two opportunities of technology to support playful experiences that are otherwise difficult to achieve without technology. First is the opportunity to support geographically separate players, playing together simultaneously (in contrast to taking turns). Second is the opportunity to scale the play experience, such as allowing three participants to play together equally.

3.1.1 The Setup

Each player uses a physical ball, a bat, and a table tennis table. The table is set up in such a way that the ball can be hit against the vertically positioned opposite half of the table (Fig. 1). This setup is familiar to table tennis players who practice alone by playing the ball against the board. This backboard has projected images of eight large "bricks" on it. These bricks are identical for all players, and they are synchronized across all three stations (Fig. 2). A projector mounted to the ceiling projects the bricks in a semi-transparent fashion on top of two video streams of the other players in the game. One player's videoconference is positioned on the left of the backboard, and the other is on the right. Each table has a set of loudspeakers

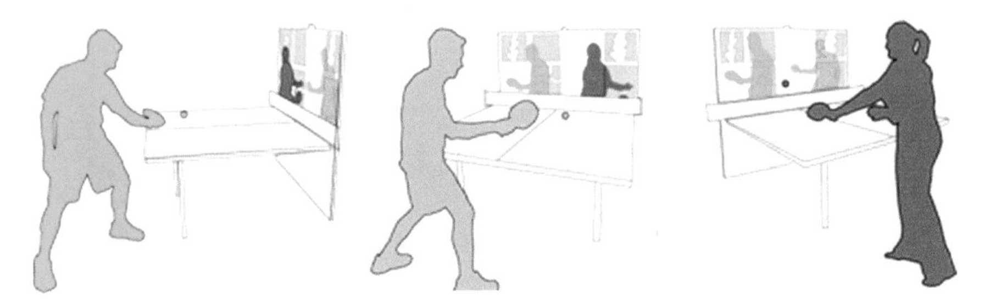

Fig. 2 The bricks are shared across the three stations, a hit is visible to all players

and each player wears a microphone so the three participants can converse with each other in a three-way audio conference during the gameplay.

The backboard is equipped with sensors mounted on the back that detect when and which brick the players hit. These piezoelectric sensors detect the impact of the ball hitting the wooden table tennis table, with the impact permeating through the wood. The sensors are located in a manner similar to the one suggested by Ishii et al. (1999) in order to determine the exact impact location based on the time it takes the impact to travel to each sensor: the sensor that receives the impact first is triggering the impact location to the software system using a digital acquisition board that samples the impact data at a high-speed.

The bricks "break" when hit by the ball as a result of the sensors registering the location of the impact. All three players see the same brick layout and the same brick status layered on top of the videoconferencing streams. If a brick is hit once, it cracks a little. If it is hit again (regardless of by which player), it cracks more. The crack appears on all three stations. If hit three times, the brick "breaks" and is removed from play, revealing more of the underlying videoconference: the player "broke" through to the remote player. However, only the player that hits the brick the third and final time receives the point. This offers players a number of strategies for winning the game. The players can either try to crack as many bricks as possible by placing the ball quickly or they can wait for the opportunity to snatch away points from other players through hitting bricks that have been already hit twice by the others.

Each brick that is completely broken scores one point, and the running score is displayed along the top end of the projection. Play continues until all bricks have been cracked three times and been removed from play. At this point the player who has scored the most points is announced as the winner and after a delay of 15 s, the game resets all the bricks and play can recommence.

3.2 Studying Playing Table Tennis for Three

The data used for the analysis of playing Table Tennis for Three comes from video recordings of participants playing Table Tennis for Three and videotaped interviews of all three participants together in one room. Notes were also taken during the interviews. The interviews contained open-ended questions about the players' experiences and their interactions with one another. Furthermore, we asked participants to answer a short questionnaire to examine specific questions we had on experience and gameplay (more on this later).

We recorded the participants during play with a video camera. Only one player was recorded at a time. If notable actions occurred on the remote end, this was observed through the videoconference. The interviews were also videotaped.

Each group played between 30–60 min. The players were brought together into one room after the game, where the interviews were conducted with all three of them together. The interviews lasted from 20 to 60 min.

3.2.1 Participants

We recruited 42 participants through personal contacts, email lists, and word-of-mouth. The volunteers were asked in the advertising material to organize themselves preferably in teams of three. If they were unable to do so, we matched them up randomly with other participants in order to have always three people participating at the same time.

We report on 14 teams of three. One participant played twice due to a last-minute cancellation (which was considered in the analysis). In total there were therefore 41 participants. The participants were between 21 and 55 years old (mean 31.6 years), and consisted of 27 males and 14 females. It is acknowledged that prior social relationships between participants can affect the social play interactions within a game (Salen and Zimmerman 2003), this aspect was taken into account by asking additional questions about how their existing relationships affected the way they played. The participants' prior exposure to table tennis was varied. One had never played before, 14 participants played less than five times, 18 players between five and 100 times and eight volunteers played more than 100 times before.

3.2.2 Data Analysis

The video data was analyzed by the authors using an iterative coding process, drawing on sensitizing ideas that relate to the research question and the relevant literature. The coding process was used to identify important concepts. The data was coded to locate concepts in an attempt to condense the data into categories. We created annotations on segments of the data we found particularly interesting, and wrote down memos during the coding process as reflective commentaries for deeper analysis.

The list of concepts we created from the first pass helped identifying emerging themes, and informed the next coding step when we viewed the data again. In this second iteration, the focus was more on the concepts and the initial codes were reviewed and examined before checked if some of them could be combined.

In a subsequent viewing of the data, the focus was on instances that illustrated concepts. We tried to make comparisons and identify contrasts to support the conceptual coding categories we have developed so far. Then, the emerged themes were grouped into logical groups and hierarchical categories were created. We then sorted the annotations, memos, and codes into groups to identify the key themes, looking out for whether they would be specific for Table Tennis for Three or exertion games in general.

4 Results

We begin by providing evidence that both social and physical exertion play occurred when engaging with Table Tennis for Three. What follows is a description of themes identified from the data.

4.1 Social Play

The players encouraged themselves and each other during the game.

> Lets' go for it! [P3, play]
> P18, come on! [P20, play]

They were not shy of engaging in competition, often expressed in statements such as:

> I really wanted to beat her. [P13, interview]

If a player snatched away the last brick, there were statements such as

> You b***! [P21, play]

Although players interacted verbally with their partners during the game, some players also commented that they felt more like they were playing against bricks:

> [...] not at each other. When playing, you are more focusing on the blocks than on the other player. [P15, interview]

P21 described vividly how she recognized that she could have focused on the bricks, but realized that playing against "a person" suited her better. In the interviews she made the following comment to the third player P4 in her group:

> [...] you are playing against yourself, sometimes, because you are very competitive against the screen, whereas I was watching P13, and it was like, when I was telling her off for cheating, then I actually aimed the ball at P13's head, at the screen! (All laughing).

Fig. 3 During the interview: "…and you just wait for the third one [crack] and try to get in there quick!"

4.2 Exertion Play

Participants were investing physical effort, and as a result, were exhausted. They commented on how quickly the game made them tired, in particular because they never had to wait for their partner, as in conventional table tennis. Participants also showed visible signs of exertion, such as sweat on their foreheads.

Next, an examination follows that investigates what aspects of the design facilitated both social and physical exertion play.

4.3 Movements Responding to Partners' Movements

The participants aimed at hitting bricks that have been hit before (see Fig. 3):

> And you just wait for the third one [crack] and try to get in there quick (makes a smash arm movement) [P11, interview]
> [You] wait for someone to break the blocks and go for it [P10, interview].

Fig. 4 Player's reaction from playing against another player

However, their ability to anticipate which brick will be hit next was limited:

> [There is a limited] sense where the other player is [...] [or] where he is playing [P12, interview]
> [...] so I didn't know this was where you were throwing the ball [P13, interview]

Anticipating the other players' movements mostly only played a role in initiating the game. P14 said during the interview:

> "You are waiting, and as soon as you are ready" [both P14 and P1 initiate a gesture for a serve but stop just before executing it], P1 interrupts: "You are ready to go" [both finishing their movement, both smiling].

4.4 Being Expressive

Users demonstrated affective expressions during play, but predominantly outside immediate play: in between points, while having a break, and after the games (see Fig. 4). Most verbal interactions were of emotional nature and not in relation to play directly, such as swearing, yelling, or dismissing the other player, but sometimes also words of encouragement were exchanged:

> You are not trying hard enough! [P5, play],
> I'm going to beat you next time [P7, play], and
> You guys are hysterical! [P8, play].

Most of these verbal exchanges were of a joking nature, with elements of mocking, teasing or "fooling around," characteristic of social sports games (Weinberg and Gould 2006).

Some users chose to supplement their verbal comments with a gesture, such as throwing their hands in the air to indicate they won. A player jokingly made a fist to the other players; another participant put her tongue out. Players often applauded others on their performance, and some performed little winning dances. This was often accompanied by laughter, facilitating a humorous atmosphere.

The exertion activity served as a starting point for social interaction, but it also hindered social interaction when attempted verbally: Players sometimes wanted to say something, but were not able to because they were too exhausted or too involved in their exertion activity. A player made this explicit by saying

> hang on… [P5, play]

when the beginning of a new game interrupted his dialog. He continued his verbal exchange in the next break of the game after he recovered from his exhaustion.

4.5 Alternative Ways of Playing

Players adapted rules from traditional table tennis to suit the interactions afforded by Table Tennis for Three. They also referred to table tennis and its rules in the interviews. Even though none of the teams discussed rules amongst each other before the game, they successfully engaged in gameplay by assuming the ball needs to be hit with the bat, cannot be returned volley, etc.

Nevertheless, players changed the way they played when an opportunity arose, such as when a player could not return the ball with his/her bat, but was able to catch it with her/his hand to increase the chance of winning. They also grabbed spare balls and played them successively, and hid under the table before serving (see Fig. 5).

The players knew that this was "breaking" the rules, because they pointed it out if they caught someone: a player yelled out loud

> You cheat, you are a cheater! (accompanied by laughter) [P9, play]

because a remote player was using her hands, but then this player used the hands herself.

Fig. 5 Player hiding to try alternative serve

4.6 Unpredictability of Physical Ball

Despite the players' best efforts, the ball often acted in unexpected ways; it hit the edge of the bat, and was diverted in the opposite direction. The ball also often bounced back quicker than expected, but a quick reaction on impulse of the player was able to deflect the ball in a manner that resulted in an unanticipated trajectory of the ball. The ball also often hit the edge of the table, being reflected off it in a surprising angle. Players looked amazed at how some of their hits returned the ball. These surprising situations are characterized by a considerably unexpected behavior of the ball that the player with his/her actions did not intend nor anticipated.

The challenging aspect of controlling the ball with the bat and the associated surprising actions that occurred contributed to the players' enjoyment, which was reflected in their verbal expressions: players often shouted short exclamations such as

> yikes!

when the ball flew off in the wrong direction, often followed by a smile (see Fig. 6). This laughter was then answered with laughs by the remote players, and

Fig. 6 Ball flew off the side

functioned as conversation starters. Players then switched their attention to the remote end if they heard such a surprise expression.

5 Discussion

We now discuss the observations in regards to social and physical exertion play and associated themes that facilitated this social and physical exertion play.

5.1 Shared Virtual Objects

On the one hand, participants appreciated being able to play with remote partners. They showed signs of playing together despite the fact that interactions were mediated. On the other hand, players reported that they sometimes felt more like playing against the virtual bricks rather than another person. We also observed incidences where players seemed to experience both, playing against another person and playing against bricks, such as described by the group around P21: they seemed to be able to navigate between these apparently contradicting experiences.

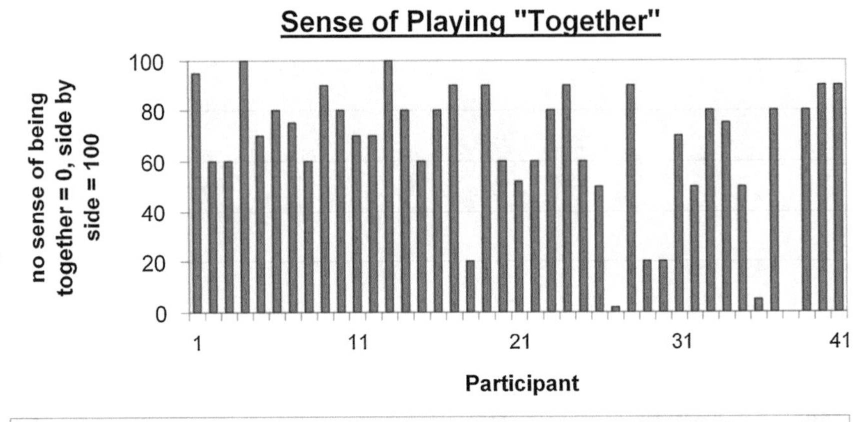

Fig. 7 Sense of playing "together" rated

We asked the participants to rate their sense of having played together on a scale from 0 to 100, with 0 being no sense of playing together, and 100 being the same extent as when playing on the same table.

The median was 70, and the arithmetic mean 65.46 (see Fig. 7). The answers were affirmative, especially in combination with the fact that most players (35 out of 40) had the feeling they were doing something together. Interestingly, this was statistically significant if correlated with "I liked the game" ($r = 0.64$, $p < 0.01$). The players who disagreed with the "together" statement were among the three that disliked the game, suggesting that there could be a link between the engagement with the game and the feeling of "doing something together."

In order to understand this further, we look at the idea of parallel and non-parallel play (Mueller et al. 2008a): by parallel play, we mean play experiences in which participants are aware of each other's exertion activity, but are not physically interfering with one another. A typical example from traditional sports is a 100-m race, in which the white lines on the track explicitly forbid any physical interference between runners. Vossen (2004) says that in parallel play, each player performs his or her exertion actions independently from one another. The players have no direct influence upon the difficulty of the task faced by other players, as they cannot directly interfere with one another.

By non-parallel play we mean play experiences in which participants can use their bodies to physically interfere with one another. An example from the traditional sports domain is wrestling, where the wrestlers contend for control of each other's bodies. Vossen (2004) reminds us that this category is distinctly different to how people can experience exertion compared to parallel play.

Table Tennis for Three facilitated parallel play in the physical world of the exertion actions, but also non-parallel play in the virtual world of the shared bricks. This explains why players expressed both, that they felt like playing against bricks but also against other players. The virtual bricks enabled a non-parallel game activity, which the players used to challenge each other in order to enrich their social experience. The players used this opportunity to engage with their remote opponents, trying to "outsmart" them to snatch points using tactical decisions to gain an advantage. As such, the virtual bricks functioned as "shared objects" that facilitated the non-parallel play. Shared objects are common in many traditional social sports; often it is a ball, such as in conventional table tennis. In Table Tennis for Three, the virtual bricks took on the role of shared object, as the physical ball was unique to each location and hence not shared.

The main exertion action of the game, however, is the hitting action of the ball, which occurs in the physical space and is of a parallel characteristic. This parallel characteristic facilitated a sensation of playing against bricks, not other players, affecting social play. The players described the parallel aspect through expressing they felt like "racing" the computer bricks. When players had such an experience of "playing against bricks" and "not at each other," the mediating design fell short in facilitating a non-parallel experience at that moment.

The affordances of both parallel and non-parallel play were used by some players to enhance their experience. Especially, the players who seemed to seamlessly navigate between playing against another person and against virtual bricks demonstrated that the boundaries of the spaces can become fluid in the context of use, and the users made them fit to meet their interactional needs.

It is also interesting to note that due to limitations in the sensor and network technologies, delays in brick updates (changing from a full brick, to a broken brick, to a disappearing brick) can occur. Network lags in online games can significantly affect the user experience (Linehan et al. 2006), and are a challenge for any online experience. Interestingly, as the physical ball required skillful control by the participants, it appeared this focused their attention on the exertion actions rather than the updates of the virtual bricks, which in turn might limit any negative effects delays in brick updates might produce. As such, any effects resulting from delays in Table Tennis for Three appeared to be tempered by the focus on the exertion actions, especially when compared to button-press computer games.

In sum, players experienced both social and physical exertion play, even though the interaction between the participants was mediated by technology. Interestingly, the social play comprised of both parallel and non-parallel play, where parallel play was occurring in the physical space, and non-parallel play in the virtual thanks to shared virtual objects. Players were able to navigate both spaces and both forms of social play, which was facilitated by the physical exertion play activity.

We now present a summary of the result and discussion, phrased in an actionable way to guide designers of future systems:

Shared virtual objects can be used to support social play. However, they might facilitate non-parallel play in the virtual world, whereas the exertion activity can be parallel in the physical world.

5.2 Anticipation

A key characteristic of exertion games is the involvement of bodily movements, and supporting bodily movements is believed to facilitate social interactions (Lindley et al. 2008). For individual activity, this support of bodily movements is often associated with self-awareness of exertion actions (Moen 2006; Consolvo et al. 2006), however, in a social setting, awareness of other bodies and their associated movements becomes also important (Fogtmann et al. 2008): an understanding of the activities of others can provide a context for one's own activities (Rettie 2003), and so-called awareness of other bodies can be an important part of how we make sense of the world through our bodies (Dourish 2001).

Knowing that awareness is an important element in supporting social play, it can now be discussed how anticipation was facilitated by the videoconferencing component. Players were contending for control of the virtual bricks. In order to gain an advantage, participants tried to anticipate their partners' next moves. Such anticipation was possible because movement was involved, and this movement was continuous. Participants could perceive this continuousness of movement through the videoconference.

The notion of continuousness draws attention to the fact that bodily gross-motor movements, as featured extensively in Table Tennis for Three, are accompanied by preparatory and follow-through movements, quite different to button presses. An exertion action involves movements that include a whole set of motions (Moen 2006): for the Table Tennis for Three players, it was a backswing, a forward swing, the contact with the table tennis ball, and the follow-through. Although only the contact with the ball counted toward the game, all motions formed part of play, and players tried to make use of it. Players were aiming to be aware of their opponents' bodily movements by attempting to "read" these pre- and post-movements to anticipate future play actions. In Table Tennis for Three, the participants made use of the continuousness nature of bodily movements in order to enhance their social play experience through anticipation.

When compared to traditional gamepad-controlled games, it can be seen that continuousness has a particular role in exertion games. Being aware of a player controlling a binary button interaction does not reveal much of that player's intention behind the button pressing action. Anticipating a future move is very difficult with buttons, as they do not require visible sequential movements (a simple flick with the thumb is sufficient) that participants could use as cues to initiate an action. Usually, this support for continuousness is assumed in co-located experiences, however, the mediation in Table Tennis for Three brought this point to the fore, as the videoconference's technical shortcomings hindered the opportunities for anticipation.

Continuousness is a direct result of "movements [being] situated in time and space" (Griffin 2005). That is why animations representing player actions in computer games such as fighting games display pre- and post-movements in response to a single button-press. For example, a simple button-command in Tekken (Namco 2010) results in the character preparing for the kick by leaning backwards and moving the arm back, before propelling forward, hitting the opponent and then getting the arm back into the original position. As a consequence, if game designers want to include anticipation in conventional computer games, they need to support this anticipation by implementing it in the virtual world. In exertion games, players can utilize anticipation due to the continuousness characteristic of movement; this means that designers have an additional way to support anticipation.

Interesting to note is that players in Table Tennis for Three utilized both, the continuousness afforded by bodily movement as well as computer game play elements to anticipate future states of play. The players used the state of the bricks to anticipate which of them will be hit next and adjusted their actions accordingly. The players anticipated future actions not only based on movement actions over the videoconference, but also based on the movements in the virtual space.

Any mediating technology can affect how anticipation is supported in exertion games. In Table Tennis for Three, the capture area of the camera limited the space in which continuousness movements could have been captured and hence transmitted, and the videoconferencing quality failed to consider very fast movements and hence anticipation was limited. Also, the conical shape of the camera did not always adequately capture the ball's flight path, hindering any anticipation of future moves based on the location of the ball. However, anticipation was further supported through the virtual objects in the game. In contrast to sports (anticipation occurs based on continuousness in the physical space) and computer games (anticipation occurs based on actions in the virtual space), anticipation in Table Tennis for Three occurred based on continuousness over the videoconference as well as based on actions in the virtual space.

Even though Table Tennis for Three supported anticipation, it did not support bodily reciprocity as, for example, a conventional game of table tennis. Although players in Table Tennis for Three were anticipating their partners' next move, they were also dealing with the task of controlling their own ball, which was acting independent from any other player. In conventional table tennis, every return hit is a response to the other player. Players in Table Tennis for Three could choose to respond to the remote players, but they could also choose to engage in parallel play, just aiming to hit bricks in a random fashion as fast as possible. So the ball was not the mediating object between the players, but the bricks, where the ball was the mediator to the bricks. In addition, due to the conical shape of the videoconference, it was difficult to link partners' actions to specific bricks: although one could anticipate when a player is going to execute a hit due to the continuousness of movement, anticipating exactly which brick will be hit was difficult to anticipate once the ball left the capture area of the camera.

Supporting the continuousness of bodily movements allows anticipating a player's next action, contributing to social play.

5.3 Secondary Performance

Participants performed using their bodies as a way to communicate, in particular outside the game, as a form of metagaming. Metagaming is a social play phenomenon that refers to the relationship of a game to elements outside of the game. One way that metagaming occurs "during a game other than the game itself [...] are social factors such as competition and camaraderie" (Salen and Zimmerman 2003). Our participants engaged in a metagame experience by socially interacting with one another beyond the immediate game play. In particular, through the exertion nature of the game, they did so by using their bodies to communicate: first, exhaustion made verbal exchange more difficult, second, the bodily focus of the game lent itself to communication behavior through the body. The participants embraced this by nonverbally commenting on other players' performances and turning the game into a bodily spectacle beyond the gameplay itself. Larssen et al. (2004) found the notion of "expressive latitude" particularly fitting in the context of exertion games to describe such behavior that is not directly influencing the game outcome, but can have communicative aspects. We call this secondary performance.

Interestingly, aspects of secondary performance also appeared in the interviews, in particular when participants were "retelling their experience." The retelling of what happened in a game is an important part of a "lived experience" (McCarthy and Wright 2004). Players predominantly used their exertion skills in the games, so they drew on these skills again during the reliving of the experience. This reliving of a "pleasurable kinaesthetic stimulation" can re-trigger the associated pleasurable emotions (Iso-Ahola and Hatfield 1986). Re-enacting the exertion movements can also support the player's cognitive processes, helping them remembering certain parts of the game (Lindley et al. 2008). Players gave further meaning to these exertion actions by reliving and sharing them with others, the support for secondary performance appeared to contribute toward a social play experience. These exertion actions supporting secondary performance are missing in conventional gamepad-controlled computer games, and their players have to rely on their cognitive skills to remember their lived experience and associated affective responses.

Supporting people in expressing themselves using their bodies—in and outside the game—can contribute to social play and facilitate metagaming.

5.4 Movement Variety

The participants exhibited bodily movement in many ways while striving to achieve their goals; we consider this a form of self-expression. The participants played with the limitations of technical mediation, using it as resource for "bending the rules." This cheating was used to enhance the experience, a phenomena which has been previously observed in traditional computer game play (Consalvo 2007). However, the inclusion of exertion in Table Tennis for Three afforded a new approach: players explored their bodily movements within the space of opportunities the sensing technology allowed (or not allowed, as not all actions were sensed), and what was communicated over the videoconference. Benford et al. used the terms "expected, sensed and desired" to differentiate different interactions within sensing systems (2005); using his words we could say that the participants were playing with the various categories that the sensing systems afforded. As the context was a game, the participants explored their movement variety within the sensing space in order to enhance their experience. The players were not so much trying to "break" the rules as to rather "bend" them, exploring alternative ways to achieve the goal in order to catch up with an advanced player, make the other player laugh, and so forth. This has been described as transformative social play, in which players actively engage with the rule system in order to shift or extend their relations with other players (Salen and Zimmerman 2003), therefore constituting a form of social play. The mediated aspect appeared to contribute to this, as it enabled a variety of opportunities for bending the rules.

The limitations of the technology enabled these opportunities in three ways:

1. The sensing system of the table tennis table afforded players exploring different ways to execute a hit: if I throw the ball, is the hit also registered? What if I throw it really hard, can I confuse the sensors so I break two targets? Players were asking these questions in the interviews but also trying out the different tactics during the game.
2. Players realized that they could grab another ball in their other hand to serve as backup if the first ball goes astray. This was difficult to see over the limited video quality, especially in fast-paced games, as the videoconference system communicated only a limited resolution displaying a limited framerate. Communicating subtle details in rapid actions, as often the case in exertion games, is still a challenge for sensing and networking technologies. Players are aware of this and they play with it to their gaming advantage.
3. There was a mismatch between the space the player occupied and what the videoconference camera saw. Participants played with this mismatch (for example by hiding below the table), contributing to a joyful atmosphere, but it also affected social play negatively. In co-located exertion games, the physical space and what the other player sees are usually interconnected. In distributed games, however, these spaces might be disparate due to technical limitations. In non-exertion games this mismatch might not be important, as players might not move much. In exertion games on the other hand, players' activities involve many large-scale movements, which might conflict with conventional awareness technologies that are aimed at supporting focused awareness cues such as facial expressions. When one of the participants stepped out of the view of the camera, she became "unsensable" by the camera. By doing so, she also left the "magic circle of play" (Salen and Zimmerman 2003). This shows how

movement variety in mediated environments can lead to problems, but also opportunities for finding alternative ways to reach the game's goal.

We agree with Salen and Zimmerman that this aspect of rule-breaking is more likely to occur in exertion games; they attribute it to the "athletic nature" of the game (Salen and Zimmerman 2003). Salen and Zimmerman compare this effect to a chess game, in which a player will not gain an advantage by having a little corner of his/her rook peek into an adjacent square. "But in the infinitely granular space of the real world, milliseconds and millimeters can mean the difference between winning and losing" (Salen and Zimmerman 2003).

Movement variety supports creativity through finding alternative strategies—including cheating—to play, including the exploitation of technological limitations. This can facilitate social play ranging from disclosing to showcasing these alternative strategies.

5.5 Uncertainty

Table Tennis for Three exploits the affordances of tangible objects, which includes the ball, bat, and table. For our players, the tangibility of the play objects contributed to an uncertainty of play, creating opportunities for excitement and surprise. These opportunities for excitement and surprise appear to have fueled social interactions between players. The notion of surprise has been previously recognized in physical play (Czajkowski n.d.) and augmented mixed reality games (Sharp et al. 2007). The results of uncertainty contribute to Gaver's claim that the physical environment can provide affordances for social interaction in games (1996), and Hornecker and Buur's suggestion that "the richness of bodily movement" in combination with tangible interfaces is particularly beneficial for social interactions (2006). This also supported the leveling of the playing field between participants of different athletic abilities: a player who was behind in points could all of a sudden receive an advantage due to a surprise event, adding to the excitement of the game. Exertion amplified the chances and outcomes of tangibility's uncertainty: tangibility can support uncertainty without exertion; however, the diverse, fast, and forceful movements exhibited in exertion play facilitated these surprising moments for the players. Also, involving the body and the "real world" has been pointed out to add an element of uncertainty in location-based mobile phone games, which designers need to be aware of (Benford et al. 2003).

In non-exertion digital game play, these chance encounters need to be artificially introduced as an element of chance is inherent in most computer games. Game creators have to take special care in finding a balance between believable chance and randomness for the players (Salen and Zimmerman 2003). For example, in an exertion game such as the Nintendo Wii Sports Tennis, the ball on the screen might also be controlled by an element of chance; but it will be generic, as the ball will never bounce off the furniture that surrounds the player. The ball will also not bounce off the racquet's frame in much unexpected ways, but if it does, the experience will be "fundamentally different," as players might not believe the probability by which it occurred, but rather assume a bug in the software (Gaver 1996).

Utilizing the uncertainty that arises in physical exertion play, in particular when the body interacts with physical objects, can add an element of surprise that facilitates social play.

6 Contributions

This work has contributed toward an understanding of social and physical exertion play through the following contributions.

Shared object: Shared virtual objects can be used to support social play. However, they might facilitate non-parallel play in the virtual world, whereas the exertion activity can be parallel in the physical world.

Anticipation: Supporting the continuousness of bodily movements allows anticipating a player's next action, contributing to social play.

Secondary performance: Supporting people in expressing themselves using their bodies—in and outside the game—can contribute to social play and facilitate metagaming.

Movement variety: Movement variety supports creativity through finding alternative strategies—including cheating—to play, including the exploitation of technological limitations. This can facilitate social play ranging from disclosing to showcasing these alternative strategies.

Uncertainty: Utilizing the uncertainty that arises in physical exertion play, in particular when the body interacts with physical objects, can add an element of surprise that facilitates social play.

7 Limitations

We made the assumption that the user data gathered in a lab environment is representative to data that would have been collected in the field. Prior research has investigated the use of exertion games in people's homes in order to investigate the impact of their day-to-day surroundings on the experience, such as the limited space issues when playing Dance Dance Revolution (Sall and Grinter 2007). The authors found that the living room is an unfamiliar and often unsuitable space for physical exercise, hence the lab environment might be just as suitable to investigate an exertion game. Furthermore, we have tried to set up the table tennis tables in ways that resembles spaces in which such tables could be encountered, for example public areas in corporate environments. Also, traditional exercise is usually performed in a gym or on an outdoor field to which the participants generally need to travel to, just as they had to travel to participate in a game of Table Tennis for Three. There are also dedicated spaces in which commercial entities offer exertion game experiences (XRtainment), sometimes described as interactive gyms, and these places also require participants to leave their familiar surroundings and travel first before they can participate. These examples demonstrate that a dedicated environment for conducting the study that is unfamiliar for the participants is not very unusual in the context of exertion games.

The participants were located in different parts of the building. This is not the same as being in different locations across the world, separated by significant distance that requires effort to overcome by travel. As the participants knew they were able to join one another physically after the game with ease, their social behavior might have been different than if they would have been geographically very far apart. The social implications that come with simulated distance have been put aside in this study.

It is also acknowledged that scaling the system from a two-player version to a three-player system is only offering limited opportunities for investigating scaling effects of mediated social play. Online computer games have pushed the envelope of how massive scaling can be supported in gaming, with some titles supporting millions of players. Supporting three players in Table Tennis for Three seems meager in comparison. However, the system only served as vehicle to investigate social play beyond two players, and represents the first attempt to support player constellations that are otherwise hard to achieve in traditional settings: allowing three players to play together equally while investing physical effort.

8 Future Work

Similar to other studies who identified a performative aspect afforded by physical exertion play including audience participation that entices social play, we also observed how physical exertion play can "turn the body into a spectacle" (Sheridan and Bryan-Kinns 2008). Our current work on Table Tennis for Three did not include the consideration of an audience aspect. Therefore, we recommend future work to investigate an audience's role in social and physical exertion play, furthering our understanding of such play experiences.

Other possible avenues for future work are investigations into the applicability of the findings in co-located experiences. Furthermore, exploring how useful the themes are for describing other play interactions and analyzing existing games might also be beneficial future work. In particular, the notion of secondary performance has been investigated in other research work on social and physical interaction around games (Segura et al. 2013) since presented in this article, as such, an exploration of the themes' potential to influence new game ideas might also be a fruitful area for future work.

9 Conclusion

We have presented a qualitative analysis of player observations and interviews from an exertion game to understand the facilitating role of design in the interrelationship between social and physical exertion play. First, we have found evidence that exertion games can facilitate social play, even in mediated environments. Second, we have identified the salient themes Shared Object, Anticipation, Secondary Performance, Movement Variety, and Uncertainty that contribute to the link between social and physical exertion play and discussed how specific design elements can facilitate (and hinder) this link. Our hope is that these themes are used to analyze existing as well as create future physical exertion games. Furthermore, we hope our work also has implications for theory that articulates the interrelationship between social and physical exertion play. As such, our work might also offer guidance for future work that aims to include exertion aspects into social play as well as for exertion games that are currently not supporting social play. However, we also note that design alone cannot guarantee social play, it is after all the players who create social play, that design features can only facilitate (Salen and Zimmerman 2003).

In sum, our goal is to contribute to a better understanding of social and physical exertion play and their interrelationship, advancing this research area featured within this book, in order to facilitate the many benefits of social and physical exertion play.

Acknowledgments The authors wish to thank everyone who has helped with the project and the writing of this article. The first author would also like to thank the University of Melbourne, with which he was affiliated while the majority of this work was done.

References

Bekker T, Sturm J, Eggen B (2010) Designing playful interactions for social interaction and physical play. Pers Ubiquit Comput 14(5):385–396. doi:10.1007/s00779-009-0264-1

Benford S, Anastasi R, Flintham M, Drozd A, Crabtree A, Greenhalgh C, Tandavanitj N, Adams M, Row-Farr J (2003) Coping with uncertainty in a location-based game. IEEE Pervas Comput 2(3):34–41

Benford S, Schnädelbach H, Koleva B, Anastasi R, Greenhalgh C, Rodden T, Green J, Ghali A, Pridmore T, Gaver B (2005) Expected, sensed, and desired: a framework for designing sensing-based interaction. ACM Trans Comput Hum Interact (TOCHI) 12(1):3–30

Bikeboard (n.d.). http://nyx.at/bikeboard/Board/showthread.php?threadid=61242

Consalvo M (2007) Cheating: gaining advantage in videogames. The MIT Press, Cambridge

Consolvo S, Everitt K, Smith I, Landay JA (2006) Design requirements for technologies that encourage physical activity. Paper presented at the proceedings of the SIGCHI conference on human factors in computing systems, Montreal

Consolvo S, Klasnja P, McDonald DW, Avrahami D, Froehlich J, LeGrand L, Libby R, Mosher K, Landay JA (2008) Flowers or a robot army?: encouraging awareness and activity with

personal, mobile displays. Paper presented at the proceedings of the 10th international conference on Ubiquitous computing, Seoul
Czajkowski Z (n.d.) The essence and importance of timing (sense of surprise) in fencing. http://www.mat-fencing.com/Akademia16.html
de Kort YAW, IJsselsteijn WA (2008) People, places, and play: player experience in a socio-spatial context. Comput Entertain(CIE) 6(2)
Dourish P (2001) Where the action is: the foundations of embodied interaction. MIT Press, Boston
Fogtmann MH, Fritsch J, Kortbek KJ (2008) Kinesthetic Interaction—revealing the Bodily Potential in Interaction Design. Paper presented at the OZCHI '08 conference of the computer-human interaction special interest group (CHISIG) of Australia on computer-human interaction, Cairns
Gaver WW (1996) Affordances for interaction: the social is material for design. Ecol Psychol 8(2):111–129
Griffin S (2005) Push. Play: an examination of the gameplay button. Paper presented at the proceedings of DiGRA 2005 conference: changing views—worlds in play, Vancouver
Hornecker E, Buur J (2006) Getting a grip on tangible interaction: a framework on physical space and social interaction. Paper presented at the proceedings of the SIGCHI conference on human factors in computing systems, Montreal
Ishii H, Wisneski C, Orbanes J, Chun B, Paradiso J (1999) PingPongPlus: design of an athletic-tangible interface for computer-supported cooperative play. Paper presented at the SIGCHI conference on human factors in computing systems
Iso-Ahola SE, Hatfield BD (1986) Psychology of sports: a social psychological approach. Wm. C. Brown Publishers, Dubuque
Knoerlein B, Székely G, Harders M (2007) Visuo-haptic collaborative augmented reality ping-pong. In: International conference on advances in computer entertainment technology, 2007, pp 91–94, ACM Press, New York
Larssen A, Loke L, Robertson T, Edwards J, Sydney A (2004) Understanding movement as input for interaction–a study of two eyetoy games. Paper presented at the proceedings of OzCHI '04, Wollongong
Lawn M, Takeda T (1998) Design of an action interface with networking ability for rehabilitation. In: IEEE Engineering in Medicine and Biology Society, Hong Kong, 1998
Lin J, Mamykina L, Lindtner S, Delajoux G, Strub H (2006) Fish'n'Steps: encouraging physical activity with an interactive computer game. Paper presented at the UbiComp 2006: ubiquitous computing conference
Lindley SE, Le Couteur J, Berthouze NL (2008) Stirring up experience through movement in game play: effects on engagement and social behaviour. Paper presented at the proceeding of the twenty-sixth annual SIGCHI conference on human factors in computing systems, Florence
Linehan C, Roche B, McLoone S, Ward T (2006) Network latency in on-line gaming: an engineering or a psychological problem? Paper presented at the CGAMES 2006—9th international conference on computer games: AI, animation, mobile, educational and serious games, Dublin Institute of Technology, Dublin
McCarthy J, Wright P (2004) Technology as experience. The MIT Press, Boston
Merleau-Ponty M (1945) Phenomenology of perception (Routledge Classics). Routledge, New York
Moen J (2006) KinAesthetic movement interaction: designing for the pleasure of motion. Dissertation, KTH, Numerical Analysis and Computer Science, Stockholm
Mueller F, Agamanolis S, Picard R (2003) Exertion interfaces: sports over a distance for social bonding and fun. Paper presented at the SIGCHI conference on human factors in computing systems, Ft. Lauderdale
Mueller F, Edge D, Vetere F, Gibbs MR, Agamanolis S, Bongers B, Sheridan JG (2011) Designing sports: a framework for exertion games. Paper presented at the CHI '11: proceedings of the SIGCHI conference on human factors in computing systems, Vancouver

Mueller F, Gibbs M, Vetere F (2008a) Taxonomy of exertion games. Paper presented at the OzCHI '08: proceedings of the 20th Australasian conference on computer-human interaction, Cairns

Mueller F, Gibbs M, Vetere F (2009) Design influence on social play in distributed exertion games. Paper presented at the CHI '09: proceedings of the SIGCHI conference on human factors in computing systems., Boston

Mueller F, Gibbs M, Vetere F, Agamanolis S (2008b) Design space of networked exertion games demonstrated by a three-way physical game based on Table Tennis. Comput Entertain 6(3):1–31. doi:http://doi.acm.org/10.1145/1394021.1394029

Mueller F, Gibbs MR, Vetere F (2010) An exploration of exertion in mixed reality systems via the "Table Tennis for Three" game. In: Dubois E, Gray P, Nigay L (eds) Engineering of mixed reality systems. pp 165–182

Namco (2010) Tekken. Namco. http://tekken.com

Rettie R (2003) Connectedness, awareness and social presence. Paper presented at the presence 2003, 6th annual international workshop on presence, Aalborg

Salen K, Zimmerman E (2003) Rules of play: game design fundamentals. The MIT Press, Boston

Sall A, Grinter RE (2007) Let's get physical! in, out and around the gaming circle of physical gaming at home. Comput Support Co-op Work (CSCW) 16(1):199–229

Segura EM, Waern A, Moen J, Johansson C (2013) The design space of body games: technological, physical, and social design. Paper presented at the proceedings of the SIGCHI conference on human factors in computing systems, Paris

Sharp H, Rogers Y, Preece J (2007) Interaction design: beyond human computer interaction. Wiley, West Sussex

Sheridan J, Bryan-Kinns N (2008) Designing for performative tangible interaction. Int J Arts Technol Special Issue Tangible Embed Interact 1(3/4):288–308

Vossen DP (2004) The nature and classification of games. Avante 10(1):53–68

Weinberg RS, Gould D (2006) Foundations of sport and exercise psychology. Human Kinetics, Champaign

Winograd T, Flores F (1986) Understanding computers and cognition: a new foundation for design. Ablex Publishing Corporation, Norwood

Woodward C, Honkamaa P, Jppinen J, Pykkimies EP (2004) Camball-augmented virtual table tennis with real rackets. Paper presented at the proceedings of the 2004 ACM SIGCHI international conference on advances in computer entertainment technology, Singapore

XRtainment (n.d.) XRtainment—where working out is all play. http://www.xrtainmentzone.com/

Designing Games to Discourage Sedentary Behaviour

Regan L. Mandryk, Kathrin M. Gerling and Kevin G. Stanley

Abstract Regular physical activity has many physical, cognitive and emotional benefits. Health researchers have shown that there are also risks to too much sedentary behaviour, regardless of a person's level of physical activity, and there are now anti-sedentary guidelines alongside the guidelines for physical activity. Exergames (games that require physical exertion) have been successful at encouraging physical activity through fun and engaging gameplay; however, an individual can be both physically active (e.g. by going for a jog in the morning) and sedentary (e.g. by sitting at a computer for the rest of the day). In this chapter, we analyse existing exertion games through the lens of the anti-sedentary guidelines to determine which types of games also meet the requirements for anti-sedentary game design. We review our own game designs in this space and conclude with an identification of design opportunities and research challenges for the new area of anti-sedentary game design.

Keywords Energames · Exergames · Sedentary behaviour · Cognitive benefits · Exercise · Games

1 Introduction

Regular physical activity has many benefits, including to a person's physical (Garber et al. 2011; Pate et al. 1995; U.S. Department of Health and Human Services 1996), emotional (Hassmén et al. 2000) and cognitive well-being (Etnier et al. 2006; Hillman et al. 2008). The Canadian Society of Exercise Physiologists recommends that adults achieve 150 min of moderate-to-vigorous-intensity physical activity per week (Tremblay et al. 2011); however, only 15 % of adults

R. L. Mandryk (✉) · K. M. Gerling · K. G. Stanley
Department of Computer Science, University of Saskatchewan,
Saskatoon, S7N 5C9 Saskatchewan, Canada
e-mail: regan@cs.usask.ca

A. Nijholt (ed.), *Playful User Interfaces*, Gaming Media and Social Effects,
DOI: 10.1007/978-981-4560-96-2_12,

meet these guidelines in at least 10-min bouts, and only 5 % of adults meet these guidelines on at least 5 days per week (Colley et al. 2011a, b). To encourage physical activity, researchers and developers have created a variety of exergames that encourage people to exercise, by integrating exercise into the game mechanics (e.g. (Ahn et al. 2009; Berkovsky et al. 2012; Gao and Mandryk 2011; Hernandez et al. 2012; Mueller et al. 2010; Stanley et al. 2011; Xu et al. 2012)). For example, in GrabApple (Gao and Mandryk 2011) (a Kinect-based digital exergame), the player has to move, jump and duck to collect apples. Because the player's body weight acts as resistance in GrabApple, playing the game yields moderate to vigorous exertion levels, but it is rated as fun as a sedentary mouse-based version of the game (Gao and Mandryk 2012).

Recent work among health researchers has shown that alongside the benefits provided by physical activity, there are also negative consequences associated with sedentary behaviour (Garber et al. 2011; Tremblay et al. 2010). For example, sedentary behaviour has been shown to influence carbohydrate metabolism (Chilibeck et al. 1999), reduce bone mineral density (Zwart et al. 2007) and affect vascular health (Hamburg et al. 2007). Interestingly, the physiological changes that result from sedentary behaviour are distinct from those that result from a lack of physical activity (Tremblay et al. 2010). A lack of exercise changes the body in different and unique ways from an overall sedentary lifestyle (Hamilton et al. 2008). Although this may seem surprising, physical activity and sedentary behaviour are not mutually exclusive; as Fig. 1 shows, even if a person is physically active (e.g. goes for a jog first thing in the morning), she can also be sedentary (e.g. by primarily sitting for the remaining waking hours); the effects of too much sitting are physiologically distinct from too little exercise (Owen et al. 2010). Thus, a physically active individual could be susceptible to the negative effects of a sedentary lifestyle (Tremblay et al. 2010). Because exercise and sedentary behaviours influence the body in different ways, the benefits of meeting the physical activity guidelines could be undone if people spent the remaining hours of the day engaging in largely sedentary behaviours (Hamilton et al. 2008). Because of the potential negative effects on health, many groups are now exploring the need for anti-sedentary guidelines to exist alongside guidelines for physical activity. For example, the Canadian Society of Exercise Physiologists now has sedentary guidelines for children and youth (Tremblay et al. 2011); however, more foundational research on the detrimental effects of sedentary behaviour on physiology is needed to establish evidence-based guidelines for all populations (Tremblay et al. 2011).

As researchers who design digital game-based interventions to promote health, we have been focused on designing games to promote physical activity; however, these exergames may or may not also work to combat sedentary behaviours. For example, a game designed to encourage a jogger to commit to and follow through with a daily jog will help a player meet the physical activity guidelines, but will not help to combat sedentary behaviour over the remaining waking hours. There simply exists no analysis of the design requirements for anti-sedentary games to

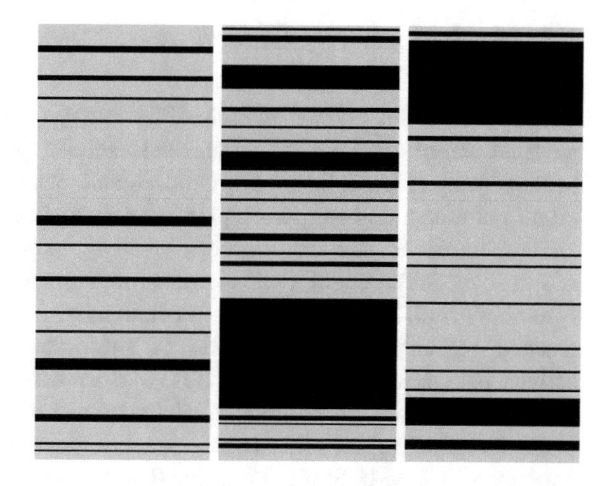

Fig. 1 Daily activity charts (*grey* = sedentary time, *black* = active time, time passes from *top* to *bottom*) representing three activity profiles. *Left* sedentary person, *Middle* Physically active person, *Right* Physically active person who is also sedentary

help inform researchers how the designs of anti-sedentary exergames should differ from physical activity-promoting exergames.

In this chapter, we first present and contrast the medical guidelines for physical activity and those for sedentary behaviours. We discuss compliance with the guidelines and common barriers to an active lifestyle. We then identify how exergames have addressed these barriers by following a series of design principles, and what additional principles need to be considered for anti-sedentary game design. We dub these anti-sedentary games energames, and provide a review of examples from the literature for digital games that partially meet the criteria of principled energame design. This critical analysis is followed by a further discussion of the fundamental design principles for energames and their differentiation from traditional exergames. We conclude by considering the challenges and opportunities in this new area of designing, deploying and evaluating digital games that combat sedentary behaviours.

2 Guidelines for Physical Activity and Anti-sedentary Behaviour

Combating inactivity has frequently been addressed through the compilation of guidelines. Physical activity recommendations aim to provide guidance for setting goals in daily life to provide sufficient exercise to improve overall health. In contrast, anti-sedentary guidelines aim to combat sedentary lifestyles, shifting the focus from increasing physical exertion to encouraging frequent, low-intensity physical activity to avoid negative health effects. In this section, we present an overview of guidelines for physical activity, guidelines to combat sedentary lifestyles and adherence to these guidelines.

2.1 Physical Activity Guidelines

The Canadian Society for Exercise Physiology (CSEP) provides guidelines for physical activity to help individuals set exercise goals that will result in health benefits (csep.ca/guidelines). For children and teenagers, they recommend at least 60 min of moderate-to-vigorous physical activity per day. To achieve this goal, the recommendations suggest vigorous-intensity exercise three times a week in addition to muscle and bone-strengthening exercises on three or more days (Tremblay et al. 2011). For adults, recommended levels of physical activity can be reached by engaging in 150 min of moderate-to-vigorous-intensity physical activity per week (Tremblay et al. 2011), which can be broken up into chunks of as small as 10 minutes. Additionally, adults are recommended to include muscle and bone-strengthening activities twice a week. Similar guidelines are provided by the American College of Sports Medicine (Garber et al. 2011) and the American Heart Association (see heart.org).

Despite the wide availability of guidelines that can help inform individuals about the benefits of regular physical activity, research reports that many people only partially achieve activity goals (e.g., they do exercise at required intensity levels, but do not engage in activity frequently throughout the week), with particularly poor compliance rates among teenagers (Pate et al. 2002). As a result, many people do not reach recommended levels of physical activity (Colley et al. 2011a, b) and health risks remain, particularly among children and teenagers (Sothern et al. 1999).

2.2 Anti-sedentary Guidelines

As a response to the growing body of evidence showing that there are distinct physiological responses to a lack of physical activity and to sedentary behaviour (Tremblay et al. 2010), the CESP has recently released guidelines to combat sedentary lifestyles among children and youth. In contrast to existing physical activity guidelines aiming to encourage physical exertion, these anti-sedentary guidelines focus on re-introducing physical activity into daily routines. Based on an analysis of behaviours that can lead to sedentary lifestyles, such as using motorized transportation, watching television, playing sedentary video games and using computers, the CSEP suggests that families introduce activity by replacing sedentary means of transportation and introducing active family time. Their guidelines suggest limiting sedentary transport, prolonged sitting and time spent indoors. Furthermore, they recommend limiting sedentary leisure activities including television and computer use to two hours per day (Tremblay et al. 2011).

The approach suggested in this chapter—applying video games to combat sedentary lifestyles—seems to contradict the CSEP recommendations. However, it is important to make a distinction between sedentary and active video games. In the context of our work, we believe that engaging with video games that encourage physical activity can help fight sedentary lifestyles despite increasing the overall time that is spent playing video games, by combining a popular leisure activity with physical activity, potentially facilitating a transition into a more active lifestyle. In this chapter, we outline how physical activity can be designed into games that support the goal of combating sedentary lifestyles, thereby encouraging players to be more active and helping them to adopt healthier lifestyles.

3 Barriers to Healthy Behaviour

As noted in previous sections, many people do not get the recommended levels of physical activity per day (Colley et al. 2011a, b). Researchers and governmental institutions have identified common barriers to physical activity to provide recommendations for persons wishing to transition from a sedentary lifestyle into more active daily routines.

3.1 Common Barriers to Physical Activity

Research results (Salmon et al. 2003) suggest that environmental issues (e.g. bad weather), the cost of exercise and individual aspects of personality (e.g., one's intrinsic motivation and perceived self-efficacy) have an impact on whether a person gets sufficient physical activity. Many initiatives aiming to foster physical activity and combat sedentary lifestyles discuss these barriers in detail. According to the Centres for Disease Control and Prevention (CDC, cdc.gov), common barriers to participating in physical activity are a lack of time, low interest in activity, low self-efficacy, a lack of social support and access to suitable facilities. Likewise, Healthy Families BC (healthyfamilies.bc) mentions that being busy, lacking exercise partners, not knowing how to approach physical activity and not wanting to sweat or feel hot are common barriers. These identified barriers can be roughly categorised into three main types: *psychological barriers* (i.e. physical activity is often boring, and antisocial, with significant skill, physical fitness and perceived capability barriers to entry), *temporal barriers* (i.e. physical activity is often disruptive to modern schedules and often has significant start-up and recovery times) and *physical barriers* (i.e. physical activity often requires access to specialised hardware or locations, or is constrained by physical processes outside the player's control such as the weather).

3.2 Barriers to Non-sedentary Lifestyles

In addition to the lack of physical activity, many people simply spend too much time sitting, and sedentary behaviours can result in negative health outcomes that are physiologically distinct from those associated with a lack of physical activity (Hamilton et al. 2008; Tremblay et al. 2010). Technological innovations have a lot to do with the amount of time spent sitting. People drive or ride buses to work, engage in computer-based jobs (where sitting is the norm) and indulge in screen-based leisure activities, such as watching television and playing video games. Consider the average day for many people—it begins with riding in a vehicle to work, sitting throughout the day, commuting home, a sit-down dinner and a few hours spent watching television, playing games or reading a book. Going to the gym or playing a sport can displace some of the time spent sitting, but modern routines themselves form a significant barrier to non-sedentary behaviour.

In addition, the aforementioned psychological barriers to exercise are also barriers applicable to people who wish to change their sedentary lifestyle. Taking breaks from sitting by climbing stairs or stretching is not particularly compelling, whereas physical fitness might prevent a person from choosing to cycle to work, rather than ride in a vehicle. Physical barriers are also still relevant; cycling to work or going for a walk over your lunch break can be difficult in many climates due to seasonal weather. However, if planned correctly, the temporal barriers to exercising—start-up and recovery (e.g. shower and change) times and finding time in a busy schedule to exercise—are not as relevant in the context of combating sedentary behaviour as they are to promoting physical activity. If designed correctly, anti-sedentary games could slot into a player's day in opportunistic moments, while the short duration and light activity of a game to combat sedentary behaviour does not require a player to change or shower. As such, there is an opportunity for anti-sedentary games to easily address the temporal barriers, while specifically targeting the physical and psychological barriers to an active lifestyle, thus improving overall health.

4 Design Principles for Exergames and Energames

To help people meet the recommendations for physical activity, and to overcome the barriers to exercise, researchers have created a variety of exergames, usually by replacing a regular game input device with controllers that require physical effort to foster activity among players and thus encourage healthier lifestyles. Mueller et al. define exergames as games "in which the outcome is predominantly determined by physical effort" (Mueller et al. 2011) and that "demand intense physical effort from players" (Mueller et al. 2010), highlighting that such games focus on providing sports-like exertion, which goes beyond simply encouraging players to be more active. The term "exergame" generally implies that physical activity resulting from game play is of moderate or vigorous intensity. In contrast, games that combat sedentary behaviour do not require intense physical effort or sport-like exertion, but encourage movement throughout the day to combat long periods spent sitting. In contrast to exergames, we dub these games *energames*, i.e. *games that reduce sedentary time by requiring frequent bursts of light physical activity throughout the day*.

In this section, we first present the guiding principles for exergame design to investigate whether they can be leveraged to facilitate the design of energames. This is followed by several examples of commercial and research-based exergames. We then define new guiding principles for energame design, followed by examples of several exergames that could be adapted to meet these principles and thus successfully combat sedentary behaviours.

4.1 Guiding Principles for Exergame Design

Research has previously addressed the design of effective exergames by providing design principles for integrating physical activity into games while fostering player motivation. A meta-level approach is provided by Consolvo et al. (2006), where they highlight the importance of providing feedback on activity levels, drawing awareness to past and current activity levels and providing feedback regarding goal achievement. Furthermore, the authors underline the importance of social influence for long-term user engagement, particularly social pressure that can be increased by sharing users' levels of activity and social support that can be achieved through the connection of users. Finally, they point out that accounting for users' lifestyles is crucial to facilitate the integration of activity-motivating technologies in daily life, which is particularly important when designing games to combat sedentary lifestyles. In the remainder of this section, we present additional considerations for exergame design by compiling guiding principles from a variety of sources and categorising them into five core areas of interest.

(1) Providing an easy entry into play. Lowering the barrier to foster physical activity can be accomplished by offering players an easy entry into play (Väätänen and Leikas 2009) using accessible core game mechanics (Campbell et al. 2008). Providing tips or hints can support entry along with advice to new players (e.g. tutorials) and coaching mechanisms that help players grasp the physical dimension of the game, e.g. learning gestures and the development of motor skills (Thin and Poole 2010).

(2) Implementing achievable short-term challenges to foster long-term motivation. To engage players over a longer period of time, many guiding principles comment on the inclusion of achievable short-term goals in order to foster long-term player motivation (Campbell et al. 2008; Thin and Poole 2010; Yim and Graham 2007). Yim and Graham (2007) refer to the concept of self-efficacy—the degree to which people attribute change in their lives to their own actions—to underline the importance of achievable goals.

(3) Providing users with appropriate feedback on their exercise effort. Providing players the opportunity to review their exercise efforts, for instance through progress charts that can be accessed after play or in-game feedback that informs players about their current performance (Thin and Poole 2010) can improve performance. Likewise, it is recommended to hide players' fitness levels in multi-player environments to avoid direct competition between players, which might discourage novice users or players with lower fitness levels (Yim and Graham 2007).

(4) Implementing individual skill-matching to keep players engaged. Adapting in-game challenges to match players' individual skill levels is one of the most important aspects of exergame design. Campbell et al. (2008) recommend the inclusion of marginal challenge to address this issue: providing the player with challenging, yet achievable in-game tasks. This is not only relevant to adapt games to the skill level of players; balancing between different players to provide enjoyable multi-player experiences is another important factor. Mueller et al. (2012) elaborate on this issue and provide a list of design tactics: to balance between players, they recommend facilitating empathy by creating awareness of other players' workout intensities and allowing players to negotiate the duration of physical activity. Stach et al. (2009) recommend that exergames be balanced for people of different fitness levels by driving play mechanics by a player's exertion relative to their own fitness level (e.g. through percent of target heart rate), rather than through absolute metrics of effort (e.g. through cycling revolutions per minute).

(5) Supporting social play to foster interaction and increase exercise motivation. Supporting social play and fostering interaction between players is a core component when trying to increase long-term exercise motivation. Campbell et al. (2008) distinguish between internal (specific to the game context) and external (brought into the game from outside) social relations that have to be accommodated by exergames. Mueller et al. (2009, 2010) provide a detailed analysis of social interaction in exergames, and offer additional design recommendations for social play, including considerations regarding the importance of meta-gaming (game-related activity that occurs outside of actual gameplay) between sessions to foster social bonding.

Because many of these principles focus on motivational aspects of exergame design, they hold valuable information for game designers, and can help overcome some of the barriers to exercise presented in the previous section. However, additional considerations are necessary to create games that fully address all design challenges that go along with combating sedentary lifestyles, rather than fostering physical exertion. In the following section, we further investigate how exergame design can help inform the creation of energames: we provide an overview of successful exergame examples, and investigate how their core mechanics can be leveraged to help inform the design of energames that can help address the barriers to an active lifestyle.

4.2 Successful Exergame Examples

To explore how games can help individuals be more active, we analyse currently available exergames, and discuss how principles applied in these games can be applied in the design of games to combat sedentary behaviour. In our analysis, we do not provide an exhaustive overview of currently available exergames; we choose successful exemplars spanning from commercially available exergames to games that were developed as research tools.

Commercially available exergames. A very popular platform for commercial exergames is Nintendo's Wii console (http://wii.com), which features the Wii Remote controller that uses different buttons and accelerometer information to track user input; an increasing number of games that require physical player input have been released for the console. One of the most successful commercially available games on that platform is Wii Sports (http://nintendo.com/games). The game consists of four different mini games that implement the Wii Remote controller in different ways, Wii Sports Bowling, Tennis, Golf, and Boxing. Research has shown that except for Wii Sports boxing, the games do not cause significant energy expenditure among players, and none of the games provide activity levels similar to the actual sport (Graves et al. 2008). However, games like Wii Sports show how integrating physical input into video games can shift sedentary playing time to more active behaviour. While levels of energy expenditure may not be sufficient to replace traditional physical activity, such games may be suitable to combat sedentary behaviour by reaching out to gaming audiences that exhibit sedentary behaviour by nature (high amounts of daily screen time) and replacing sedentary play with active alternatives.

Higher levels of energy expenditure are achieved by video games that use music and simulate dancing, and they are among the most successful commercially available exergames. A prominent example is Konami's Dance Dance Revolution (DDR) (konami.com/ddr). DDR uses a custom controller—a mat equipped with sensors to detect the player's steps—and invites players to dance along with different songs, displaying the necessary steps on screen. Because of the fast pacing and increasingly difficult dance moves, the game encourages higher levels of energy expenditure than many other exergames, partially reaching recommendations of intensity levels of physical activity by the American College of Sports Medicine (Unnithan et al. 2005). Similar to DDR, Dance Central (DC) by Harmonix integrates music and physical activity to engage players in the game (dancecentral.com). Players perform Kinect-tracked dance moves along with music, requiring complex physical input sequences. By integrating the whole body in play, the game has the potential of providing higher levels of physical activity than other systems such as Nintendo Wii Sports. When designing games to combat sedentary lifestyles, the motivational pull of music can be leveraged to encourage individuals to become more active. However, an issue that designers must address is the difference in energy expenditure caused by currently available games. Some motion-based games require so little activity that there is little benefit over playing with a standard game controller (sedentary game); however, the energy required to play some dancing games may be too high to engage players used to sedentary play (i.e. the fitness barrier may be insurmountable). Energames targeted at sedentary players need to tune energy expenditure requirements to promote active screen time, while not discouraging people accustomed to sedentary play.

Exergames in research. Research on exergame design has approached the topic from two sides. Games such as Jogging over a Distance by Mueller et al. (2010), where persons in remote locations are connected to allow them to go on runs together, aim to bring game elements into the realm of traditional exercise. Another slightly different approach towards augmented outdoor sports experiences is the skateboarding game Tilt 'n' Roll by Anlauff et al. (2010), which requires users to ride a skateboard that is equipped with sensors to detect board movements and tricks. This set-up is extended by a mobile application that keeps track of user achievements. Such sports-like exergames focus on augmenting the real-world experience with technology to motivate players, and to provide an engaging player experience. Sensor-based approaches provide an example of how gaming technologies can be applied to overcome barriers to physical activity, which may be an interesting design opportunity for games to combat sedentary lifestyles: connecting persons in remote locations can help provide social support for exercise, and adding game elements to sports can help increase their motivational pull. Likewise, research on exergames has addressed their design from the perspective of game development. Projects such as Heart Burn (Stach et al. 2009)—a racing game that is controlled using a recumbent stationary bike—and Swan Boat (Ahn et al. 2009)—a multi-player game where two players' hand and arm gestures combined with treadmill input to collaboratively steer a boat on its way down a virtual river—integrate sports equipment to implement physical activity into games. These projects demonstrate that it is possible to create games with engaging game mechanics that have the added benefit of physical activity, with greater creative freedom than when trying to combine existing sports with entertainment technologies. Such games may provide the opportunity of encouraging people who are not interested in sports in different kinds of physical activity, contributing to their overall activity levels. In addition, the use of custom hardware can introduce people who are unable to participate in traditional sports to exertion-based play, such as Liberi's adapted bicycle that allows children with Cerebral Palsy to play a multi-player open world game (Hernandez et al. 2012). Finally, to better fit physical activity into a person's busy day, and to address the barrier of users feeling too hot or too sweaty (cf. Sect. 3), the casual exergame GrabApple by Gao and Mandryk (2011), shown in Fig. 2 is based on the idea of providing players with short, 10-minute chunks of exercise to help them obtain the recommended levels of exercise by making it easier to fit physical activity into daily schedules.

Summary. Commercially available and research-based digital exergames integrate physical activity into video games by drawing from aspects of sport and game design. The aforementioned examples of successful exergames show that certain aspects of such games can encourage physical activity. This potential may also be leveraged for the design of energames; however, further considerations are

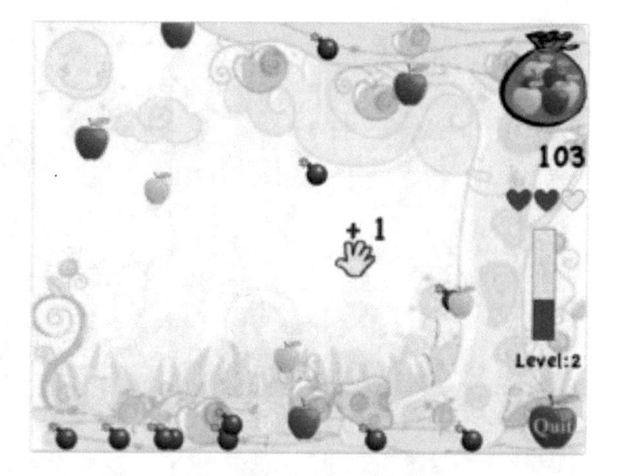

Fig. 2 GrabApple, a casual exergame

necessary to account for differences when encouraging individuals to integrate frequent, light-intensity physical activity into daily routines as compared to encouraging them to participate in moderate-to-vigorous physical exercise.

4.3 Guiding Principles for Games that Combat Sedentary Behaviour

Exergames that encourage physical activity have seen some success in providing individuals with moderate-intensity exercise in an engaging play session. The goal of energames, however, is not to provide moderate-intensity physical activity, but to reduce the amount of sedentary time. Whether a game is built to promote moderate-intensity sustained exercise, or to reduce the amount of time spent sitting, the aforementioned principles of good exergame design still apply. Games should still provide easy entry, implement achievable challenges to foster long-term motivation, provide feedback, offer adequate and balanced challenge, and support social play. However, for successful energames, there are additional requirements related to providing low-intensity activity, multiple times over the course of the day.

Casual interaction. GrabApple was the first exergame engineered specifically as a *casual exergame*, which is defined as "computer games that players can learn easily and access quickly, using simple rules and special game mechanics, to motivate them to exercise at a moderate intensity for short periods of play" (Gao and Mandryk 2011). The goal of GrabApple was to create an exergame that could be played in 10-min bursts multiple times throughout the day to meet the recommended guidelines for physical activity. By applying the principles of casual game design to the design of an exergame, Gao and Mandryk created a game that was easy to access, produced moderate-to-vigorous-intensity exercise similar to

Fig. 3 Gemini: an accumulated context exergame. Screenshots of phone interface (*left*) and RPG game (*right*)

running on a treadmill, but was as fun to play as a sedentary version of the game (Gao and Mandryk 2012). Although GrabApple was designed to promote physical activity, the concepts behind casual play are important when considering games that combat sedentary behaviour. Anti-sedentary guidelines promote lower intensity activity frequently throughout the day, thus game-based interventions require accessible games with short set-up times, and the use of readily-available equipment. Therefore, the first additional principle for energame design is:

(6) Keeping interaction casual. To support users playing the game multiple times per day, start-up interfaces, play time and equipment must be kept casual. Following the guide for casual exergame design (Gao and Mandryk 2011, 2012) will help to create energames that can be accessed quickly and easily.

Pull to repeat. The intention of GrabApple was that it should be played multiple times per day to meet the recommended levels of physical activity; however, the design of the game did not explicitly encourage repeated play sessions. To combat sedentary behaviour, energames should ideally motivate users to play frequently, with play sessions spread throughout the day. There are many examples of games that encourage repeated play through the use of social play mechanics (e.g. Zynga's Farmville (http://Farmville.com), persistent worlds (e.g. Blizzard's World of Warcraft (http://battle.net/wow)), or simple but addictive interactions (e.g. Rovio's Angry Birds (http://angrybirds.com)). Although some of these games allow for the short play sessions of a casual game (e.g. Farmville, Angry Birds), others are designed to be more immersive and thus promote longer play sessions (e.g. World of Warcraft). In an alternate approach to promote frequent play, some exergames have been designed to decouple the physical activity from game play. Gemini (Stanley et al. 2011) (shown in Fig. 3) is a role-playing game that allows users to collect their activity over the course of a day and integrate these real-world behaviours for in-game rewards in a standard immersive play session. This approach (described in more detail in the next section) helps to encourage physical activity over the entire course of a day, and not just during the play session. Because anti-sedentary guidelines promote lower intensity activity frequently throughout the day, the second additional principle for energame design is:

(7) Motivating repeated play sessions throughout the day. To break up long periods of sitting, energames should be played frequently over the course of the day. Using social games mechanics, persistent worlds, or accumulated activity could motivate players to repeat play sessions multiple times in a day.

Persuasive games. Although playing an energame that follows the principles of casual game design multiple times a day could help to decrease sedentary behaviours, helping a user change their habits and routines to decrease sedentary time will also have a positive impact on a user's health. Persuasive games (game-based persuasive technologies that aim to bring about desirable change in attitude and behaviour without using coercion or deception (Fogg 2002)) could help users to replace sedentary behaviours with active ones. For example, a persuasive game that encourages users to commute via bicycle (rather than by car) to reduce carbon emissions also has potential to reduce sedentary behaviour, and thus has value as an energame. Persuasive games could also be designed with the specific goal of reducing sedentary time (e.g. by encouraging cycling or walking to work instead of driving or taking the stairs instead of the elevator). A key idea behind persuasive games for behaviour change is that they scaffold new routines—unlike some other approaches, the new behaviour should remain after the game intervention is removed. These games are not outside of the context of sedentary game design, but simply represent a specific approach to reducing the time spent sitting. The third energame principle is:

(8) Persuading players to change their routines and habits. To scaffold routines that better fit the guidelines for non-sedentary behaviour, principles from persuasive game design can be used to help users make small changes with big impact.

Summary. These new guiding principles for energame design suggest that video games can be applied to reduce sedentary behaviours and thus improve player health. In the following section, we review examples from the exergame literature that either fit the principles for energame design or could be adapted to fit the principles and reduce sedentary behaviours.

4.4 Successful Energame Examples

Exergames have been traditionally designed to increase physical activity; there has been little direct intention to create games that combat sedentary behaviour. However, there are several games in the larger space of ubiquitous games, casual exergames, and accumulated activity games that overcome some of the barriers to non-sedentary behaviour identified in this chapter.

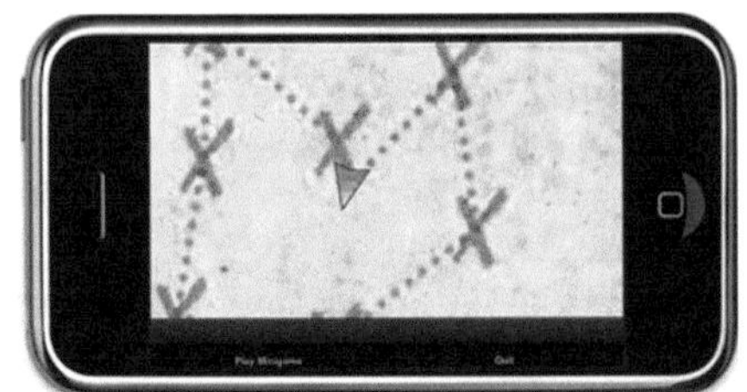 

Fig. 4 PiNiZoRo, a low-intensity ubiquitous exergame. Screenshots of (*left*) orienteering interface and (*right*) one of several minigames

Ubiquitous games. Ubiquitous games in general (Magerkurth et al. 2005) and mixed reality games in particular (Lindt et al. 2007) utilise real-world context as a game mechanic or input, often focusing on the player's location in the world (see (Magerkurth et al. 2005) for a review). Because these location-based games are driven by a player's movement through the world (e.g. (Bell et al. 2006; Benford et al. 2006; Stanley et al. 2010)), they can be considered exergames so long as the player is propelling themselves (as opposed to riding in a vehicle). Fast-paced ubigames such as Can You See Me Now (Benford et al. 2006), which pitches virtualand real players against each other in a game of team tag or Zombies, Run! (http://www.zombiesrungame.com), which has the player performing interval training to escape a zombie hoard, can fit the traditional definition of an exergame; whereas slower-paced walking games inspired by geocaching like Feeding Yoshi (Bell et al. 2006)—in which the player moves through the world to plant seeds and gather fruit for their virtual character—and PiNiZoRo (Stanley et al. 2010)—an orienteering game for families that has players 'walk a beat' in their neighbourhood (Fig. 4)—tend to provide lower-impact activities. These games that require low-intensity physical activity could be construed as early energames; however, they do fall short of addressing all the energame design requirements. Games like PiNiZoRo have a not-inconsequential set-up time as minigame rewards must be placed sensibly around the local environment by a game designer, often a parent. Likegeocaching, levels are typically designed to take 20–30 min to complete, too long to be played frequently, multiple times per day. Designing casual walking games that promote activity throughout the day could be a promising first step in energame design.

Casual exergames. Casual exergames such as GrabApple (Gao and Mandryk 2012) do fit the temporal requirements for energames, in that they are designed to be played in short bursts. However, as an academic effort, GrabApple has a lower replay value than commercially-designed games. Although a leaderboard or other competition-based incentives could motivate more frequent play, GrabApple requires fixed hardware (i.e. a Kinect sensor) and thus is not accessible at any time to any individual. As smartphone activity sensing improves, and reduces cheating in accelerometer-based play (e.g. shaking a pedometer), the use of compelling game mechanics, such as those demonstrated in Angry Birds (http://angrybirds.com) or Temple Run (http://imangistudios.com), could create low-barrier, highly-repeatable games. Although attempted in a number of commercial applications such as Teemo (http://goteemo.com), the design and implementation of smartphone-based casual exergames is subject to 'cheating' accelerometer-based input and the social awkwardness that results from performing the game mechanics (e.g. squatting) in public places (Gao and Mandryk 2012).

Accumulated activity games. Accumulated activity exergames decouple exercise and play to provide asynchronous in-game rewards (often in a traditional sedentary game) for previously-completed activity in the real world. Gemini (Stanley et al. 2008, 2011), Neat-o-Games (Fujiki et al. 2008), Move2Play (Bielik et al. 2012), American Horsepower Challenge (Xu et al. 2012), Play, Mate! (Berkovsky et al. 2009), and Neverball (Berkovsky et al. 2012) are examples of this concept in the academic literature. Pokewalker from Nintendo (http://nintendo.com/consumer/systems/ds/pokewalker.jsp), where pedometer-enhanced pokemon battle for supremacy, is the best-known commercial instantiation of an accumulated activity game. Interaction can be designed to provide players with directly mapped in-game benefits such as more powerful pieces (Stanley et al. 2008) or companions (Stanley et al. 2011), in-game currency to purchase mechanic-impacting (Fujiki et al. 2008) or cosmetic (Xu et al. 2012) virtual items, or unlocks for in-game capabilities (Berkovsky et al. 2012) or new games or minigames (Bielik et al. 2012). This class of games sidesteps some of the design requirements for short duration, and ease of accessibility in energames by measuring activity continuously. Whenever activity occurs, it is measured, accumulated, and credited for digital rewards in the offline game. Players are implicitly encouraged to fit in small bouts of activity whenever possible to increase their benefit in the sedentary portion of the game. However, the reward structure for these games does not perfectly match the requirements for energames. In a typical example, exercise is accumulated over the day, but the in-game impact of physical activity may be scaled based on individual factors, such as fitness level or historical activity levels, to encourage continued play and prevent disengagement (Berkovsky et al. 2012). However, the sequence of activities is not generally prescribed. In simple accumulated activity games, a single bout of walking for an hour is given the same weighting as six 10-minute walks. While both activities are desirable, according to anti-sedentary guidelines, the six shorter walks spread throughout the day would be preferable to the single long walk. Finally, accumulated activity games are often linked to sedentary gaming experiences, so the non-sedentary portion of the game is facilitating the sedentary activity. Careful cost-benefit balancing is required to combat this issue.

Summary. While no existing game or game genre meets all the requirements for energames, analysing how different genres do and do not match the principles can provide some insight into how future games should be designed to counter sedentary behaviour. Classic exergames often have too high a barrier to entry and too long a duration for use as energames. Casual exergames overcome the duration and accessability barriers, but have typically not provided sufficiently rewarding experiences to entice players into multiple bouts through a day, week in, week out. Accumulated activity games completely remove the barriers to entry by always

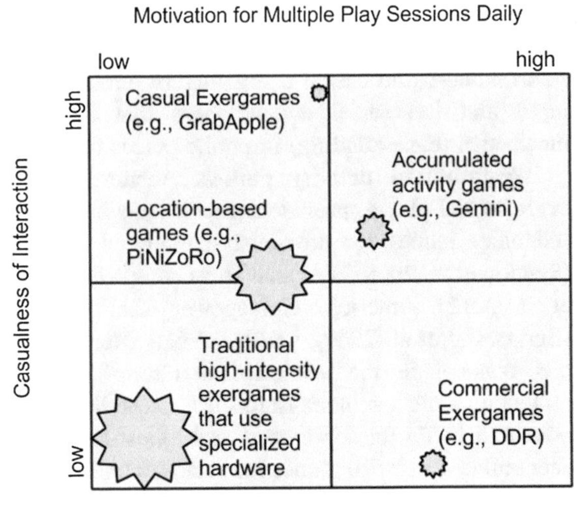

Fig. 5 Exergames plotted on a matrix of casualness and motivation for multiple play sessions. Energames should appear in the upper right quadrant, which is largely unexplored

measuring activity level, but do not currently distinguish whether activity is occurring frequently throughout the day, or in a single burst (although this could be incorporated into gameplay). In addition, these games also require a delayed gratification by decoupling exercise and play, which may inhibit transfer of game-based routines to general lifestyle choices through scaffolding.

If we consider the design of energames, and where existing exergames fail to combat sedentary behaviour, it seems apparent that energames should possess a low barrier to entry in terms of physical or game ability, access to specialised equipment or locations and required recovery time, but also have a high replay ability factor to allow users to continue to repeat the experiences multiple times per day, every day of the week. If we plot existing games on a matrix with casualness (low-to-high) on one axis, and motivation for multiple play sessions (low-to-high) on the other, we demonstrate how applying the principles of energame design should result in games in the upper right quadrant (see Fig. 5). We also find that most traditional academic games fall within the lower left quadrant, having a generally high barrier to entry (i.e. due to specialised hardware), a design for moderate-intensity activity and no pull to repeat, either due to a game design that targets a single session of play, or having mechanics that simply are not compelling enough to engage players frequently throughout the day. The casual exergame GrabApple (Gao and Mandryk 2011) fits in the upper left quadrant, because although high in terms of the requirement for casual play, it has little to compel a player to repeat play multiple times throughout the day. The commercial exergame DDR sits in the lower right quadrant matrix, having a mechanic that encourages replayability in multiple play session, but has significant barriers to entry (requiring specific hardware). Smartphone-based (or pedometer-based) accumulated activity games have exceptionally low barriers to entry, being always on, but have a more diffuse pull to repeat due to the delayed gratification of a single play session. The current instantiations of accumulated activity games sometimes integrate with a non-casual game engine (e.g. as in the role-playing game Gemini based on Neverwinter Nights); however, the smartphone-based portion of the game that encourages activity throughout the day is very casual. It is not surprising that current exergame types do not also meet the principles for energame design; they were generally designed to promote moderate-intensity exercise for a dedicated and sustained period of play, not to reduce sedentary time. As is apparent from the figure, the upper right quadrant—which should be inhabited by energames—is largely unexplored, providing fertile ground for additional research. The properties of the upper quadrant and its future potential are explored in the following section.

5 Opportunities and Challenges

5.1 Towards Energames

Casual games and accumulated activity games address some, but not all, of the requirement for energames. These games experience temporal shortcomings as energames: casual games are not sufficiently compelling in the long-term, and accumulated activity games do not address the timing of activity with sufficient resolution. Being compelling both in short bursts and over the long term is a daunting design challenge. However, certain game genres, such as MMORPGs or casual games, can provide compelling examples of how games can have a high pull to repeat, encouraging players to come back multiple times a day, several times a week over prolonged periods of time. Analysing the pacing and mechanics of examples such as the MMORPG World of Warcraft (http://battle.net/wow) or casual games such as Farmville (http://farmville.com) or Bejeweled (http://bejeweled.popcap.com) can provide additional insight into the creation of energames. For example, people spend 300 million minutes a day playing the casual game Angry Birds (McGonigal 2012), showing that a compelling game mechanic has a lot of motivational pull. The wide variety of highly-motivating sedentary casual games and their broad appeal from hardcore gamers to casual players shows how carefully designed energame concepts can be tailored towards different audiences, potentially encouraging children, teenagers and adults to adopt more active lifestyles.

5.2 Research Challenges and Opportunities

As previously discussed, one of the main barriers to non-sedentary behaviours is the modern professional's need to be sitting at a desk or using a computer. While it may be easier to fit several small bouts of physical activity—rather than a prolonged high-intensity activity—into a sedentary-office worker's workday, it is still difficult to incorporate multiple bouts of physical activity within a workday context. Because non-sedentary tasks must be performed repeatedly, the player cannot be left to play only outside of work, as this could lead to long sedentary periods at work. However, game-based interruptions to sedentary behaviour must be sensitive to both larger routines, and specific individual schedules. Asking the player to go for a walk in the middle of their carpool commute, or to perform jumping-jacks in the middle of a scheduled meeting would be inappropriate and potentially unsafe. Research advances in context-sensitive interruption from the field of ubiquitous and pervasive computing could determine if the player is in an interruptible state, prior to triggering a reminder to encourage the player to engage with the game.

Even if the player has the physical space during a meeting, performing jumping-jacks while the CEO is addressing the company would be considered socially unacceptable. Even performing callisthenic exercises outside of work contexts (e.g. waiting at a bus stop) could be socially awkward. Novel mixed reality game designs might be a potential solution to this problem. Context-appropriate missions could be triggered to allow the player to engage in a short period of movement without seeming to be socially inappropriate. Examples of such tasks include pretending to deliver a file to another floor in an office environment in an "OfficeDash"-style game, or walking briskly, but unobtrusively for five minutes without anyone noticing you in a spy-themed game, or using the social lubricant of children or pets to perform socially acceptable short bursts of free-play.

Another challenge is associated with encouraging physical activity among special populations. For example, people with certain motor disabilities or chronic diseases, people who use wheelchairs, and institutionalised older adults are at a higher risk of sedentary behaviour as they often depend on the assistance of others to be able to engage in physical activity, and because side-effects of their condition (e.g. range of motion, gross motor control) may limit their ability to be physically active. Research has recently addressed this issue by studying how motion-based controls can be introduced in ways that lower accessibility barriers, for instance by integrating wheelchair-based game input (Gerling et al. 2013) that can turn sedentary video games into motion-based games, and by analysing how motion-based controls should be designed to be suitable for older adults (Gerling et al. 2013) to successfully encourage physical activity (Gerling et al. 2012). Likewise, research on the design of exergames for children with cerebral palsy (who have gross motor impairments) has shown that exergames are a great opportunity to encourage children with cerebral palsy to be more active; however, it was necessary for the researchers to design a specialised bike ergometer to enable play (Hernandez et al. 2012). Many game projects for populations with reduced motor ability resemble energames—requiring some physical effort, but focusing on gaming accessibility, and providing modest physical activity in order to reduce the negative effects of sedentary lifestyles. However, like most exergames, these academic efforts generally are intended to be played in a single dedicated session, and not multiple times per day.

Even if the scheduling can be appropriate managed, and suitable incentives and difficulty levels for a variety of gamer types and populations can be created, energames suffer from a difficult feedback problem—the health benefits are not immediately and clearly visible to the user. Many of the benefits of non-sedentary behaviour are preventative—reducing the risk of serious conditions from appearing later in life (Garber et al. 2011; Pate et al. 1995; Tremblay et al. 2010), and thus little immediate benefit is evident to a player. Changing lifestyle to reduce the risk of future coronary disease can be less compelling than the more immediately apparent gains—such as losing weight, or increased fitness—that are associated with traditional exercise. People may become disengaged with energames if they see no immediate or tangible return on their investment. Contributions from visual analytics that are built into game play could help address this problem. Providing players with visualisations of how their activities are affecting their long-term health, and how not pursuing these activities could be detrimental to their long-term health, could potentially help people understand why playing the energame has benefit. Integrating persuasive strategies (Oinas-Kukkonen and Harjumaa 2008) (such as simulating the long-term effect of short-term changes) into a game design could also make the future benefits of behaviour change more tangible for a player. In addition, highlighting the temporary benefits of short bursts of activity (e.g., improved concentration, working memory and mood; see (Gao and Mandryk 2012)) may help players to see that their efforts are paying off in immediate, although acute, benefits to their well-being.

6 Summary

In this chapter we motivate the concept of an energame—a *game that reduces sedentary time by requiring frequent bursts of light physical activity throughout the day*—to combat the negative health effects of sedentary behaviour. Because these effects are distinct from those due to a lack of physical activity, it is important to consider possible technological interventions for both classes of health behaviours. We arrive at the definition of an energame through scrutiny of the health guidelines for physical activity and non-sedentary behaviour, and consideration of the barriers to an active lifestyle. We employ the results of this investigation to extend existing exergame design principles to energames, particularly with respect to the temporal barriers. Energames must be played frequently for short bursts of time, requiring aspects of casualness and motivational pull not necessarily required of traditional exergames. While our analysis of existing games from both industrial and academic sources did not produce any examples that met all of the suggested design principles, several titles—particularly from the casual and accumulated activity genres—were approaching viable solutions. Drawing design wisdom from sedentary titles such as MMORPGs, casual games and social games, game designers and researchers should be able to build novel energames that manage the temporal, social and personal constraints imposed by the design principles. Sedentary behaviour is linked to many of the lifestyle diseases prevalent in the developed world, making this new class of digital game important for individual well-being.

References

Ahn M, Kwon S, Park B, Cho K, Choe SP, Hwang I, Jang H, Park J, Rhee Y, Song J (2009). Running or gaming. In: Proceedings of advances in computer entertainment (ACE), pp 345–348

Anlauff J, Weitnauer E, Lehnhardt A, Schirmers S, Zehe S, Tonekaboni K (2010) A method for outdoor skateboarding video games. In: Proceedings of advances in computer entertainment (ACE), pp 40–43

Bell M, Chalmers M, Barkhuus L, Hall M, Sherwood S, Tennent P, Brown B, Rowland D, Benford S, Capra M, Hampshire A (2006) Interweaving mobile games with everyday life. In: Proceedings of the SIGCHI conference on human factors in computing systems (CHI), pp 417–426

Benford S, Crabtree A, Flintham M, Drozd A, Anastasi R, Paxton M, Tandavanitj N, Adams M, Row-Farr J (2006) Can you see me now? ACM Trans Comput-Hum Interact (TOCHI) 13(1):100–133

Berkovsky S, Bhandari D, Kimani S, Colineau N, Paris C (2009) Designing games to motivate physical activity. In: Proceedings of the 4th international conference on persuasive technology, pp 37–40

Berkovsky S, Freyne J, Coombe M (2012) Physical activity motivating games: be active and get your own reward. ACM Trans Comput-Hum Interact (TOCHI) 19(4):32

Bielik P, Tomlein M, Krátky P, Mitrík Š, Barla M, Bieliková M (2012) Move2Play: an innovative approach to encouraging people to be more physically active. In: Proceedings of the 2nd ACM SIGHIT international health informatics symposium (IHI), pp 61–70

Campbell T, Ngo B, Fogarty J (2008) Game design principles in everyday fitness applications. In: Proceedings of the ACM conference on computer supported cooperative work (CSCW), pp 249–252

Chilibeck PD, Bell G, Jeon J, Weiss CB, Murdoch G, MacLean I et al (1999) Functional electrical stimulation exercise increases GLUT-1 and GLUT-4 in paralyzed skeletal muscle. Metabolism 48(11):1409–1413

Colley RC, Garriguet D, Janssen I, Craig CL, Clarke J, Tremblay MS (2011a) Physical activity of Canadian adults: accelerometer results from the 2007 to 2009 Canadian health measures survey. Health Rep 22(1):1–8

Colley RC, Garriguet D, Janssen I, Craig CL, Clarke J, Tremblay MS (2011b) Physical activity of Canadian children and youth: accelerometer results from the 2007 to 2009 Canadian health measures survey. Health Rep 22(1):15–23

Consolvo S, Everitt K, Smith I, Landay JA (2006) Design requirements for technologies that encourage physical activity. In: Proceedings of the ACM conference on human factors in computing systems (CHI), pp 457–466

Etnier JL, Nowell PM, Landers DM, Sibley BA (2006) A meta-regression to examine the relationship between aerobic fitness and cognitive performance. Brain Res Rev 52(1):119–130

Fogg BJ (2002) Persuasive technology: using computers to change what we think and do. Ubiquity, 2002(5)

Fujiki Y, Kazakos K, Puri C, Buddharaju P, Pavlidis I, Levine J (2008) NEAT-o-Games: blending physical activity and fun in the daily routine. Comput Entertain (CIE) 6(2):21

Garber CE, Blissmer B, Deschenes MR, Franklin BA, Lamonte MJ, Lee IM, Swain DP (2011) American college of sports medicine position stand. Quantity and quality of exercise for developing and maintaining cardiorespiratory, musculoskeletal, and neuromotor fitness in apparently healthy adults: guidance for prescribing exercise. Med Sci Sports Exerc 43(7):1334–1359

Gerling KM, Dergousoff KK, Mandryk RL (2013) Is movement better? comparing sedentary and motion-based game controls for older adults. In: Proceedings of graphics interface 2013, pp 133–140

Gerling KM, Kalyn MR, Mandryk RL (2013) KINECTWheels: wheelchair-accessible motion-based game interaction. In: Extended abstracts of the ACM conference on human factors in computing systems (CHI), pp 3055–3058

Gerling KM, Livingston IJ, Nacke LR, Mandryk RL (2012) Full-body MOTION-based game interaction for older adults. In: Proceedings of the ACM conference on human factors in computing systems (CHI), pp 1873–1882

Gao Y, Mandryk RL (2011) GrabApple: the design of a casual exergame. In: Entertainment computing (ICEC), LNCS, Springer, Heidelberg, pp 35–46

Gao Y, Mandryk R (2012) The acute cognitive benefits of casual exergame play. In: Proceedings of the ACM conference on human factors in computing systems (CHI), pp 1863–1872

Graves L, Stratton G, Ridgers ND, Cable NT (2008) Energy expenditure in adolescents playing new generation computer games. Br J Sports Med 42(7):592–594

Hamburg NM, McMackin CJ, Huang AL, Shenouda SM, Widlansky ME, Schulz E et al (2007) Physical inactivity rapidly induces insulin resistance and microvascular dysfunction in healthy volunteers. Arterioscler Thromb Vasc Biol 27(12):2650–2656

Hamilton MT, Healy GN, Dunstan DW, Zderic TW, Owen N (2008) Too little exercise and too much sitting: inactivity physiology and the need for new recommendations on sedentary behaviour. Curr Cardiovasc Risk Rep 2(4):292–298

Hassmén P, Koivula N, Uutela A (2000) Physical exercise and psychological well-being: a population study in Finland. Prev Med 30(1):17–25

Hernandez HA, Graham TC, Fehlings D, Switzer L, Ye Z, Bellay Q, Hamza MA, Savery C, Stach T (2012) Design of an exergaming station for children with cerebral palsy. In: Proceedings of the ACM conference on human factors in computing systems (CHI), pp 2619–2628

Hillman CH, Erickson KI, Kramer AF (2008) Be smart, exercise your heart: exercise effects on brain and cognition. Nat Rev Neurosci 9(1):58–65

Lindt I, Ohlenburg J, Pankoke-Babatz U, Ghellal S (2007) A report on the crossmedia game epidemic menace. Comput Entertain (CIE) 5(1):1–8(article 8)

Magerkurth C, Cheok AD, Mandryk RL, Nilsen T (2005) Pervasive games: bringing computer entertainment back to the real world. Comput Entertain (CIE) 3(3):4

McGonigal J (2012) Building resilience by wasting time. Harv Bus Rev 90(10):38

Mueller F, Gibbs MR, Vetere F (2009) Design influence on social play in distributed exertion games. In: Proceedings of the ACM conference on human factors in computing systems (CHI), pp 1539–1548

Mueller F, Gibbs MR, Vetere F (2010a) Towards understanding how to design for social play in exertion games. Pers Ubiquit Comput 14:417–424

Mueller F, Vetere F, Gibbs MR, Agamanolis S, Sheridan J (2010) Jogging over a distance: the influence of design in parallel exertion games. In: Proceedings of the SIGGRAPH Sandbox, pp 63–68

Mueller F, Edge D, Vetere F, Gibbs MR, Agamanolis S, Bongers B, Sheridan JG (2011) Designing sports: a framework for exertion games. In: Proceedings of the ACM conference on human factors in computing systems (CHI), pp 2651–2660

Mueller F, Vetere F, Gibbs MR, Edge D, Agamanolis S, Sheridan JG, Heer J (2012) Balancing exertion Experiences. In: Proceedings of the ACM conference on human factors in computing systems (CHI), pp 1853–1862

Oinas-Kukkonen H, Harjumaa M (2008) A systematic framework for designing and evaluating persuasive systems. In: Proceedings of persuasive technology'08, LNCS, Springer Berlin Heidelberg, pp 164–176

Owen N, Healy GN, Matthews CE, Dunstan DW (2010) Too much sitting: the population health science of sedentary behaviour. Exerc Sport Sci Rev 38(3):105–113

Pate RR, Pratt M, Blair SN, Haskell WL, Macera CA, Bouchard C, Wilmore JH (1995) Physical activity and public health. A recommendation from the centers for disease control and prevention and the American college of sports medicine. JAMA: J Am Med Assoc 273(5):402–407

Pate RR, Freedson PS, Sallis JF, Taylor WC, Sirard J, Trost SG, Dowda M (2002) Compliance with physical activity guidelines: prevalence in a population of children and youth. Ann Epidemiol 12(5):303–308

Salmon J, Owen N, Crawford D, Bauman A, Sallis JF (2003) Physical activity and sedentary behaviour: a population-based study of barriers, enjoyment, and preference. Health Psychol 22(2):178–188

Sothern MS, Loftin M, Suskind RM, Udall JN, Blecker U (1999) The health benefits of physical activity in children and adolescents: implications for chronic disease prevention. Eur J Pediatr 158(4):271–274

Stach T, Graham N, Yim J, Rhodes RE (2009) Heart rate control of exercise video games. In: Proceedings of graphics interface (GI), pp 125–132

Stanley KG, Pinelle D, Bandurka A, McDine D, Mandryk RL (2008) Integrating cumulative context into computer games. In: Proceedings of the international academic conference on the future of game design and technology, pp 248–251

Stanley KG, Livingston I, Bandurka A, Kapiszka R, Mandryk RL (2010) PiNiZoRo: a GPS-based exercise game for families. In: Proceedings of the international academic conference on the future of game design and technology (FuturePlay), pp 243–246

Stanley KG, Livingston IJ, Bandurka A, Hashemian M, Mandryk RL (2011) Gemini: a pervasive accumulated context exergame. In: Entertainment computing (ICEC) 2011, pp 65–76

Thin AG, Poole N (2010) Dance-based exergaming: user experience design implications for maximizing health benefits based on exercise intensity and perceived enjoyment. In: Pan Z, Cheok AD, Müller W (eds) Transactions on edutainment IV. LNCS, Springer, Berlin, pp 189–199

Tremblay MS, Colley R, Saunders TJ, Healy GN, Owen N (2010) Physiological and health implications of a sedentary lifestyle. Appl Physiol Nutr Metab 35(6):725–740

Tremblay MS, Warburton DE, Janssen I, Paterson DH, Latimer AE, Rhodes RE, Kho ME, Hicks A, LeBlanc AG, Zehr L, Murumets K, Duggan M (2011) New Canadian physical activity guidelines. Appl Physiol Nutr Metab 36(1):36–46

Unnithan VB, Houser W, Fernhall B (2005) Evaluation of the energy cost of playing a dance simulation video game in overweight and non-overweight children and adolescents. Int J Sports Med 27(10):804–809

U.S. Department of Health and Human Services. (1996). Physical activity and health: a report of the surgeon general. Atlanta, Georgia: Centers for Disease Control and Prevention, National Center for Chronic Disease Prevention and Health Promotion

Väätänen A, Leikas J (2009) Human-centered design and exercise games. In: Kankaanranta M, Neittaanmäki P (eds) Design and use of serious games. Springer Science and Business Media B.V, Heidelberg, pp 33–47

Xu Y, Poole ES, Miller AD, Eiriksdottir E, Kestranek D, Catrambone R, Mynatt ED (2012) This is not a one-horse race: understanding player types in multiplayer pervasive health games for youth. In: Proceedings of the ACM conference on computer supported cooperative work (CSCW), pp 843–852

Yim J, Graham TCN (2007). Using games to increase exercise motivation. In: Proceedings of future play, pp 166–173

Zwart SR, Hargens AR, Lee SM, Macias BR, Watenpaugh DE, Tse K, Smith SM (2007) Lower body negative pressure treadmill exercise as a countermeasure for bed rest-induced bone loss in female identical twins. Bone 40(2):529–537

Part V
Learning by Creating

Playing in the Arcade: Designing Tangible Interfaces with MaKey MaKey for Scratch Games

Eunkyoung Lee, Yasmin B. Kafai, Veena Vasudevan and Richard Lee Davis

Abstract Most tools for making games have focused on-screen-based design and ignored the potentially rich space of tangible interface design. In this chapter, we discuss how middle school youth (ages 10–12 years) designed and built their own tangible game interfaces to set up a game arcade. We conducted two workshops in which students used the MaKey MaKey, a low-cost tangible interface construction kit, to build touch-sensitive game controllers using everyday conductive materials for games they remixed in Scratch. We address the following research questions: (1) What types of tangible interfaces do youth create for their games? (2) How do youth designers deal with the complexities of coordinating the design of tangible interfaces with online Scratch games? (3) What do young users have to say about their tangible interface designs? We found that youth designers mostly replicated common controller designs but varied in their attention to either functionality or esthetics. An unexpected finding was how these different approaches followed traditional gender lines, with girls more focused on esthetics and boys more focused on functionality. These findings might point toward different expectations and informal experiences that need to be taken into consideration when bringing tangible design activities into educational settings. During the arcade, the youths' perspectives on their games and controllers changed as they observed other people playing their games. They expressed pride in their creations and saw ways to refine

E. Lee (✉)
Korea Institute for Curriculum and Evaluation, Seoul 100-784, Korea
e-mail: eklee76@kice.re.kr

Y. B. Kafai · V. Vasudevan
University of Pennsylvania, Philadelphia, PA 19104, USA
e-mail: kafai@gse.upenn.edu

V. Vasudevan
e-mail: veenav@gse.upenn.edu

R. L. Davis
Stanford University, Stanford, CA 93405, USA
e-mail: rldavis@stanford.edu

A. Nijholt (ed.), *Playful User Interfaces*, Gaming Media and Social Effects,
DOI: 10.1007/978-981-4560-96-2_13, © Springer Science+Business Media Singapore 2014

their designs in order to improve usability. In our discussion, we address how the inclusion of tangible interface design can extend game making activities for learning. Ultimately, we want youth to move beyond and experiment more with conventions, not just to increase their technological understanding and flexibility but also as a way to more critically approach the design of everyday things.

Keywords Tangible interfaces · Design · Interface construction kits · Game controllers · Scratch games · MaKey MaKey · Gaming arcade

1 Introduction

Over the past decade, new game controllers like the Nintendo Wii Remote, Microsoft Kinect, and the Sony Playstation Move have transformed console gaming by responding to gesture, body movement, or touch. In the rapidly growing arena of mobile gaming, such motion-responsive and touch-based controls have fast become the norm, not the exception. However, most youth never get the chance to build their own controllers because hacking or building custom physical interfaces typically requires access to costly tools, technology, and understanding that lie beyond the reach of most K-12 students. The recent development of low-cost tangible interface construction kits such as the Lego WeDo®, the PicoBoard, and the MaKey MaKey have made it substantially easier for amateurs to engage in interface construction (Millner 2010). These kits have opened the door to extending game making activities into the physical realm and added new opportunities for learning about topics like coding, circuitry, and interface design.

In this chapter, we report on what we learned in two workshops where middle school students' remixed Scratch games, designed tangible interfaces with the MaKey MaKey, and set up an arcade (Davis et al. 2013; Vasudevan et al. 2013). In the first workshop, students remixed a Scratch game and created a custom controller with the MaKey MaKey, a small USB device that connects to conductive materials and transforms them into touch-sensitive buttons that can control and move objects on the computer screen (Silver et al. 2012). In the second workshop, we expanded the social contexts by including a game arcade as a culminating public event in which other students from the school were invited to try out games and controllers. The following questions guided our analyses: What kind of tangible interfaces would youth create for their games? How would beginning programmers deal with the complexities of coordinating the screen and tangible designs of their games? How does framing the workshop within the context of an arcade impact participants' understanding of their work? We discuss what the findings from the two studies tell us about youths' creations, approaches, and perceptions as well as the associated challenges and opportunities that arise when game design activities move into the tangible realm.

2 Background

While much research has explored the learning benefits of playing video games (Gee 2003; Squire 2010), the focus has recently shifted to also consider the learning benefits of making games (Kafai 1995; Kafai and Peppler 2011). Involving learners in game design activities can have numerous learning benefits (Hayes and Games 2008). Designing games can foster computational thinking (Repenning et al. 2010), provide motivation for learning programming (Fowler and Cusack 2011), and increase technological fluency (Peppler and Kafai 2010). In addition, these production-oriented approaches have successfully broadened interest in gaming and computing (Denner et al. 2008; DiSalvo and Bruckman 2011).

Based on these successes, numerous platforms have been developed for novice game designers, ranging from specialized tools to open-ended programming languages (Burke and Kafai in press). For instance, Sploder is a game design platform that restricts the types of games users can create to four genres: platforms, puzzles, shooters, or algorithms. Although these more specific tools limit the variety of game projects users can create, they also provide a lower barrier of entry that is attractive to designers with very little experience. In contrast, Scratch is an example of a platform with wide walls that allows beginning designers to create many genres of interactive media, including stories, animations, and games. Even Microsoft has released their own design platform called Kodu, bringing game design activities to anyone with an Xbox 360 (MacLaurin 2009).

While game making activities have become quite popular, there have been few efforts to include the design of tangible controllers such as joysticks, touch pads, or other devices, most likely because the technical and material components are not easily accessible. This is a conspicuous omission because controllers are an essential part of the gameplay experience (Bayliss 2007) and designing and building controllers could bring additional benefits to game design activities (Marshall 2007). Benefits might include offering opportunities for collaborative activity and providing ways of making abstract ideas more concrete (Antle 2007), in addition to promoting more active, physical engagement in learning activities (Marshall et al. 2003). When Horn et al. 2012 compared learning with tangible interfaces to more traditional methods, they found that tangible interfaces are not only more inviting but also are better at supporting active collaboration, and have broader appeal across genders.

With the development of easy-to-use tangible interface construction kits, it is now possible to investigate these learning benefits in more detail. What can youth learn by designing and constructing tangible interfaces to go along with their games? Though there are a handful of studies that describe interface-design courses (Martin and Roehr 2010), few of them examine the benefits and challenges of this activity. Most relevant to our work is Millner's pioneering research with the Hook-Ups tangible interface construction kit that enabled children to become creators of interactive tangible experiences by minimizing programming and providing lower cost access (2010). His research illustrated how youth can learn about electricity, design, and programming while crafting tangible interfaces from found materials. We built on this research by expanding it into game design activities using a commercially available tangible construction kit called MaKey MaKey (Silver et al. 2012). We wanted youth to design custom physical controllers to go along with their Scratch games. Such an activity provides a good introduction to software and hardware design, in particular for middle school students, because it builds on their prior experience with popular gaming platforms and can draw from the large repertoire of games available on the Scratch site. One of our research goals was to examine the opportunities for learning programming and tangible interface designs with MaKey Makey in a school context. For that reason also we focused on students remixing rather than designing games from scratch (no pun intended!) as to counterbalance their limited experience with programming.

In addition to designing tangible game interfaces and remixing Scratch games, we also wanted to better understand the social contexts in which these design activities can be situated. When Papert (1980) described successful learning environments, he drew on Brazilian samba schools that have the annual Carnival festival as a public, culturally relevant event to work toward. As we know from other research, having an explicit audience in mind helps groups of students to focus their efforts (Zagal and Bruckman 2005). One such example is the recent development of online game competitions (Kafai et al. 2012) that, much like robotics competitions (Manseur 2000), provide a high level of motivation and broader audience. With a gaming arcade as our culminating, public event, we chose a design that would be more about the social experience and less about winning or losing. We invited younger students in the school to come play the games, test the controllers, and provide us with feedback.

3 Methods

3.1 Participants and Settings

The two workshops took place at a K-8 neighborhood school in a metropolitan area in the northeastern United States. Students in 6–8th grades (11–12 years old) opted to participate in the game design workshops as part of their elective (or choice) time. These elective courses are offered throughout the school year and students can choose how they want to spend their time for two periods a week. While a total of 18 youth participated in the two game design workshops (workshop 1 had four boys and five girls while workshop 2 had five boys and four girls), only 13 assented to participate in the research (six in Workshop 1 and seven in Workshop 2). Four of the students who participated in the second workshop also participated in Workshop 1. The first game design workshop was co-taught by

Table 1 Workshop schedules

Sessions	Workshop 1	Workshop 2
1	Introduction to Scratch and MaKey MaKey	Introduction to Scratch and MaKey MaKey
2		
3		
4	Reusing and remixing Scratch games	
5		Reusing and remixing Scratch games
6		
7	Building and crafting game controllers	Building and crafting game controllers
8		
9		
10	-	Playing in the arcade

Fig. 1 Screenshots of Scratch programming interface (*left*) and the Internet Portal (*right*)

three of the authors (Davis, Lee, and Vasudevan) while two authors (Davis and Vasudevan) taught the second workshop. Each workshop met nine times, with each session lasting about 50 min each.

3.2 Game Design Workshops and Arcade Setup

The first workshop focused on game and controller design, and the second added the arcade experience (for an overview of workshop activities, see Table 1).

In the first game design workshop, youth were asked to remix existing Scratch games and design their own tangible interfaces. In the initial three sessions the youth were introduced to the Scratch environment, the basics of creating circuits, and working with MaKey MaKey construction kits (Figs. 1, 2).

In the next three sessions they spent time modifying (remixing) their games. After selecting a specific game to remix for their final projects, youth designed physical interfaces using the MaKey MaKey, Play-Doh (a modeling compound that is nontoxic and comes in several different colors, similar to clay), pipe cleaners, and

Fig. 2 The MaKey MaKey board (*left*) and a hooked up version (*right*)

other materials available in the classroom. In the final three sessions, the youths developed and tested their interfaces with their Scratch games. The MaKey MaKey (see Fig. 2) is a small USB device that connects to conductive materials and transforms them into touch-sensitive buttons that can control and move objects on the computer screen (Silver et al. 2012). It requires no drivers, no specialized software, and no knowledge of programming (Collective and Shaw 2012).

In the second workshop, youth also created video games and designed a controller (or interface) using a MaKey MaKey. However, with more scaffolding in using Scratch, youth also had the chance to create their own simple games. The main difference between the two workshops was the addition of a culminating arcade to showcase the games in the second workshop. To accomplish this, youth were introduced to the Scratch environment in the first four sessions during which they spent time creating a simple Scratch game through guided practice in class. In the next two sessions they spent time modifying (remixing) their games. After selecting a specific game to remix for their final projects, youth designed physical interfaces using the MaKey MaKey and Play-Doh. In the final three sessions youth developed and tested their interfaces and modified their games. Finally, on the day of the arcade, youth made minor tweaks, set up their games, designed signs for their respective arcade stations, and hosted younger students at the arcade.

3.3 Data Collection and Analysis

We documented student design work and group interactions in observation notes, photographs, and video recordings for both workshops. In addition, we collected all final Scratch programs and used a framework developed by Brennan and Resnick (2012) to analyze the complexity of computational concepts (such as loops and conditionals) and use of design practices (remixing and debugging). We captured the progress of game controllers designed with the MaKey MaKey and Play-Doh with photos after each session and analyzed final designs in respect to functionality and esthetics. We also conducted post interviews with youth in which they reflected on the their approaches and experiences. These interviews were coded using a two-step process that identified two themes: audience considerations when designers referred to players and device creation when designers reflected on

Fig. 3 Example remixed Scratch games

the challenges and benefits for creating their own controllers. In addition, we asked players during the arcade to comment on the games and controllers in terms of what they liked and what they would improve.

4 Findings

In the following sections, we present findings on how youth designers remixed or constructed Scratch games and built and crafted tangible game controllers. We are combining the results from the two workshops. In addition, we present what youth learned about their games and controllers from other students who came to the arcade, an event that only took place in the second workshop.

4.1 Software Design: Remixing of Scratch Games

The youths in both workshops found different ways to remix games. The remixes included both functional (e.g., updating, removing, or tweaking code to change functionality) and esthetic (e.g., changing the appearance or other effects) changes. All of the participants spent significant time and effort on the esthetic features of their games by drawing new characters, designing custom backgrounds, modifying existing images, and adding background music. We also found that the youths used a wide range of computational concepts such as sequences, loops, conditionals, event handling, operators, and variables in their final games. Figure 3 shows different students remixes.

Many youth added graphic and sound effects in their remixes in addition to changing game mechanics. One example is Ishita, who added a shrinking piece of ice to her game because she wanted the penguin to eat all the fish before the ice melted (see Fig. 4). She also modified the code to include sound effects whenever

Fig. 4 Ishita's game screen and code for melting ice

Fig. 5 Amani's game screen with features for usability

a fish was eaten. Another youth, Amani, added good brains (pink) and bad brains (green) to her updated version of Zombie Attack. Each brain was labeled with point values so players could distinguish the good brains from the bad ones. She also linked the size of the brain to the point value, so brains with higher point values were larger and vice versa, and added a total score and a final win and lose screen (see Fig. 5).

In contrast, Marcus focused on the game mechanics of winning and losing in his remixed game. To win his game, the main sprite, a hungry fish, needed to reach a size of 105. Each time the hungry fish ate a smaller goldfish, the physical size of the fish incremented and the score increased. However, if the hungry fish touched the seaweed before it reached the size of 105, then the player would lose the game. To accomplish these changes, Marcus remixed the original game in the following ways: switching control of the fish to the arrow keys, adding sound and animation effects each time a goldfish was eaten, increasing the size of the fish, setting up a limit for when the fish exceeded a certain size, and establishing conditions for losing/winning game. The only graphic change he made in his remix of the game was to add a new seaweed sprite (see Fig. 6).

When reflecting on their design decisions in remixing games, audience consideration was a key factor in students' thinking and coding. In the case of James, this meant that he was aware that someone else might play his game: "I think it's

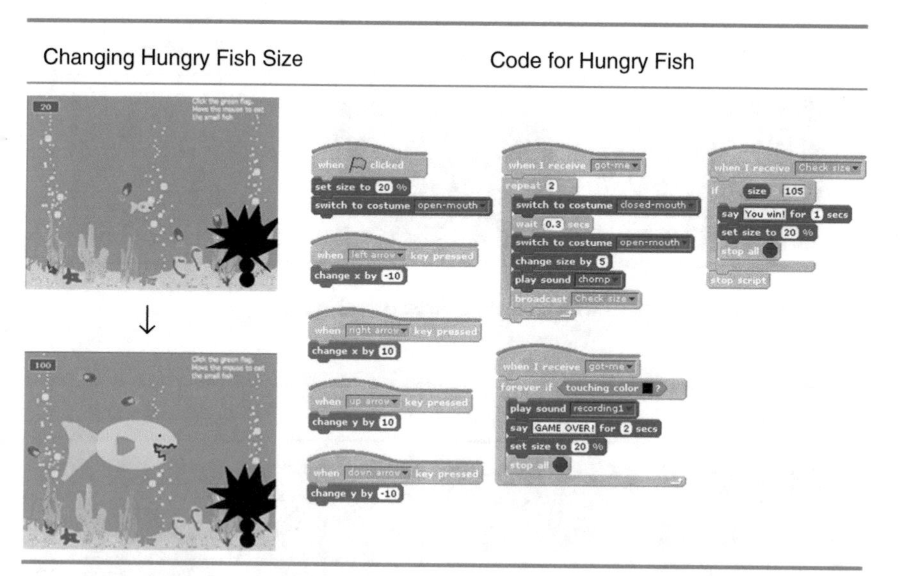

Fig. 6 Marcus' game screen and code for hungry fish

good, because it's pretty basic, if uh, like, like, in most games, like, there's like, in the arrow keys and space bar and that kind of stuff. So I, didn't really want to confuse the player, the person that's going to play." Others wanted to make their games harder and thus more interesting to play by adding levels or complexity such as a bad guy or a distractor. While Isabel mentioned that she would add more functionality to make enemies more difficult to chase, Amani intended to add more levels: "I think I would change, is for it [her game] to have levels. And then there would be more like enemies. Like, we said in the presentation the enemies would be thrown in there in the harder levels. And the levels would just get harder, and the monkey would be moving faster and faster and faster." These reflections illustrate how youth began to assume the role of game designers who focus on making games that are both playable and challenging enough to keep players' interest (Gee 2003).

4.2 *Tangible Design: Building and Crafting Game Controllers*

In addition to designing the screen interface in Scratch, youth also designed tangible game controllers for their games. Some of these tangible designs emphasized functional elements by making buttons large enough to touch while others focused more on esthetic elements such as matching colors to Scratch remixes or creating specific shapes.

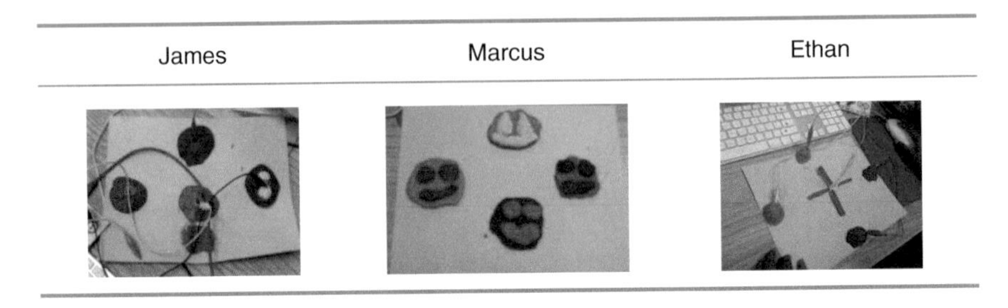

Fig. 7 Functional controllers with minimal esthetic considerations

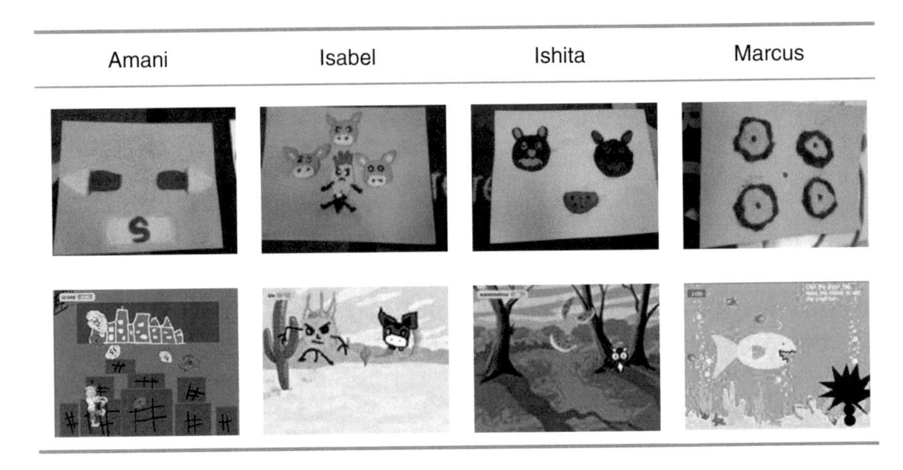

Fig. 8 Examples of controllers and screens that balance esthetics and functionality

In the first workshop, we found that one group of students designed controllers that used directional arrow keys (up, down, right, and left), so that the main characters or sprites in the games could move. Many of these youth did not match their controller designs to the topical focus of their Scratch game design (see Fig. 7). For instance, James and Marcus created three or four round buttons which were large enough for a user to place their hand on and, more importantly, easy enough to play their games. In contrast to these controllers, Ethan's design stands out. Instead of creating directional touch pads, he used metal wire to create a hand-held joystick and Play-Doh mounds as the touchpoints to complete the circuit for his game controller. Ethan mentioned in the post-interview that he was proud of his design because he felt it was most unique when compared to others.

In the second workshop, the interfaces varied from the extremely detailed, with Isabel and Ishita matching their controller components to the sprites (characters) in their Scratch games (see Fig. 8), to the less nuanced, with Amani, Marcus, and Jonathan matching the colors or themes of their games to their controllers.

In reviewing trends across both workshops, we found that girls focused more on the esthetic elements whereas boys focused mostly on functional elements in the design of their controllers. Only one of the boys (Marcus) aligned the esthetics of his game controller with his Scratch game. Most girls referenced 'usability' as their design rationale. For instance, Isabel noted that she wanted to make sure that the fish in her game was represented on her game controller. Ishita explained that she "tried to like match it with the theme, like, on my picture. Because if someone was trying to play my game, I thought that umm, if I used like, different kind of, umm buttons, it would be hard for them to understand like, which part is which." When asked about his second controller design for his Fish Chomp game, Marcus stated, "…what do you find, in this sea that actually can be put into a flat thing, you could touch?… So I finally thought oysters with pearls, and that will be perfect because then the MaKey MaKey clips could be touched to the pearls and would make more sense." The change in Marcus' design was prompted by one of the teachers who asked him think more about his design in the second workshop.

Others referenced not only usability but also personal preferences. For instance, Amani thought her initial interface would be easier for her first-grade sister than a regular keyboard because "she can use her whole hands, at some point her hands will be big enough to do it, but sometimes, right now, her hands are a little bit too small. So, with the interface, she can just put her whole hand on it. And it would just be fun and it wouldn't get her as frustrated." But she referred to the matching as a personal value saying that, "it wouldn't make sense if I had like a princess keyboard with a zombie game… that's kind of my thing, like I have to be matching." In addition, some girls, unlike any of the boys, created extraneous, decorative Play-Doh pieces on their controllers (see Fig. 9). Amani built four arrow keys despite the fact that her game only needed two, and Ishita also built extra pieces to match her controller and game screen design. However, they removed these extraneous pieces when they created controllers again in the second workshop.

4.3 Arcade Design: Reflections of Designers and Players

In contrast to the first workshop that ended with youth designers playing each others' games, the second workshop closed with an arcade in which younger students in the school were invited to come to the computer lab and try out the newly created Scratch games and controllers (see Fig. 10).

As the younger students moved between games, the youth designers explained their game and controller designs and provided help when players ran into challenges. These challenges ranged from alligator clips falling out of controllers, to controller pieces falling on the ground, to designers realizing they needed to adjust their game designs by adding a score so that more students could play their game. Three youths made on the spot changes to their Scratch code while four youths also made changes to their controller designs. For instance, Amani changed her

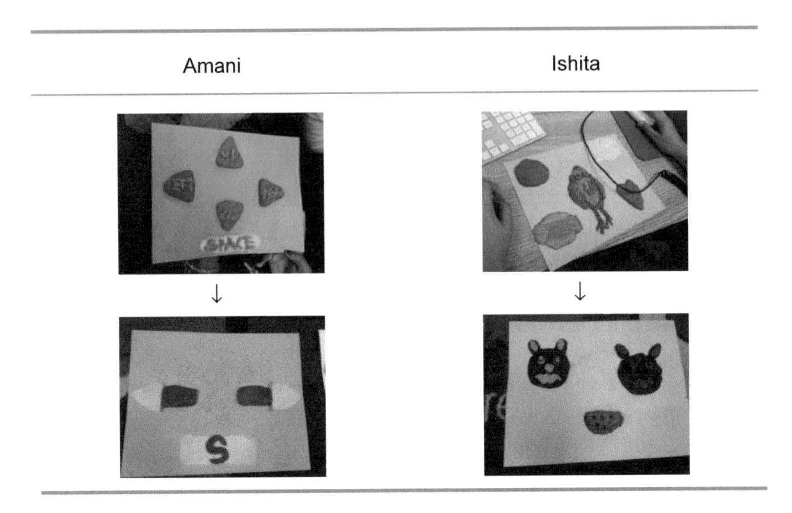

Fig. 9 Changes in controller designs from the first to second workshop

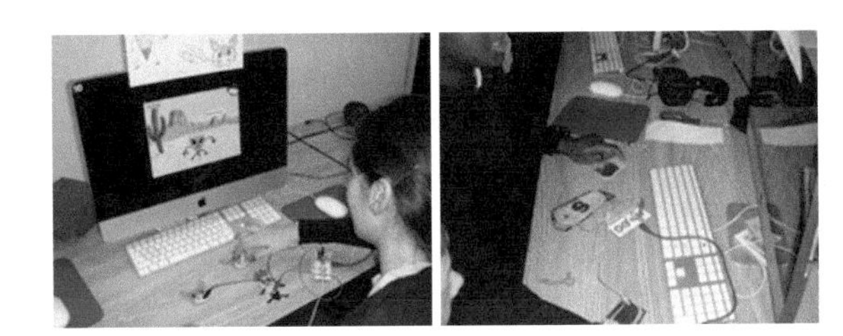

Fig. 10 Playing Isabel and Amani's game in the arcade

Scratch code to reduce the difficulty of the game by reducing the goal score value and increasing the speed and jumping height of the main character. These changes were made in response to players' feedback while playing her game. She also added a final information screen to show players if they won or lost the game. One of the major challenges of the controllers was that many players did not realize they needed to touch the earth clip. To address this issue, the designers quickly designed and added conductive bracelets and touch pads to make this connection more obvious. This rapid prototyping happened during the breaks between visiting fourth and fifth grade classes.

Coming to the arcade and playing the games was a rich experience for all of the players. The players openly expressed their likes and dislikes about the Scratch games and controllers and generated many ideas for changes in designs. The large majority of them (70 %) wanted designers to make the games more complex by adding more enemies or by making it harder to get points, while others wanted to make the games easier by increasing the main character's ability to jump higher or go faster, or by reducing the threshold score for leveling up. While players loved the tangible controllers because they were easy to control, they also had recommendations for changes. For example, players wanted the touch pads to be closer together in Isabel's controller, and similarly for Amani's controller, one player suggested connecting the touch pads together because the pieces of her controller were separated and often fell off the table.

The designers also learned a great deal from watching others interact and play with their games in the arcade. Some youths felt an increased sense of confidence when they observed younger students having fun with their games. Ishita explained that "spending the time with those kids, it was kinda fun cause they really like my game." Another participant, Earl, said that he originally thought his game was boring but then "when the fifth graders just played it a lot… they played it, said it was fun and that made me think that okay, it's… it's good, fun." Youth also gained some valuable insights about design and usability from seeing others playing their game and having to make real-time adjustments. For example, Isabel mentioned one of the improvements she thought of while observing game play: "The keyboard (game controller) have more Play-Doh because it so thin. And the MaKey MaKey go out all the time," referring to the alligator clips slipping out of her thin component pieces. Throughout their feedback, youths explicated that watching others play their games provided insights and gave them ideas that they hadn't otherwise considered.

5 Discussion

In this chapter, we examined different ways of combining software game design with tangible interface design and situating the design experience in a social context, the arcade. Our goal was to understand youths' creations, approaches, and perceptions as well as the associated challenges and opportunities that arise when game design activities move into the tangible realm. We observed that youths' tangible interface designs replicated common controller designs. While the functional variations in the controller designs were minor, what did vary was the attention to esthetics. We saw striking differences in how youth mapped out their physical designs as controllers ranged from unformed heaps of Play-Doh to meticulously designed sculptures that mapped tightly to on-screen elements.

An unexpected observation was how the attention to esthetics in game controller designs was linked to gender, with girls paying more attention to balancing esthetics and functionality. Making tangible interfaces for games revealed that the girls, unlike the boys, were able to combine both technical functionality and graphical esthetics in their controller designs and thus, one could argue, created more user friendly designs. While these are preliminary findings based on a small group of participants, they point towards promising directions in expanding computing activities. Research on the users of textile computational construction kits has demonstrated that broadening the range of materials and activities in computing can also help broaden participation in computing (Buechley and Hill 2010). In these cases, tangible technologies seem to provide a promising way to bridge the gender gap in computing.

Designing interfaces with Play-Doh and the MaKey MaKey exposed participants to technical concepts like conductivity and electrical circuits in new and imaginative ways. Although we provided participants with aluminum foil, metal tape, pipe cleaners, wire, and Play-Doh, all the participants chose to use Play-Doh exclusively, a choice that determined particular creative opportunities and challenges. The benefits of using Play-Doh are that it is safe, easy to access in schools, and doesn't require special tools to mold, shape or build. However, once participants built an initial prototype the moisture started to dissipate, which caused short circuits and other malfunctions. Despite these challenges, we found that the creative opportunities offered by the Play-Doh outweighed these problems. Bringing the MaKey MaKey into the classroom also revealed some of the potential design improvements that could be made. Several youths mentioned that they found working with the alligator clips challenging, because the holes on the MaKey MaKey were small and the clips were difficult to open. In addition, three youth participants mentioned they would remove the requirement to connect to the Earth section on the MaKey MaKey because it was not intuitive.

As mentioned earlier, the vast majority of the controller designs replicated common gamepads. One explanation for this is that the open-ended design of the MaKey MaKey gave participants too much freedom, and that reasonable constraints could lead to more unique designs. One way to increase the diversity of designs would be to incorporate periodic challenges throughout the workshop. Challenges could include asking participants to work with conductive materials that aren't as easily crafted as Play-Doh, or asking them to generate different layouts than the traditional directional arrow keys. Another way could be to provide participants with different sensors and materials so they could create a wider variety of interfaces: soft interfaces using conductive fabrics, pressure-sensitive interfaces using strain gauges, or motion-sensitive interfaces using photoresistors.

The addition of the arcade provided a meaningful context in which youth designers were able to test and view their work. It also provided them with an authentic audience of younger students and teachers who played with their games and controllers. This experience led to insights about the quality, complexity, and usefulness of their game and controller designs. It also helped youths see their game in an interactive context, where they could make connections between their games, their peers' games, and the larger experience they were helping to develop. There was anecdotal evidence that the arcade led to changes in the participants' perception of their games, though this was not something we were looking for initially.

In future studies, we would consider interviewing the participants before and after the arcade to see how the experience changed their view of themselves and their designs. In addition to the learning benefits, we also gathered insights about working with construction materials and the MaKey MaKey, increasing diversity of controller designs, and expanding the arcade contexts and collaborations. While setting up the arcade provided youths with rich context to design, evaluate, and improve their tangible interface designs, it also provided us with insights on how to design future environments for authentic audiences. Game design workshops could draw inspiration from youth-created spaces like Caine's Arcade, an arcade created entirely out of cardboard by an eight-year-old boy. This example could encourage youths to create interactive environments from found materials beyond those provided. Imagine children creating interactive playgrounds with musical slides and light-up ladders, immersive arcades that use computers to provide audio and visual effects, and houses or classrooms with door alarms and pet detectors.

6 Conclusion

The inclusion of tangible game controller designs and the social context of an arcade provided compelling insights into how game making activities can be expanded to create richer learning opportunities. Tangible construction kits like the MaKey MaKey provide novel ways and create authentic, interactive environments for youth to learn about programming and design.

Acknowledgments The research reported in this chapter was supported by a collaborative grant from the National Science Foundation to Yasmin Kafai (NSF-CDI-1027736). The views expressed are those of the authors and do not necessarily represent the views of the Foundation or the University of Pennsylvania. We also would like to thank the teacher Peter Endriss and his students for participating in the study.

References

Antle AN (2007) The CTI framework: informing the design of tangible systems for children. In: Proceedings of the 1st international conference on Tangible and embedded interaction, ACM Press, Baton Rouge, pp 195–202

Bayliss P (2007) Notes toward a sense of embodied gameplay. Situated play. In: Proceedings of the digital games research association conference, Tokyo, Japan, pp 96–102

Buechley L, Hill BM (2010) LilyPad in the wild: How hardware's long tail is supporting new engineering and design communities. In: Proceedings of the designing interactive systems (DIS), Aarhus, Denmark, pp 199–207

Burke Q, Kafai YB (in press). A decade of game-making for learning: from tools to communities. In Agius H, Angelides MC (eds) The handbook on digital games. Wiley, New York

Brennan K, Resnick M. (2012). New frameworks for studying and assessing the development of computational thinking. Paper presented at the annual meeting of the American Educational Research Association, Vancouver, BC, Canada

Collective BSM, Shaw D (2012) MaKey MaKey: Improvising tangible and nature-based user interfaces. In: Proceedings of the sixth international conference on tangible, embedded and embodied interaction, Kingston, ON, Canada. pp 367–370

Davis RL, Kafai YB, Vasudevan V, Lee E. (2013). The education arcade: crafting, remixing, and playing with controllers for scratch games. In: Proceedings of the 12th international conference on interaction design and children. New York, United States

Denner J, Werner L, Bean S, Campe S (2008) The girls creating games program: Strategies for engaging middle school girls in information technology. Front J Women's Stud 26(1):90–98

DiSalvo B, Bruckman A (2011) From interests to values. Commun ACM 54(8):27–29
Fowler A, Cusack B (2011) Kodu game lab: improving the motivation for learning programming concepts. In Proceedings of the 6th international conference on foundations of digital games, Bordeaux, France, pp 238–240
Gee J (2003) What video games have to teach us about learning and literacy. Palgrave Macmillan, New York
Hayes ER, Games IA (2008) Making computer games and design thinking: a review of current software and strategies. Game Cult 3(4):309–322
Horn MS, Crouser RJ, Bers MU (2012) Tangible interaction and learning: the case for a hybrid approach. Pers Ubiquit Comput 16(4):379–389
Kafai YB (1995) Minds in play: computer game design as a context for children's learning. Lawrence Erlbaum, New Jersey
Kafai YB, Burke Q, Mote C (2012) What makes competitions fun to participate?: the role of audience for middle school game designers. In: Proceedings of the 11th international conference on interaction design and children, Bremen, Germany, pp 284–287
Kafai YB, Peppler K (2011) Youth, technology, and DIY: developing participatory competencies in creative media production. Rev Res Educ 35:89–119
MacLaurin M (2009) Kodu: end-user programming and design for games. In: Proceedings of the 4th international conference on foundations of digital games, 2 Port Canaveral, FL, USA
Manseur R (2000) Hardware competitions in engineering education. In: Proceedings of the 30th annual frontiers in education, FIE, p F3C/5–F3C/8
Martin FG, Roehr KE (2010) A general education course in tangible interaction design. In: Proceedings of the fourth international conference on tangible, embedded, and embodied, and embodied interaction, Cambridge, Massachusetts, USA, pp 185–188
Marshall P (2007) Do tangible interfaces enhance learning? In: Proceedings of the 1st international conference on Tangible and embedded interaction, Baton Rouge, Louisiana, USA, 15–17 Feb 2007
Marshall P, Price S, Rogers Y (2003) Conceptualising tangibles to support learning. In: Proceedings of the interaction design and children, ACM Press, Preston, pp 101–109
Millner A (2010) Computers as Chalk: Cultivating and Sustaining Communities of Youth as Designers of Tangible User Interfaces. Unpublished Doctoral Dissertation. Massachusetts Institute of Technology, Cambridge
Papert S (1980) Mindstorms: children, computers, and powerful ideas. Basic Books, New York
Peppler K, Kafai YB (2010) Gaming fluencies: pathways into a participatory culture in a community design studio. Int J Learn Media 1(4):1–14
Repenning A, Webb D, Ioannidou A (2010) Scalable game design and the development of a checklist for getting computational thinking into public schools. In: Proceedings of the 41st ACM technical symposium on computer science education, Milwaukee, WI, USA, pp 265–269
Silver J, Rosenbaum E, Shaw D (2012) Makey Makey improvising tangible and nature-based user interfaces. In: Proceedings of the ACM tangible embedded and embodied interaction, Kingston, Ontario, Canada, pp 367–370
Squire K (2010) Video games and learning: teaching and participatory culture in the digital age. Teachers College Press, New York
Vasudevan V, Kafai YB, Lee E, Davis RL (2013) Joystick designs: middle school youth crafting controllers for scratch games. In: Proceedings of games, learning, and society 9.0, Madison, WI, USA
Zagal JP, Bruckman AS (2005) From samba schools to computer clubhouses: cultural institutions as learning environments. Convergence 11(1):88–105

Playful Creativity: Playing to Create Games on Surfaces

Alejandro Catala, Javier Jaen, Patricia Pons and Fernando Garcia-Sanjuan

Abstract Creativity is of vital importance for human development since it allows individuals and ultimately society to successfully overcome new challenges. Besides social factors, the environment can also influence the development of such an important skill. We therefore considered it of interest to explore this capacity in the context of new information technology and game-based learning. Tabletop systems greatly facilitate the characteristics behind creative processes such as communication, the exchange of ideas, and collaborative interaction between individuals. This chapter explores the suitability of interactive surfaces in collaborative creative tasks carried out by teenage students using software to create 2D game worlds for tabletops.

Keywords Creativity · Games · Tabletops · Surface · Collaboration

1 Introduction

Many different childhood activities are focused on both learning and entertainment. They rely on creating artifacts from everyday materials that children can (learn to) handle. Two typical examples can be seen in Fig. 1, concerned with

A. Catala (✉) · J. Jaen · P. Pons · F. Garcia-Sanjuan
Grupo ISSI, Universitat Politècnica de València, Camí de Vera s/n,
46022 Valencia, Spain
e-mail: acatala@dsic.upv.es

J. Jaen
e-mail: fjaen@dsic.upv.es

P. Pons
e-mail: ppons@dsic.upv.es

F. Garcia-Sanjuan
e-mail: fegarcia@dsic.upv.es

A. Nijholt (ed.), *Playful User Interfaces*, Gaming Media and Social Effects,
DOI: 10.1007/978-981-4560-96-2_14,

Fig. 1 Traditional approaches regarding play and learning

paper and pencil drawings and modeling clay with simple tools. Although these can be seen as entertaining activities, they are an essential part of children's development and encourage the development of physical motor skills as well as the mental skills related to creativity and imagination by involving children in creative activities. The active role of children in such tasks and their ability to keep children motivated make these activities very effective.

The common element in these sample activities is that they are creative. This means that we are not just consuming knowledge from pre-existing sources, but are actually forming, reproducing, and improving ideas and/or concepts, which requires greater cognitive effort.

Computer-based educational alternatives could provide other advantages, such as introducing players to a process of digital literacy (Gros 2007) and the development of computational thinking skills (Wing 2006), which are especially important for creativity in the information society (Resnick 2002). So, a pertinent question here is whether some engaging digital activities similar to the traditional or physical ones can be delivered in terms of development and creativity.

The video games in electronic entertainment systems, although interactive and with a lot of potential to learn from them, are more focused on consuming information rather than creating things, confining creativity to a set of selection commands in a limited controller. The potential of technology to address creativity therefore seems limited by the current user interfaces and controllers.

However, new possibilities for entertainment seem to be more promising, with the arrival of more natural user interfaces, intended as those providing more natural user behavior and feeling during the experience (Wigdor and Wixon 2011, p. 10), such as motion sensing input devices (e.g., Kinect), graspable or tangible interfaces, interactive tabletops, and computational input devices, which go further than the current gamepad controllers. These interfaces will bring innovative interactions to address creative tasks, overcoming the previous input limitations. However, the way in which these input technologies must be involved to be effective is still an open question. Will they allow users to become more focused on the task in hand? Will they allow activity's designers to deliver improved experiences? Or will they simply remain as cool interfaces without actually improving human processes?

In order to explore this and other related issues, we conducted a 3-year research project known as CreateWorlds. From the current range of available technology, we focused on exploring how interactive tabletops can be used to foster creativity by involving users in active group tasks. As we will see in this chapter, the primary idea is to provide a tool rather than a specific videogame to be consumed, with the aim of putting creative and social aspects into action, aspects that were difficult to include in earlier interface technology.

2 Creativity and Entertainment

There is nowadays a general belief that creativity is an important asset for the competitive global market. In fact, supranational entities such as the EU Commission have recognized creativity as a key driver in economic development (EU 2009) and try to encourage creativity and innovation help bring about long-term changes in our society.

Although this may sound easy, it really is not, as the term *creativity* has always been subject to discussion, for example, the wide range of definitions for everyday creativity (Aleinikov et al. 2000). According to Amabile, creativity arises as a result of the combination of three factors: knowledge, creative thought, and motivation (Amabile 1983). Knowledge consists of all the information possessed by individuals to solve a problem. Creative thinking can be summed up as the presence of the individual abilities such as the ability to disagree with others and experiment with different solutions to proposals chosen by the majority, the ability to persevere in difficult or problematic situations, and, finally, the ability to gestate ideas during periods in which we alternately forget the problem and return to it with a new perspective. Finally, Amabile stresses the prevalence of intrinsic motivation, which is directed by interest, satisfaction, and the challenges presented in order to solve a problem, over "external pressures" or extrinsic motivation.

In one way or another, creativity is essentially related to the idea of originality, novelty, unusual, innovation, etc. Hence, given the conceptual complexity of the term, it is mostly considered as a multidimensional concept by creativity theories, which focus on different aspects, such as products, processes, persons, or places (Runco 2010). Although people usually think about masterpieces or great inventions when they are asked about creativity, everyday creativity is more important (little-c and mini-c) (Richards 2007; Beghetto and Kaufman 2007) rather than eminent creativity (big-c and pro-c) (Beghetto 2007). Such an eminent creativity is more related to professional careers (e.g., painters, publicists, writers, engineers or architects), whereas everyday creativity is concerned with quotidian situations in which people can find different or novel improvements for their own benefit (e.g., self-organization and daily planning or home cooking). The idea is that any improvement in the creativity of ordinary people will produce benefits to our society in the long term, as it is related to divergent thinking and other skills concerned with generating and exploring multiple ideas, which will allow people to face new challenges in the future.

The study of creativity in context of Information and Communications Technologies (ICT) has been rather limited to date. Some research has focused on evaluating creativity in Software Engineering issues (Wang et al. 2010), the capability of a collaborative distributed tool in a shared writing task to support creativity by improving awareness (Farooq et al. 2007), or even a mobile tagging application shown to effectively support participant awareness and coordination as well as facilitating the exploration of artifacts in the creative process of design work (Vyas et al. 2010a, b; Vyas and Nijholt 2010). Meanwhile, other research work has focused on supporting brainstorming sessions and the creation of concept maps as in Forster (2009), Friess et al. (2010), Buisine et al. (2007), and Geyer et al. (2010). All these explorations are a big step forward in understanding and supporting creativity in professional or technical contexts, although they do not pay much attention to developing everyday creativity and are more focused on professional contexts.

In such professional dimension, there are some outstanding successful tools for creative expression. For example, the *reacTable* (Jordà et al. 2007) is a tabletop-based musical instrument which allows users to experience music visually by means of touch and tangible input. Musicians can experiment with sound in real-time, changing and controlling parameters in a direct way, facilitating refreshing creative improvisation, and even allowing collaborative multiuser performances. Another example is *IntuPaint* (Vandoren et al. 2008), which is a tangible interface for a digital paint easel, using an interactive surface and specifically designed electronic brushes. The system consists of a powerful model-based paint simulation that is able to render realistic paint results by capturing the subtle paint nuances of the artists through their paint strokes. It is a big step forward to bridge the gap between physical and digital painting, and therefore supporting the creative expression in the form of drawings.

A different approach is addressing creativity from a less strict or formal point of view, by including some kind of creative flavor in an entertainment environment to enable creativity to be developed. Entertainment has traditionally been considered as a facilitator and effective tool in informal learning. Videogames, which are not only for fun but also for learning, allow active participation and high task engagement. Some general benefits, as described in Michael and Chen (2006), are the ability to model more complex systems, a higher engagement with learning materials, interactivity and quick testing and evaluation of answers, proximity to learning strategies founded on constructivism, and cost savings by reducing training times and using virtual environments rather than expensive real settings.

For two decades there have been a series of game-based entertainment systems that have focused on the creation of different artifacts to foster creativity. They range from systems provided with a programming language for non-experts to tangible and smart devices that allow creating and controlling simulated worlds.

For example, *LogoBlocks* is a graphical programming language to support programming for the LEGO programmable brick (Begel 1996). The language uses a drag-and-drop methaphor in a Windows-Icons-Mouse-Pointers (WIMP) user interface. The *brick* is a small computer that can be embedded and used in LEGO creations by reading from sensors and controlling engine activations. In this way, children and teenagers can create worlds where robots are physical programmable entities. It showed how physical devices can successfully be mapped to digital elements and be successfully manipulated to specify robot behavior.

The well-known tool, *Scratch*, is a graphical programming environment that allows children to program interactive stories, games, and animations composed of a set of sprite-based objects (Maloney et al. 2004). The programming language to specify behavior is also based on a drag-and-drop metaphor composed of virtual blocks representing instructions. The main screen of the tool shows the stage and the sprite representing the entities, allowing program debugging, and iterative testing of new ideas. The environment is a single-user application based on WIMP interaction. It has led a worldwide online community fostering creativity by means of sharing and exchanging projects, and it is successfully used at secondary schools to introduce teenager students to basic computational concepts.

Similarly, *Agentsheets* is a tool based on agents that allow users to create simulations and interactive games (Repenning et al. 2000). Users can create simulations of sprite-based agents in a 2D world arranged in a rectangular array. The users are responsible for designing the visual aspect of the agents by drawing icons, so that these agents are actually sprite-based entities. Their behavior is based on event-based rules which are edited following a visual approach to the rewriting rule paradigm.

Topobo is a 3D constructive assembly system that allows the creation of biomorphic forms like animals and skeletons (Parkes et al. 2008). This is achieved by means of pieces embedded with kinetic memory. *Topobo* is designed to be a user interface that encourages creativity, discovery, and learning through active experimentation with the system. It can help students to learn about several physical concepts such as balance, center of mass, coordination, and relative motion.

Previous works can alternatively be used to support interactive stories as some behavior can be programmed to control their progress. However, there are also approaches that specifically support storytelling processes by focusing on the performance and the enactment by the user rather than following programmatic schemas. A good example of such approach is *TOK*, a tangible interface for children to create their own stories (Sylla et al. 2011). Children use cards that represent story elements such as characters, settings, and actions and put them on an electronic surface that is able to identify the cards. The system renders on a screen the animated story according to the card composition on the surface.

Another interesting work is *ShadowStory* (Lu et al. 2011), which is a storytelling system inspired by traditional Chinese shadow puppetry. Children use a Tablet PC to create digital animated characters, and then they are allowed to act out stories on a back-illuminated screen, controlling the characters with simple movements by means of handheld orientation sensors.

As can be seen from all these samples, many different technologies are being used to support the creation of games or interactive media, although there is a clear trend toward more tangible interfaces. Even though some of these approaches can be used with groups, they are mostly designed to support single-user interaction. Interactive tabletops or surfaces would be an appropriate interface technology for such small groups, since they could also consider other important dimensions of creativity, such as collaboration, reflection, and divergent thinking in face-to-face group scenarios.

An interactive tabletop is a computer interface that usually allows multiple users to interact with the system using hands and fingers and also by manipulating physical objects on its surface. It seems to be a suitable target platform for experimental learning environments because it has a "socio-constructivist flavor" as pointed out in Dillenbourg and Evans (2011). Provided with the appropriate software, it could be a facilitator or catalyst of typical creative learning processes for small groups. Starting from these ideas, in this chapter we present our experiences in developing and evaluating AGORAS, which is the tabletop-based platform for creating games within our CreateWorlds project.

3 Playing to Create Games

It is no coincidence that the approaches for informally fostering creativity seen in the previous section rely on entertainment and creating interactive simulations. Although a good (video)game properly used in a learning environment is a strong source of motivation, creating artifacts is a more rewarding learning activity, as it is more demanding than just consuming a story through playing a game.

Because it is not always possible to design games to deal with a specific content and is not a cost-effective strategy, the use of commercial videogames has been explored in combination with traditional learning activities. Some studies using this approach have been conducted in the context of formal learning settings, as in Ellis et al. (2006), Gros (2007), and McFarlane et al. (2002), but their learning activities are normally focused on traditional tasks on paper and are based completely on predefined videogame content. These studies have concluded that videogames traditionally support three types of learning. The first is learning from the tasks required from the games themselves. The second is learning from the content of the game, although this does not usually coincide closely with the educational curriculum. The third is learning by practicing the skills and abilities required for the game. These works normally introduce a cyclical process considered important for learning, and consists of several phases including experimentation, reflection, activity, and discussion and are not usually supported entirely by digital games. From these observations, we can say that it would be more effective to design a learning environment consisting of both traditional and digital activities, and therefore an effort should be made to provide flexible tools adaptable to the teaching methodology.

However, the use of games as a way of learning, as suggested by Clark Abt, even before the digital age in the 1960s, is one of the most effective strategies in fostering creativity (Abt 1970). In his book on serious games, Abt proposed that the game-creation process should be considered an important learning activity. This is a perspective that serious digital games do not seem to have seriously considered. He points out that the first learning phase, design and preparation, can be divided into two parts: (a) the relatively passive preparation activity, and (b) the actual design of the game itself. The first one involves studying the background of the rules, roles, concepts, etc. The second one is more satisfying and involves inventing a game model of the process to be re-created, during which the different important variables involved with their interrelationships and dynamic interactions are controlled. If this is to be done satisfactorily, pupils must understand the process involved in the game or simulation being created, and in this way, increase not only their factual knowledge, but also the interactions and processes involved. As all this activity is to be performed in a group, it is also important from the point of view of social learning, which considers knowledge emerging from interaction and communication between individuals who pursue shared objectives.

Our approach is not about building a creative problem-solving application in a specific domain with predefined behavior and pre-established reactions at the users' disposal. Following the ideas of Abt in the pre-digital age, we aim at "the construction of videogames" as the creative activity to be carried out. The interest therefore does not lie in the game itself, but in the design activities, which will allow the group to acquire in-depth and in-breadth knowledge, put critical, convergent and divergent thinking into practice, and provide high doses of intrinsic motivation or flow (Csikszentmihalyi 1991) while these activities are being carried out. What we propose is more concerned with the creation of a platform that will profit as much as possible from traditional non-technological gaming activities, while allowing the pupils to create their own games according to the rules they themselves lay down. It will also use digital technologies to provide a stimulating environment in which the subjects will be able to experience interactively the results of their design decisions.

4 Touchable Digital Worlds: Creation and Play

A direct consequence of Abt's work is that the focus is not on what can be learned by playing games but what is learned by playing at creating games. The type of game and the complexity of the building blocks will therefore determine the possibilities of success of the platform. The types of games to be supported are those with a 2D world inhabited by reactive entities capable of being simulated.

Interactive tabletops enable direct manipulation by means of fingers and tangible objects. They are highly effective when there is a meaningful spatial relationship among the digital objects, as people are used to distributing the objects across the surface. Surface-based user interfaces should therefore take this consideration into account at the design stage. In this respect, the tabletop is like a canvas on which the stage can be organized, the entities designed, and the behavior specified. The game world consists of stages which contain graphic elements as decoration. Entities can be included in stages from the beginning (i.e. by definition), or can be added as needed at runtime. The entities have a distinctive look and can exhibit two types of behavior. They combine physics-based simulation with logical rule-based behavior and thus provide a higher level of customization and make possible a wider range of activities.

4.1 From Bones to Skin

As in the systems that rely on entity simulations, entities are the first-order citizens of our system. The main difference is their composition. In the previous approaches, they were based on single sprites or icons, which are the basic visual representation. In our approach, the entities can be composed of several components connected by joints. In order to facilitate the use and reuse of multiple costumes in such multi-component entities, the entity model considered a skeleton-based approach.

In this way, the internal structure or skeleton of complex entities can be first defined and then dressed by covering the skeleton with a costume skin which will give it its distinctive look.

These internal components, which are like architectural units in the same way that a skeleton is composed of bones, are actually simple polygons. For convenience, a set of predefined shapes (e.g. circles, rectangles, triangles) can be used. Figure 2 shows several entity samples, with their internal structure and external skins.

The entity model is much more than multiple components and sprites. The big difference with previous systems is that the different parts of the entities form a body, have mass, and can collide with each other. According to several physical parameters (e.g. the coefficient of friction or mass), the entities can be simulated in terms of Newtonian physics, automatically producing movement and collisions without need to explicitly control these phenomena by means of additional scripts. There are also several types of joints to keep the entity's components together. The most useful joints are the pin and elastic joints, which work as ropes and springs.

In most cases, using single-sprite entities is the easiest way to bring entities to life. However, a model supporting skeleton-based entities makes it possible not only to create interactive media but also autonomously evolvable entities, which by means of inherent physical behavior can facilitate the creation of games requiring collision management. All this behavior can be disabled, so that the entities can be held or fixed to specific places on the stage, or can even be allowed to collide or not.

An additional feature of this model is that it allows more natural and realistic puppet shows to be created. It is relatively simple to design puppets, as the gravity on the stage can be adjusted to keep them hanging from their control strings. Figure 3 shows a user interacting with a string puppet while telling a story.

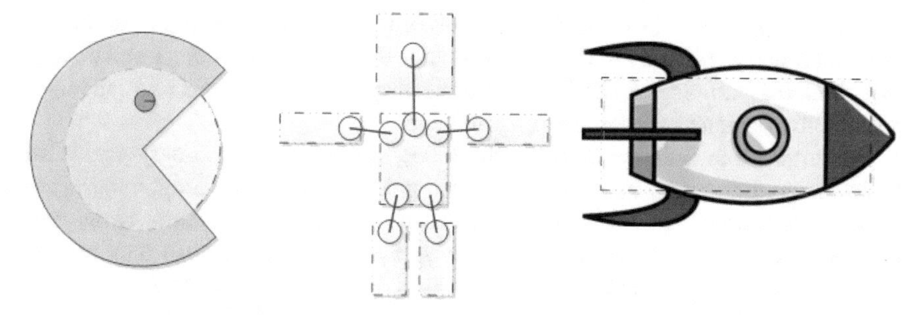

Fig. 2 Sample entities

Fig. 3 An example of string puppet

4.2 Touchable Reactive Behavior

Those born in the digital age have grown up with digital information and are familiar with a variety of computer applications, many of which consist of interactive educational games and videogames. This higher exposure to digital media and computer applications, have facilitated the introduction of complex computational concepts and some sort of programmability, which is important for the development of computational thinking (Wing 2006). It is considered a fundamental skill, since it leverages powerful abstractions and tools to solve meaningful problems in many different disciplines. Despite the higher digital literacy and early training, programming is still difficult for beginners and/or non-programmers, although possibly made easier by the environments and languages specifically designed for novice programmers.

As surveyed in Kelleher and Pausch (2005), systems whose primary goal is not to teach programming but aim at another goal by means of programming are mostly event-based. In one way or another, events and rules are the basic components needed to express behavior in these systems. Just because they are targeted at non-programmers, interfaces designs have focused on trying to create languages, programming methods, and interaction mechanisms that allow people to build those programs. An event-based approach has also been adopted in our AGORAS platform, since rules have been traditionally used in reactive environments. It seemed better suited to the expression of reactive behavior in virtual worlds, especially since rule structures are better understood and used by young people or non-programmers, as reported in Pane et al. (2001) and Good et al. (2010). This reactive behavior is an important part of the AGORAS worlds, expanding the entities' physical characteristics and therefore providing many more possibilities in building games.

When designing a tool to support for non-programmers, the main challenge is to supply all the rule expressiveness while considering textual input techniques inconvenient for tangible surface interfaces. We conducted an experiment in a local school involving 36 students enrolled in a computer science subject (16 year old) as a part of the curriculum who faced computational concepts for the first time (Catalá et al. 2013b). The study showed feasible the inclusion of a visual dataflow language to express short and medium assignment expressions. Hence, the designed rule editor deals with visual dataflows as the main conceptual artifact. Moreover they are suitable for dealing with expression programmability in environments based on tangible surface interfaces. The main rationale is that a suitable editor in this context would require co-located, cooperative, and collaborative performances from multiple users, flexibility in user interface layout and viewpoint independence, plus a set of direct touch techniques. All this can be achieved by a decomposable expression model such as the one based on visual dataflow expressions (see Fig. 4).

To facilitate the order of the interactions when editing a rule and for reasons of usability, the editor is divided into four views. The first view allows users to select the entities and the event (see Fig. 4a). The other three views are concerned with the condition (b), filter, and operation (c). Each of these assists in the creation of the corresponding data processes. The user has to include the operators and the properties needed by selecting them from the pie menus specifically designed to support parallel interactions (Catalá et al. 2012b), and all these elements are finally rearranged and connected by means of touch input (Pons et al. 2013).

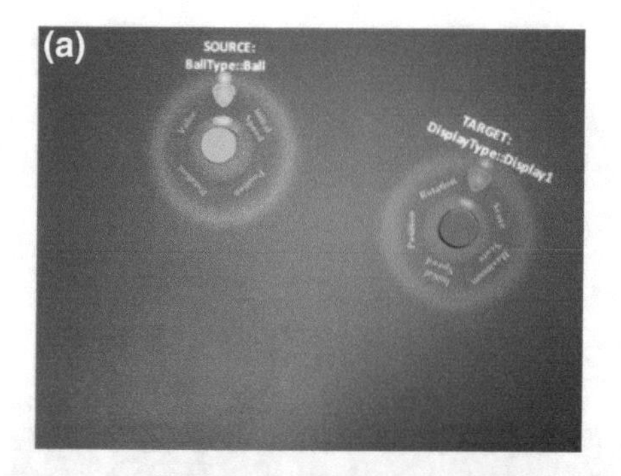

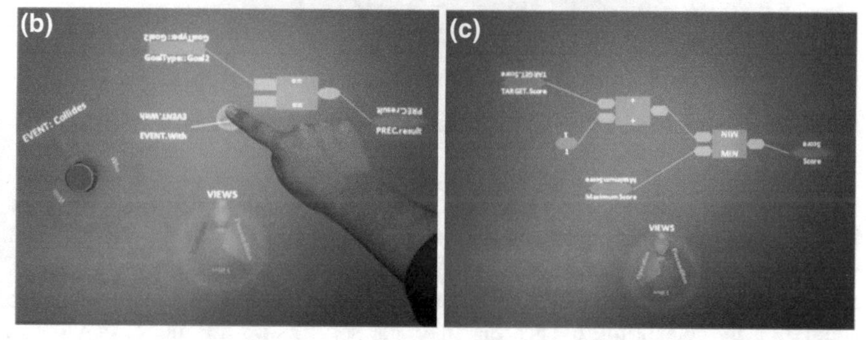

Fig. 4 Rule editor's views. **a** Selection view, **b** condition view, **c** operation view

4.3 Enactment Model and Architecture

The previously described behavioral models need a runtime environment to perform the simulation of the edited worlds. The middleware required to process and enact AGORAS worlds can be broken down into three layers (see Fig. 5). The Model layer refers to the definition and specification of the game concepts and their storage. This layer is responsible for managing the concepts that deal with the stages, entities, rules, etc., available in the game world being created, by giving support to update and retrieve all these basic elements.

The Controller layer holds the core functionality to orchestrate the simulation of a world stage. Basically, the simulator has to take a stage to be simulated and all the data from the Model layer. The simulator has an event queue for the event occurrences produced during the simulation. Three types of events are queued: those thrown by the actions when executed in this layer, those that are a consequence of the physical simulation carried out by the underlying physics engine; and those related to the gestures or interactions of the user on the surface. This queue is regularly consumed by the rule processor, which determines which rule must be triggered and then performs the execution of the action of the matched rules. Actions can involve changes in logical entity properties (e.g., increment the

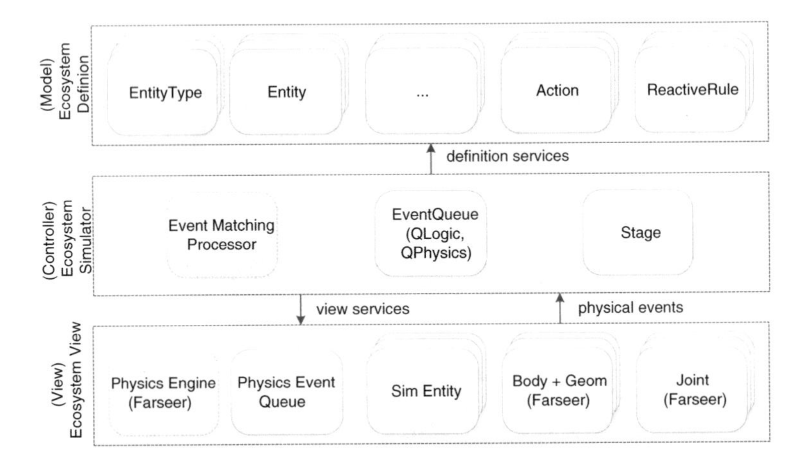

Fig. 5 Architectural logical view of the implemented runtime environment

variable "Hits" of a block in a Breakout game) or in visual properties of the game (e.g., change the visual representation or skin of an entity). The Controller layer uses the services from the Model and View layer to address these two types of runtime changes. In this way, the simulator controls the evolution of the stage simulation by consuming events and invoking services on the model or view as needed.

Finally, the View layer is responsible for visualizing the representation of entities under simulation. It offers a core set of view services that allow us to include entities and change their visual properties from the Controller layer.

The View layer is by far the most complex layer to implement. In order to perform the visualization of physics components, this layer relies on the Farseer physics engine.[1] This is an open-source physics engine simulator that allows the simulation of shapes defined internally in terms of bodies, geometries, and joints. Since Farseer does not understand the concepts in which the visual structure of entities is defined, the View layer has to translate the components of the entities in our model into Farseer primitives. It also has to maintain the correspondence between these elements in Farseer and the model, in order to be able to track which entities are producing physical events such as collisions. In this way, when two shapes collide, the middleware is able to determine which entities these shapes belong to, and is therefore able to throw a physics event occurrence associated with the entities involved that will be queued in the Controller layer for further processing. All the components of this middleware have been implemented in C# and using the Microsoft Surface SDK.

The behavioral models included in AGORAS enable users to imagine more advanced and complex games. To demonstrate the capability of the platform,

[1] Farseer Physics Engine in Codeplex: http://farseerphysics.codeplex.com/.

several classic games have been specified. Some basic AGORAS worlds inspired by *Pong,* the *Breakout* game or the *Asteroids* can be found in Catalá et al. (2012c, 2013a).

5 Workshops on Playful Creativity

We conducted an exploratory experiment involving a version of AGORAS with limited functionality. The study aimed at obtaining experimental evidence on how tabletop interfaces deal with creativity aspects, collaboration, and ownership in face-to-face interaction by teenage participants.

The 22 participants (14 male and 8 female, mean age of $M = 16.23$, $SD = 1.6$) had previously taken part in a short course on new technologies designed to motivate teenagers to study core subjects such as physics and computing. At the end of the course, the participants took part in a workshop where the experimental sessions were carried out. The course was organized by a club belonging to the Education and Culture department of the local city council.

Two tabletop platforms were developed for the study. One was the digitally augmented AGORAS-based platform, using an interactive surface that allows multi-touch and tangible input to manipulate the internal virtual blocks and the joints, whereas the other is completely physical and tangible without computer mediation, as described below.

The physical-only platform is made entirely from hardware with no software simulation. It consists of a conglomerate tabletop with a regular grid of holes. Several wooden blocks of similar size to the ones in the digital platform are available. The tabletop has four legs to keep it horizontal, and also a stand to configure it as a slanting plane to simulate similar conditions in the digital platform.

The physical blocks can be fixed and assembled as required, by using the holes drilled in them by means of screws, bolts, and other joint elements such as elastic bands and pieces of string. More complex joints and other functional components can be assembled by combining several blocks and joint elements.

In a nutshell, both platforms allow the construction of a varied set of fixed or articulated components based on basic rigid bodies and joints. The study considered two different creative assembling tasks. In the first, the main interest was focused on exploring how participants perform in a general problem whose solution is completely open, i.e. we needed a task that was only constrained by the building blocks and the participants' creativity. Hence, the task consisted of the free creation of entity-like figures, and the participants were requested to produce as many solutions as possible.

In the second task, we were more interested in exploring a creative task with a clearer objective, as in many real problems. This time, creativity would be needed in the problem-solving process, i.e. finding a solution to a specific problem. The previous experiment was involved with creating entities in similar conditions to

Fig. 6 Live in-session solution samples

those of the entity editor in AGORAS. However, the second experiment looked at the process of building a stage to fulfill a specific goal, as a consequence of the physical governance of basic components. For this, we used the inspiring scenarios provided by Rube-Goldberg Machines (RGMs), which are mechanical systems mainly composed of building blocks connected to actionable devices, normally providing a complex solution to a simple problem.

The task involved a simple problem whose solution required the construction of a machine. The participants were requested to design as many creative RGMs as possible to solve a given problem. This consisted of making a box fall from a shelf located in the center of the tabletop. See Fig. 6 for samples of both entity-like figures and the actual machines constructed by the participants.

Following previous work by McFarlane et al. (2002) and Gros (2007), which explored the use of commercial videogames in learning settings, we considered a specific set of places involving traditional activities with paper and pencil. These places correspond to individual thinking-reflection, discussion, and testing/implementation processes performed iteratively to foster creativity. This setting was expected to provide information on how the playful interfaces could be integrated into wider learning activities that could meet several sessions and involve many different teaching and learning materials apart from the digital platform. In fact, the platform is simply a tool that must be put into the teachers' hands to be flexibly used in the best way.

Three specific places were therefore considered in this iterative process: an individual thinking place, a discussion place, and a testing/implementation place.

In the individual thinking place, the subjects had to generate solutions to the problem on paper. Once each member had produced various solutions, they discussed improvements and possible new solutions and decided what solutions to implement on the testing platform. As they had discussed the ideas on paper, they already knew what parts were needed to be constructed and could collaborate on implementing them. The first two stages are thus also important, as they promote divergent thinking, which is important for creativity, since sketches from the traditional paper and pencil may encourage new ideas and also set the basis for collaboration on the experimentation platform.

The participants formed eleven groups, which were assigned to a different platform in each task, balancing the number of groups per platform. In the task of creating entity-like figures, a total of 161 proposals were generated and 91 were tested in the end, whereas 64 solutions were tested out of 122 in the RGMs task. For all these solutions, we obtained the paper sketches and the video recordings at the testing platform, showing how solutions were actually constructed in pairs. From this data, we performed an analysis in terms of creativity and collaborative interaction. Further information and more findings can be found in Catalá et al. (2011, 2012a, 2012d). Next, some of the most relevant findings are briefly presented.

5.1 Assessing Creativity

Creativity is typically intended as a componential concept, and therefore different aspects or features could be measured. Hence, our creativity assessment model consisted of a representative core set used in the psychology field: *novelty*, *fluency* and *flexibility of thinking*, *elaboration*, and *motivation*. Among them, we reported about *novelty* and *motivation*. The most important trait is undoubtedly *novelty*, which is defined as the characteristic conferring something unusual, unique, or surprising. *Motivation* was included as it is also important in human development and many other learning activities (Csikszentmihalyi 1991) and it is also included in one of the seminal definitions given by Amabile.

Novelty is difficult to assess and no clear objective measures can be found. Thus, we opted for ratings by experts following the Consensual Assessment Technique (Amabile 1982; Baer and McKool 2009). Two people with background in creativity studies were asked to rate each solution on a 5-point scale obtained as a cumulative assessment on several inner features related to novelty. Each feature was described in a single scale of three levels (+0, +0.5, +1). These features concerned how unusual the creation was, whether the idea was useful or pointless, whether there was any surprising element or not, whether there were elements better suited to represent the idea or the mechanism or not, and whether the way of assembling pieces was commonplace or unexpected but advantageous. To check whether both judges agreed on the meaning of novelty and therefore on rating consistently the solutions, an inter-rater agreement test based on Kappa statistics

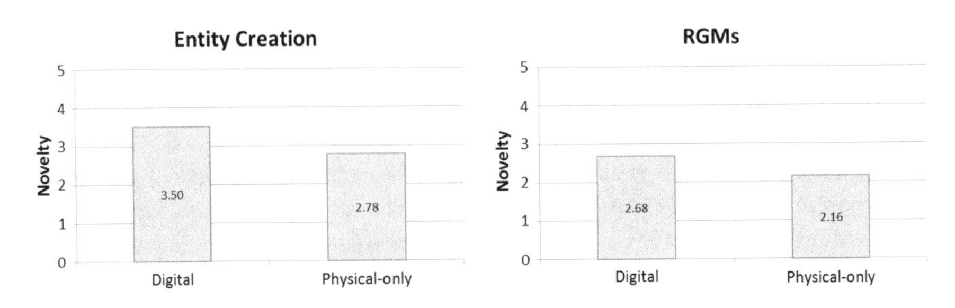

Fig. 7 Novelty mean plot by platform

was conducted. This test showed that the agreement was very good in both tasks ($K = 0.860$ for entities task and $K = 0.733$ for RGMs task). Figure 7 shows the average rates of novelty comparing both platforms, showing that solutions in the digital platform obtained significant higher rates.

Although many solutions simply complied with the problem in hand in a trivial manner, the solutions with higher rates actually made a difference in the way the components are used, especially since the digital platform only used a simplified version of the entity editor focused on structures without using skins. Figure 8 illustrates some sample solutions with higher rates.

Motivation was broadly considered by measuring the actual participation. Considering this objective approach may give us an estimation of how motivated the subjects were on using the platform, this measure was operationalized as the user manipulation time over implementation time. Figure 9 shows *motivation* by platform and task. Alternatively, motivation was also self-rated by the participants through a questionnaire. Participants rated on a 5-point scale their agreement with the statement "The platform keep me motivated to participate." This subjective user perception consistently showed that self-rated motivation in the digital platform was on average higher ($M = 3.83$, SD $= 1.030$) than in the physical one ($M = 3.22$, SD $= 0.833$).

5.2 Collaborative Interaction

For tasks that supposedly must be carried out in pairs, an interesting measure is cooperation time, which is the time that both participants in a group were effectively co-manipulating the platform, doing useful work, during the time needed to complete the solution implementation. It gives us an idea of how facilitating the platform is to support sharing and co-manipulation in the construction of structures. A priori, since both platforms are based on tabletops, an expected result would be obtaining similar cooperation profiles. However, cooperation was significantly higher in the digital platform than in the physical-only one as Fig. 10 shows.

Fig. 8 Some high rated solutions in terms of novelty

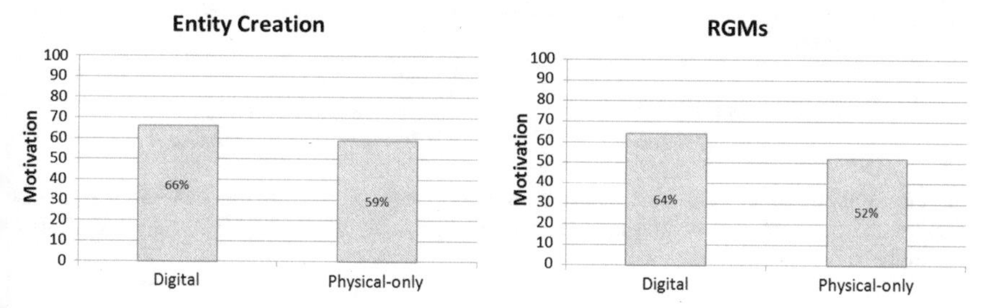

Fig. 9 Motivation mean plot by platform

Although the surface is a shareable workspace operated in pairs, the proposals were originally created by a single individual before being discussed and implemented. Psychological ownership and feelings of possession (Pierce et al. 2003) on either ideas or objects along with the characteristics of each specific interface can therefore reduce the degree in which users share objects and are willing to cooperate in the task. Moreover, although an ideal balance of interaction in group interaction could be expected when interface elements are equally accessible, group members could still dominate the construction of solutions in the platforms.

In order to study both the *ownership* and *dominance* during the implementation, we depicted the dominance profiles by task and platform as shown in Fig. 11.

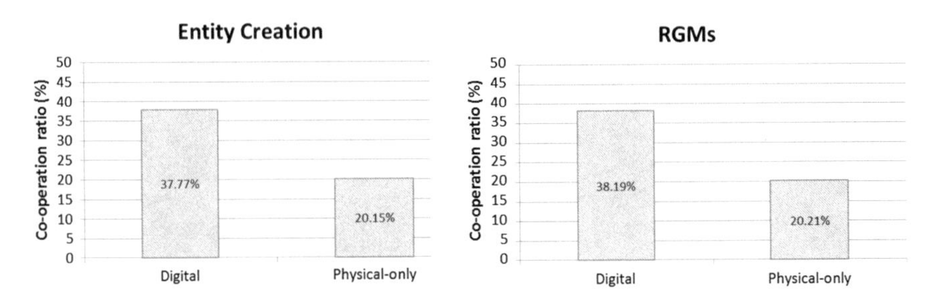

Fig. 10 Cooperation ratio by platform

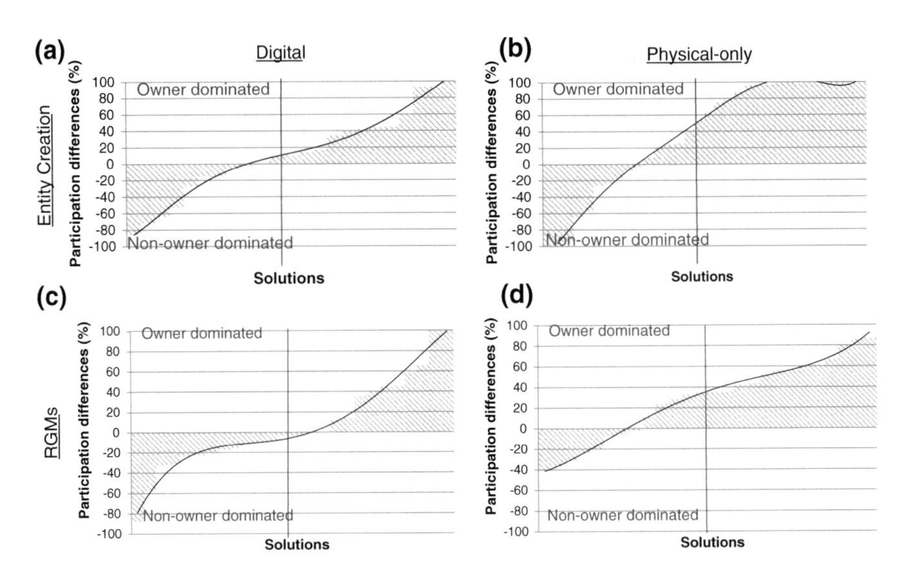

Fig. 11 Dominance by task and platform. **a** Entity creation and digital condition, **b** Entity creation and physical-only condition, **c** RGMs and digital condition, **d** RGMs and physical-only condition

The *dominance* was measured as the relative difference in the participation between the members of a group during the implementation of a single solution. For example, if a member of the group constructed alone the solution, without any active interaction of the other member, the dominance value would be 100. In turn, if both members participated in a similar degree, the dominance would be close to zero because there would be little difference on the time that they participated on this solution. Therefore, a lower dominance value means that the interaction on the platform is more balanced.

On average, for both tasks the dominance measurement was higher in the case of solutions implemented in the physical-only platform. This can be partly explained by the particular characteristics of the platforms. Although interaction on both platforms is tabletop-based and supposedly natural, manipulation still has differences favoring dominance in the case of the physical-only platform. The elements on this platform can be physically grasped and easily moved to the personal working space. Moreover, as seen in the video recordings, interaction avoidance is possible by physically pushing away a partner's hand. In the digital platform, the phenomena of territoriality and interaction avoidance or interference (Hornecker et al. 2008) are also observed but they are not as stressed as in the physical platform. In fact, the observed interferences seem better managed as blocks remain on the surface level and are therefore reachable by two participants, also facilitating equable object sharing. In this respect, the tasks under consideration are also relevant. In general, the type of solutions for the creation of entities task leads to smaller structures, taking lesser space and therefore less accessible for both members simultaneously. Instead, RGMs are usually bigger, being their blocks scattered across the surface, and then the work can be divided as soon as several components of the machine are clearly identified. We could say that participants usually take blocks to their territory, and then the co-manipulation becomes difficult when the structure is not large enough to enable each participant to create different parts collaboratively. All this suggests that the digital platform could be positive to facilitate a more balanced interaction.

On these plots, the bars arranged along the horizontal axis quantify the dominance or the participation difference on the solutions. For those solutions in which the owner dominated the implementation, the corresponding bars are plotted in the positive side whereas those dominated by the non-owners are represented in the negative side of the chart. In this way, the proportion of owner/non-owner domination can be easily assessed at a glance in order to analyze how different the profile was by platform. In general, the more balanced the interaction the smaller the area formed by the bars is.

For both tasks, the profiles by platform are quite similar, what means to a great extent that the platforms have an effect on the way that participants performed. In the case of the physical-only platform, there are many more solutions in which the owner clearly dominated than in the digital one, leading to a more overall unbalanced interaction where the non-owner hardly participated. In the digital platform, there are a significant number of solutions in which the dominance value is low. Because the differences in participation are not large, they cannot actually be considered as dominated by one or another although they are either in the positive or negative side. This result on the solutions with low dominance suggests that the digital platform allowed both owners and non-owners to equally participate in the construction process, calling for balanced interaction.

Finally, it is surprising that there are a non-negligible number of solutions in which the non-owner clearly dominated in both platforms. This suggests that the discussion stage prior to implementation is definitely profitable and that it plays a relevant role in the development of the tasks, allowing non-owners to take over the control during the implementation phase, being useful and advantageous in terms of promoting collaborative interaction.

From the video recordings some interaction patterns were identified. In the solutions with high dominance, one of the members implemented entirely (or almost) the solution. In this *individualistic* pattern, the dominant participant seemed to reject help and even avoid interaction from the other as it was clear how to proceed. A possible explanation of this behavior is that individualistic users feel that the teammate intervention would threat the efficient completion of the proposal and therefore they preferred to work alone. This behavior was more often observed in the physical-only platform,

However, there were two collaborative patterns that were more usual and associated to the solutions with low/medium dominance in both platforms. They are the *provider-constructor* and the *leader–follower*. They are closely related and normally intertwined in the construction of single solutions. The provider-constructor refers to when one participant systematically provided building elements to the other member, who eventually assembled them. The role of either provider or constructor is usually switched several times during the implementation of the solution so that both participants normally play both roles to some extent.

The leader–follower refers to the pattern in which clearly one participant played the role of leader or director during the whole construction, whereas the other remained attentive in order to participate and provide help at any time. It seemed that this alert state was more often present in the digital platform than in the physical-only, in which the followers used to keep ready a block in their hands but with no particular purpose.

6 Further Work and Future Directions

An interface's design and its technological capacities undoubtedly determine the users' capability to manipulate digital objects. While the current state-of-the-art game controllers are adequate in conventional gaming scenarios focused on consuming digital information, they are limited in supporting the production of digital objects. Thus, more research and development is needed to successfully support creative tasks, which seem to require more natural and tangible input devices.

With this aim in mind, this chapter has explored the use of tabletop interfaces to foster discussion, action, and reflection when supporting the creation of digital game worlds. The platform is designed for the playful creation of 2D videogames by direct finger manipulation. For this reason, both entities and rules were given a spatial dimension to be manipulated on the surface in collaboration with a partner. We have seen how such platforms can be used and combined in a wider learning setting and how the discussion process facilitates the sharing and manipulation of objects by non-authors.

Future directions on this line of research should look into how input interfaces can be improved to better support creative tasks. The idea is that with the proper use and combination of new technologies we should be able to overcome current limitations and reach a degree of tangibility and ubiquity that will allow users to actually focus on specific domain issues and not on the technological barriers.

Acknowledgments This work received financial support from the Spanish Ministry of Education under the National Strategic Program of Research and Project TIN2010-20488. This work is also supported by a postdoctoral fellowship within the VALi+d program from Conselleria d'Educació, Cultura i Esport (Generalitat Valenciana) to A. Catalá (APOSTD/2013/013). Our thanks to Polimedia/UPV for the support in computer hardware.

References

Abt C (1970) Serious games. Viking Press, New York

Aleinikov AG, Kackmeister S, Koenig R (2000) Creating creativity: 101 definitions (what Webster never told you). Alden B. Dow Creativity Center Press, New York

Amabile T (1982) Social psychology of creativity: a consensual assessment technique. J Pers Soc Psychol 43(5):997–1013

Amabile T (1983) The social psychology of creativity: a componential conceptualization. J Pers Soc Psychol 45:357–376

Baer J, McKool SS (2009) Handbook of research on assessment technologies, methods, and applications in higher education. In: Christopher GSS (ed) Handbook of research on assessment technologies, methods, and applications in higher education. Information Science Publishing, Hershey

Begel A (1996) LogoBlocks: a graphical programming language for interacting with the world. Electrical Engineering and Computer Science Department. MIT, Cambridge, MA

Beghetto RA (2007) Creativity research and the classroom: from pitfalls to potential. In: Tan AG (ed) Creativity: a handbook for teachers. World Scientific, Singapore

Beghetto RA, Kaufman JC (2007) Toward a broader conception of creativity: a case for "mini-c" creativity. Psychol Aesthetics Creativity Arts 1(2):73–79

Buisine S, Besacier G, Najm M, Aoussat A, Vernier F (2007) Computer-supported creativity: evaluation of a tabletop mind-map application. In: Proceedings of the 7th international conference on engineering psychology and cognitive ergonomics, Berlin, Heidelberg, pp 22–31

Catalá A, Jaen J, Martinez-Villaronga AA, Mocholi JA (2011) AGORAS: exploring creative learning on tangible user interfaces. In: 2011 IEEE 35th annual computer software and applications conference (COMPSAC), pp 326–335

Catalá A, Garcia-Sanjuan F, Azorin J, Jaen J, Mocholi JA (2012a) Exploring direct communication and manipulation on interactive surfaces to foster novelty in a creative learning environment. Int J Comput Sci Res Appl 2(1):15–24

Catalá A, Garcia-Sanjuan F, Jaen J, Mocholi JA (2012b) TangiWheel: a widget for manipulating collections on tabletop displays supporting hybrid input modality. J Comput Sci Technol 27(4):811–829

Catalá A, Garcia-Sanjuan F, Pons P, Jaen J, Mocholi JA (2012c) AGORAS: towards collaborative game-based learning experiences on surfaces. In: Proceedings of the international conference on cognition and exploratory learning (CELDA 2012), pp 147–154, IADIS, Madrid, 19–21 October, 2012. ISBN 978-989-8533-12-8

Catalá A, Jaen J, van Dijk B, Jordà S (2012d) Exploring tabletops as an effective tool to foster creativity traits. In: Proceedings of the sixth international conference on tangible, embedded and embodied interaction, New York, NY, USA, pp 143–150

Catalá A, Jaen J, Pons P, Garcia-Sanjuan F (2013a) Creativity and entertainment: experiences and future challenges, I. In: Spanish symposium on entertainment computing (SEED)

Catalá A, Pons P, Jaen J, Mocholi JA, Navarro E (2013b) A meta-model for dataflow-based rules in smart environments: evaluating user comprehension and performance. Sci Comput Program 78(10):1930–1950

Csikszentmihalyi M (1991) Flow: the psychology of optimal experience. Harper Perennial, New York

Dillenbourg P, Evans M (2011) Interactive tabletops in education, I. J. Comput-Support Collab Learn 6(4):491–514

Ellis H, Heppell S, Kirriemuir J, Krotski A, McFarlane A (2006) Unlimited learning: computer and video games in the learning landscape. ELSPA Entertainment and Leisure Software Publishers Association, pp 1–66

EU (2009) Progress Report: Chapter IV—enhancing creativity and Innovation, including enterpreneurship at all levels of education and training. In: Towards the Lisbon objectives in education and training, indicators and benchmarks, European Commission

Farooq U, Carroll JM, Ganoe CH (2007) Supporting creativity with awareness in distributed collaboration. In: Proceedings of the 2007 international ACM conference on supporting group work, New York, NY, USA, pp 31–40

Forster F (2009) Improving creative thinking abilities using a generic collaborative creativity support system. In: M-ICTE'09, pp 539–543

Friess MR, Kleinhans M, Forster F, Echtler F, Groh G (2010) A tabletop interface for generic creativity techniques. In: International conference on interfaces and human computer interaction (IHCI 2010)

Geyer F, Klinkhammer D, Reiterer H (2010) Supporting creativity workshops with interactive tabletops and digital pen and paper. In: ACM international conference on interactive tabletops and surfaces, New York, NY, USA, pp 261–262

Good J, Howland K, Nicholson K (2010) Young people's descriptions of computational rules in role-playing games: an empirical study. In: Proceedings of the 2010 IEEE symposium on visual languages and human-centric computing, Washington, DC, USA, pp 67–74

Gros B (2007) The design of learning environments using videogames in formal education. In: Proceedings of the the first IEEE international workshop on digital game and intelligent toy enhanced learning, Washington, DC, USA, pp 19–24

Hornecker E, Marshall P, Sheep Dalton N, Rogers Y (2008) Collaboration and interference: aware-ness with mice or touch input. In: Proceedings CSCW08, ACM, pp 167–176. http://dl.acm.org/citation.cfm?id=1460589

Jordà S, Geiger G, Alonso M, Kaltenbrunner M (2007) The reacTable: exploring the synergy between live music performance and tabletop tangible interfaces. In: Proceedings of the 1st international conference on tangible and embedded interaction, New York, NY, USA, pp 139–146

Kelleher C, Pausch R (2005) Lowering the barriers to programming: a taxonomy of programming environments and languages for novice programmers. ACM Comput Surv 37(2):83–137

Lu F, Tian F, Jiang Y, Cao X, Luo W, Li G, Zhang X, Dai G, Wang H (2011) ShadowStory: creative and collaborative digital storytelling inspired by cultural heritage. In: Proceedings of the SIGCHI conference on human factors in computing systems, New York, NY, USA, pp 1919–1928

Maloney J, Burd L, Kafai Y, Rusk N, Silverman B, Resnick M (2004) Scratch: a sneak preview. In: Proceedings of the second international conference on creating, connecting and collaborating through computing, Washington, DC, USA, pp 104–109

McFarlane A, Sparrowhawk A, Heald Y (2002) Report on the educational use of games, TEEM/DfES

Michael DR, Chen S (2006) Serious games: games that educate, train, and inform. Thomson Course Technology, London

Pane JF, Ratanamahatana CA, Myers BA (2001) Studying the language and structure in non-programmers solutions to programming problems. Int J Hum-Comput Stud 54(2):237–264

Parkes AJ, Raffle HS, Ishii H (2008) Topobo in the wild: longitudinal evaluations of educators appropriating a tangible interface. In: Proceedings of the SIGCHI conference on human factors in computing systems CHI'08, pp 1129–1138

Pierce JL, Kostova T, Dirks KT (2003) The state of psychological ownership integrating and extending a century of research. Rev Gen Psychol 7(1):84–107

Pons P, Catalá A, Jaen J (2013) TanRule: a rule editor for behavior specification on tabletops. Extended Abstracts of the ACM Tangible, Embedded and Embodied Interaction (TEI 2013), pp 1–8

Repenning A, Ioannidou A, Zola J (2000) AgentSheets: end-user programmable simulations. J Artif Soc Social Simul 3(3). http://jasss.soc.surrey.ac.uk/JASSS.html

Resnick M (2002) Rethinking learning in the digital age. In: Kirkman G (ed) The global information technology report: readiness for the networked world. Oxford University Press, Oxford

Richards R (2007) Everyday creativity: our hidden potential. In: Richards R (ed) Everyday creativity and new views of human nature: psychological, social and spiritual perspectives. American Psychological Association, Washington, DC

Runco MA (2010) Creativity: theories and themes: research, development, and practice. Elsevier Science, Amsterdam

Sylla C, Branco P, Coutinho C, Coquet E, Skaroupka D (2011) TOK: a tangible interface for storytelling. CHI'11 extended abstracts on human factors in computing systems, New York, NY, USA, pp 1363–1368

Vandoren P, Van Laerhoven T, Claesen L, Taelman J, Raymaekers C, Van Reeth F (2008) IntuPaint: bridging the gap between physical and digital painting. In: 3rd IEEE international workshop on horizontal interactive human computer systems, 2008. TABLETOP 2008, pp 65–72

Vyas D, Nijholt A (2010) Building boundaries on boundary objects: a field study of a Ubicomp tool in a design studio. Int Rep Socio-Informatics7(1):282–299

Vyas D, Nijholt A, Heylen D, Kröner A, van der Veer G (2010a) Remarkable objects: supporting collaboration in a creative environment. In: Proceedings of the 12th ACM international conference on ubiquitous computing, New York, NY, USA, pp 37–40

Vyas D, Nijholt A, Kröner A (2010b) CAM: a collaborative object memory system. Mobile HCI, pp 415–416

Wang J, Farooq U, Carroll JM (2010) Does design rationale enhance creativity? Hum Technol Interdiscipl J Hum ICT Environ 6(1):129–149

Wigdor D, Wixon D (2011) Brave NUI world: designing natural user interfaces for touch and gesture, 1st edn. Morgan Kaufmann, San Francisco

Wing JM (2006) Computational thinking. Commun ACM 49(3):33–35

Bifocal Modeling: Promoting Authentic Scientific Inquiry Through Exploring and Comparing Real and Ideal Systems Linked in Real-Time

Paulo Blikstein

Abstract The improvement of STEM education through new pedagogies and technologies has been the chief concern of policy-makers and educators for the past decades. Common threads among the proposed solutions have been to promote inquiry, discovery, and authentic scientific practices in the classroom. In this chapter, we present a novel inquiry-based framework which combines computer simulations and real-world sensing in real-time: bifocal modeling. Even though educational researchers have come to realize the potential of simulations, computer models, and probeware separately, little research and design have been done on the *combination* of these new technologies. When creating a bifocal model, students build a computer simulation and the analogous sensing apparatus, and link them in real-time, being able to validate, compare, and refine their conceptual models using data. In this chapter, I will focus on the technical and pedagogical aspects of this framework, describe several example models, and discuss four pilot studies, which suggest that the synergy between physical and simulated systems catalyzes further inquiry toward a deeper understanding of the scientific phenomena.

Keywords Computer modeling · Sensing · Constructivism · Physical computing · Bifocal modeling · Constructionism · Probeware · Scientific inquiry

1 Introduction

1.1 *Faraday's Motor*

In 1821, Michael Faraday built the first of his electromagnetic motors and demonstrated that electrical current could generate a continuous rotary movement. Eager to share the discovery with his colleagues, he wrote numerous letters to

P. Blikstein (✉)
Graduate School of Education, Stanford University, 520 Galvez Mall, CERAS 232, Stanford, 94305 CA, USA
e-mail: paulob@stanford.edu

A. Nijholt (ed.), *Playful User Interfaces*, Gaming Media and Social Effects,
DOI: 10.1007/978-981-4560-96-2_15, © Springer Science+Business Media Singapore 2014

fellow scientists across Europe. A little-known fact is that those letters were included in a package containing a "motor kit"—a set of materials for his colleagues to assemble their own motors and continue the experimentation (Baird 2006). The scientific community could not yet explain *why* the motor worked, but there was no denying *that* it worked. Philosophers of science have been using these and other examples to call into question what some term the "text bias" (Baird 2006)—a tendency of historians of science to consider literary production as the ultimate product of science (see Latour and Woolgar 1979), although throughout history many scientists exchanged products and experimental apparatuses as much as they shared papers.

The idea that scientific instruments are more than the disposable means by which platonic scientific truths are brought into being has garnered interest in recent years. Whether we speak of instrumentation in terms of scientific models, measuring devices, or working devices, these tools have come to be regarded as multifaceted objects which bring along their own epistemologies (Radder 2003). However, despite the extensive debate on such matters in the philosophy of science, relatively less research has been done in education. If scientific instruments have a crucial role in the generation of scientific discourses, their epistemology, and the unfolding of the scientific method, a focus on instrumentation might very well lead to similar benefits for the learning and teaching of science. The goal of this chapter is to investigate the design of learning environments and technologies that embrace, rather than negate, these complex relationships between instruments and theory, real and ideal systems, tangible and virtual apparatus, as well as the role of experimentation in the scientific method.

1.2 Scientific Instruments, the Scientific Method, and Model-Based Inquiry

Contemporary educators' conceptualizations of scientific instrumentation and experimentation have a profound influence on how science is taught: is science a method, a cognitive habit, a set of content topics, or all of the above? This discussion must begin in the context of the history of science education. In the second half of the nineteenth century, when the teaching of science and its methods were being progressively introduced into the curriculum, scientists (such as Faraday himself) argued for the teaching of science not primarily to train scientists, but for its civic benefits—as a way of empowering individuals to fully participate in an open, democratic society—on the assumption that a grasp of rational processes of scientific inquiry would have practical consequences for daily life and for political engagement at all levels. However, other factors were at play; as early as the 1920s, national reports were calling for the teaching of "real" scientific content to improve economic competitiveness. Throughout the twentieth century, every two decades or so, the pendulum has swung between the civic and economic sides of this debate (DeBoer 2000), but over the past 20 years, a new emphasis has emerged. New national standards and policy documents have established that students should learn the *content* of the scientific disciplines in the context of *being* a scientist through experiential engagements in the practices of the profession (National Research Council 2012; NGSS Lead States 2013). In other words, to achieve either civic or competitive advantages from science education, our current educational policy recognizes that instructional *content* must be deeply embedded in *investigatory practices*. It took nearly a century of debate for science practice, scientific content, and the application of science to daily life to be reconciled in national policy documents ("learning science, learning about science, doing science," Hodson 1998). Fortunately, the resolution of the debate has generated considerable interest in implementing new approaches to the learning of science in schools.

Within this process, more attention has been given to the kinds of methods and practices to be taught to students; if students are to experience being scientists, what is it, after all, that scientists *do*? One important development in this area has been a challenging of the idea of a uniform *scientific method* (Rudolph 2005; Windschitl et al. 2008):

> Reference to a universal scientific method is common in discourse at all levels of science education. [...] We assert, however, that the scientific method is not scientific at all when considered from an epistemic perspective, and that it subverts young learners' understandings of both the practices and the content of the discipline. (Windschitl et al. 2008, pp. 942)

These issues about scientific practices and methods are even more pronounced in relation to the connections between the science lab and the science classroom. Windschitl et al. (2008) point out that the classic version of the scientific method taught in schools leads to lab practice in which children conduct "contentless" experiments, trying out random experimental ideas without context and without any grasp of the underlying theoretical models (Windschitl 2004), reinforcing naïve ideas about the nature of scientists' practice (Hodson 1996, 1998). This separation of the cognitive and empirical aspects of the scientific process deepens the chasm between "serious," high stakes science, taught in the classroom, and "playful," hands-on science, taught in the lab, and optional (Roth and Garnier 2006). Windschitl et al. (2008) also note that the school version of the scientific method "works too well for teachers" (p. 947). These speedy, one hour, step-by-step, cookbook "experiments" packaged as a linear sequence of prescribed tasks turn the scientific method into a rote procedure and do little to inculcate this method as a characteristic mode of thought and approach to experience. Finally, in these traditional educational models, lab applications of the scientific method always focus on the systematic relationship of conditions and outcomes, but not on the mechanisms or internal rules that govern these relationships (which often are not available for inspection without a theoretical model).

These shortcomings have motivated scholars to suggest alternatives, oftentimes looking into the practice of real scientists for inspiration (Duschl and Grandy 2008; Nersessian 2005). Because models have become increasingly important for science practice, *model-based inquiry* is one of the most important frameworks to have emerged within the educational research community, and is an approach which has made considerable inroads in education (Blikstein and Wilensky 2009, 2010; Hmelo-Silver et al. 2007; Lehrer and Schauble 2006; Lesh et al. 2000; Levy and Wilensky 2008; Schwarz and White 2005; Stewart et al. 2005; Wilensky and Reisman 2006).

1.3 New Epistemic Forms

Model-based inquiry has overwhelmingly focused on students' creations of, or interactions with theoretical models of scientific phenomena. But contemporary science has gone much further than this, incorporating a variety of new, tangible instruments and sensing devices, and, indeed, new forms of tangibility. The processes and ways of thinking and engaging with the world that led to technological advances long before Faraday's "motor kit" have come full circle with the arrival of computer simulations that allow for multiple forms of sense-making beyond traditional mathematical modeling, speaking directly to our intuitions of geometry, mechanism, space, and time. In particular, alongside sophisticated forms of computer modeling, scientists are also developing complex measuring instruments and techniques, often with real-time connections to observation. These novel means of scientific inquiry which bring together computational modeling and real-time sensing to validate and refine models give rise to new kinds of *epistemic forms* (Collins and Ferguson 1993): cognitive tools that afford new lenses and modes of inquiry, and introduce new ways of knowing, learning from, and engineering systems.

This process has intensified in recent decades as a consequence of technological advancements. When modeling a chemical system, a beehive, or an economy, researchers must devise ways to collect data that match the complexity of the model and validate its causal or statistical inferences (Bryson et al. 2007; Cagnacci et al. 2010). They use sophisticated tools such as computer vision systems to track bees in real time, GPS systems to track the interaction of mammals as they happen (Wark et al. 2007), high-speed photography and 3D imaging to capture motion in space (Chen et al. 2009), and autonomous chemical sensors to understand the impact of climate change in remote locations (Johnson et al. 2007). These tools, in turn, provide new ways to understand and interact with the world. Consider, for example, the research on animal behavior by Wark et al. (2007). In conjunction with a computer model for predicting animal movement, this team employed wireless sensors for position tracking and created a system that will *optimize itself* until a satisfactory solution is reached.

Educational tools and practices that connect computational modeling and real-world sensing technologies can enable pedagogical practices better aligned with real scientific practice. In humans, mind and matter are intertwined through the sensorimotor coupling of rapid feedback loops: we see, we act, we see.

Scientific reasoning appears to build on this capacity for seeking attunement between our cognitive models and our sensory experience. But as we try to understand natural phenomena that are much more complex and sophisticated than everyday events, and as long as we need to perform each of those mental activities separately, the mind is encumbered by technological and cognitive constraints that do not enable it to operate scientifically as it does in the natural coupling of sight and action. In such cases, real-time data systems might become particularly useful through implementation in technologies that make such couplings, once again, seamless. Some practical examples include **measurement/sensing technologies** that extend human perceptual categories (big/small, far/near, fast/slow), **cognitive artifacts** that amplify human reasoning (Norman 1991) and **data archiving and representation**, which extend the memory capacity and reduce cognitive load.

Whereas the practice of using models and sensors in real-time might already be familiar in scientific circles, exploring how to bring this powerful technique to classrooms is a relatively new, but promising, endeavor (Tinker 1991, 1996). However, science classrooms and laboratories are still not well suited for these kinds of integrative activities and modes of interaction. A student examining an acid–base reaction in a laboratory might identify the chemical constituents involved and even hypothesize about their proportions and concentrations. Nevertheless, this student will not be able to directly witness the chemical mechanisms at work. Later, in the classroom, the students might learn about chemical equations and theories that bear little resemblance to the phenomenon observed in the laboratory, either in terms of scale or in the representation of the mechanisms involved. In this case data collection and modeling are disconnected.

However, the connection between data and theory can never be simple. Naïve ideas about scientific theories as uncompromised representations of truth readily available to scientists through careful examination of empirical data have been questioned by many researchers, who regard the reality to be much more dynamic, unstable, and only tentatively understood (Thagard 2007). Still, students and teachers hold naive epistemological beliefs about science—for example, that the body of scientific knowledge develops merely as a gradual accumulation of facts through testing. Challenging this naive epistemology is often a slow and challenging process (Hofer and Pintrich 1997; Ingham and Gilbert 1991). Also, students' understandings about the nature of modeling have been found problematic. Treagust et al. (2002), suggest that secondary students fail to understand that models cannot and need not be exact replicas of reality. Grosslight et al. (1991) found that students held "naive realist" conceptions of modeling in which models are believed to correspond directly to the reality they represent, while experts uniformly held the view that models are imperfect but improvable approximations.

For these reasons—cognitive, pedagogical, and epistemological—there is urgency and opportunity to explore the potential for bringing together real-time sensing and computational modeling in classrooms.

1.4 Experimentation, Labs, and Models

School science laboratories are one type of educational environment where these connections among theory, models, and real-world data can begin to be explored. However, these labs have been subject to widespread controversy in the research community (NRC 1996), especially regarding the benefits of physical, virtual, and combined laboratories (Olympiou and Zacharia 2012; Triona and Klahr 2003; Zacharia 2007). The popularity of simulation environments, such as PhET (Perkins et al. 2006), have led policy-makers and scholars to question the real value of physical labs for student learning—especially in the face of the associated costs and complicated logistics. A wave of research studies within the past 10 years has attempted to determine the relative advantages of physical labs relative to virtual labs and manipulatives, and whether the latter can replace the former (Triona and Klahr 2003), or how virtual models could simulate complex phenomena and permit student experimentation in domains that might be costly, impractical, or dangerous (Finkelstein et al. 2005; Jaakkola and Nurmi 2008; Klahr et al. 2007; Perkins et al. 2006; Resnick and Wilensky 1998).

The literature comparing hands-on or physical models (PM) and virtual models (VM) for science learning has sought to establish rules for choosing one modality over another or for sequencing them together (de Jong et al. 2013). Zacharia and Anderson (2003) found that combining physical and virtual models increased teachers' learning of content knowledge in physics. A treatment group of teachers performed a physics inquiry activity with a virtual model and, subsequently, a physical model, while a control group used only the physical model, and the authors found that the treatment group had greater conceptual understanding. Zacharia et al. (2008) recreated this result with undergraduate physics students first employing a physical model rather than a virtual one, and Jaakkola and Nurmi (2008) obtained similar results for elementary school students. Most of these early studies pointed to the advantage of virtual over physical labs, but researchers soon after began combining both types of procedures and found that the combination of the physical and virtual labs led to still greater conceptual understandings than either type singly.

For example, Liu (2006) compared two groups of female high-school students utilizing computer simulations and/or hands-on lab activities in chemistry. Controlling for time-on-task, the combination of PM and VM was more effective than either option alone. But there are interesting interactions between content learning and epistemology observed for this composite approach: there was a correlation between students' understanding of the chemistry content and a belief that the chemistry model shown was an exact replica of reality. In other words, students who understood the content better were not necessarily more epistemologically sophisticated. This finding is a preliminary indication of the importance of directly addressing epistemological issues in lab-based and model-based inquiry environments, either in virtual, physical, or combined modes.

The literature suggests that multiple representations can help students understand underlying scientific concepts, but such representations can also be overwhelming to new learners, who lack the knowledge that would permit them to focus on the appropriate elements of the process (Kirschner et al. 2006). One approach to assisting new learners' efforts to make sense of these multiple representations is to explicitly link them, so that changes in one modality will directly affect the other. Van der Meij and de Jong (2006) investigated this question in a virtual physics learning environment, employing multiple graphical representations to convey the relationships between variables in a mechanical system. In one condition, the representations were dynamically linked so that each responded to changes in the other, and they were "integrated" through their close visual proximity to each other. In a second condition, the variables were integrated but unlinked, and, in a third, they were both unlinked and unintegrated. The authors found that students who learned the most were most able to transfer their new knowledge to new problems, and that these same students reported the least difficulty with the *linked and integrated version*—a finding that expands upon the design principle of "multiple representations" offered by Blake and Scanlon (2007). However, intergroup differences emerged only with more challenging problems presented to the students. Although the groups' performances were approximately equal for the easier questions, the results suggest that for more difficult problems involving the use of many sources of information, the availability of scaffolding is increasingly important. The authors explain this finding in terms of the scaffolding's ability to reduce the working memory load involved in the tracking of multiple representations carefully enough to identify their relationships.

These promising, but nuanced, findings point to the combined use of computational tools and physical labs as an increasingly viable option for classroom science learning. Even though the potential of this combination of virtual and physical models as a tool for science learning has been documented over a wide range of ages and domains, the findings also point to a need for better design principles and theoretical frameworks to understand how this potential can be leveraged to address the cognitive, pedagogical, and epistemological issues at play.

In particular, two areas in this realm have not been researched sufficiently. First, the literature has focused almost entirely on predesigned physical and computer models or labs. Predesigned models can provide scaffolding and make students aware of the relevant information about a problem, but they fail to give students opportunities to evaluate the assumptions and limitations of the models themselves (Papert 1980). The practices of *creating* and *critically evaluating* models constitute an important part of the scientific practice and have been valued increasingly as educational goals (Blikstein and Wilensky 2009, 2010; Gire et al. 2010). Second, the literature also has not adequately explored the potential for deeper support for students' *explicit comparisons* between physical and virtual models. Smith and collaborators (2010) noted that scaffolds in virtual models or direct data-sharing between virtual and physical models could help students recognize the similarities and differences between the model and reality. Likewise, most of the research has focused either on the comparison of physical and virtual labs or on their sequencing, but not on their mutual synergies when they are connected in real-time. Most of the virtual labs employed in these studies are mere transpositions of physical labs to a virtual environment. Beakers, test tubes, and chemicals are simply made virtual in a computer-based environment and students conduct experiments in them. But when scientists use models and simulations together with real-world data, they are looking for synergies rather than replacement—they are aware that each brings different sets of information, questions, and insights. The very idea of deciding which mode is superior is problematic. Therefore, the framework that I will present in this chapter is more concerned with the exploration of these synergies than choosing which type of lab is better.

2 Mixing Sensors and Models: Bifocal Modeling

In this chapter, I present a framework that links computational models and sensors in real-time, providing continuity between observation, the physical construction of artifacts, and model building. Because this framework enables the seamless integration of theoretical/computational models and the physical world, allowing modelers to focus *simultaneously* on their "on-" and "offscreen" models, I have termed it *bifocal modeling* (Blikstein 2010, 2012; Blikstein et al. 2012; Blikstein and Wilensky 2006, 2007). Having a computer model and a data collection apparatus connected in real-time changes the epistemic game—instead of the traditional cycle of devising a theory and fitting empirical data to a curve, the entire system [theory + data] constitutes a new type of cognitive object, allowing learners to run "what-if" scenarios, hypothesize alternative solutions, and debug them, effectively blurring the division between data (the measured) and theory (the predicted) ("instrument epistemology," Baird 2006). In building a bifocal model, students have three main activities:

1. Students build, modify, or interact with a computational model of a phenomenon using computer modeling platforms such as NetLogo (Wilensky 1999, updated 2006), Algodoo, Scratch, Logo, or Processing. In building or interacting with the model, students should be able to express their hypotheses and initial theories about the phenomenon.
2. Students use electronic sensors and low-cost analog-to-digital interfaces such as Arduino or the GoGo Board (Sipitakiat and Blikstein 2010; Sipitakiat et al. 2004) to build their own sensor-equipped "science lab" and collect data about that same phenomenon. At times, this equipment is given to students entirely or partly assembled.
3. Finally, students run both systems, which are connected in real-time, examining the data (real and virtual) side-by-side, to validate, refine, and debug their hypotheses. The computer screen becomes a display for the computer model, whose programming allows it to proceduralize equations, text, or other

representations of scientific content, as well as the actual phenomenon, which is discretized and measured using sensors and other laboratory apparatuses.

Because the bifocal computer models are carefully constructed to imitate a phenomenon's visual language, the bifocal methodology could minimize interpretive challenges. The observed and the hypothesized results are displayed in a way that makes the perceptual differences between them less obvious, while making their procedural differences more visible. Thus bifocal modeling constitutes a tool for students to conduct scientific research that reduces the interpretive and the menial burdens of scientific practice, freeing cognitive, discursive, and material resources for the validation of hypotheses. In the next section, I present a framework for categorizing different types of uses of these technologies for the better framing of bifocal modeling with regard to the other modeling modalities that involve sensor/model interaction.

3 Three Modalities for Combining Sensors, Tangible Interfaces, and Models: Augmentation, Mimicking, and Modeling

In my effort to create a taxonomy of the possible modes for merging sensors, actuators, and models for science learning, I define three broad categories: Augmentation, Mimicking, and Bifocal Modeling. *Augmentation* encompasses all forms of enhancing a computer-based system, or a real experiment, through the use of sensors and actuators. Either sensor data may be augmented with the help of computational tools, or a computer model may be augmented with physical input and output devices. *Mimicking* refers to the use of tangible technologies to exhibit physically a particular mechanism of a phenomenon typically invisible to unassisted human vision. Bifocal Modeling, the main topic of this chapter, goes beyond simple augmentation and mimicking and connects two models in real-time. The process "duplicates" the models in the physical and virtual worlds, as I show in the examples in this section. We begin with an explanation of the two modes of augmentation (data augmentation and model augmentation), mechanism mimicking, followed by an explanation of bifocal modeling.

3.1 Data Augmentation

What I term "data augmentation" is a collection of techniques employed by designers to understand sensor data through the utilization of computational tools (Fig. 1). Augmenting physical sensors with computational tools is commonly used in schools to analyze or visualize raw sensor data for matching to a scientific

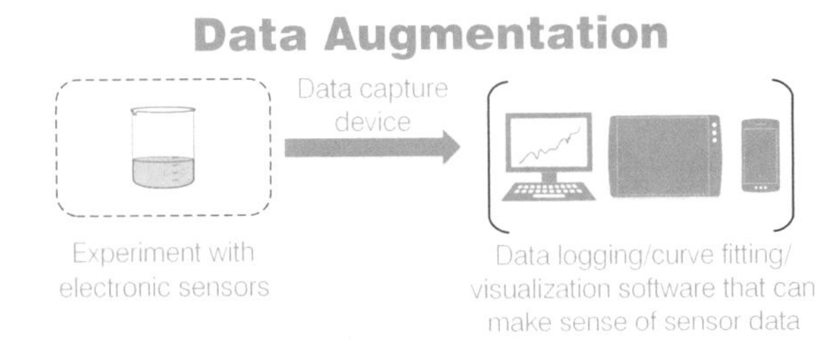

Fig. 1 Augmentation of physical sensor data using computational tools

formula. For example, an electronic sensor measuring pH in a solution can be augmented with software that either graphs the sensor values, fits curves, or generates multidimensional visualizations of the data. This is probably the most common form of augmentation and is found in many probeware-based educational software products ("PASCO Scientific"; "Vernier Software and Technology") and mobile apps. Many schools have commercial kits with sensors and data loggers for this purpose. The common forms of data augmentation include plots, heat maps, curve-fitting tools, automated statistical tools, and geotagged maps.

3.2 Model Augmentation

Augmentation can also take place on the simulation or model side (Fig. 2), in which case a computer simulation can be augmented with sensors (as input devices) or actuators (as output devices). Augmentation on the model side is not very common in mainstream education, but is an intensively researched topic in the interaction design community. Researchers have been advocating its use when, for example, it would be beneficial for students to physically feel, hear, or experience events that are too small, too large, or transpire in inaccessible timescales (Birchfield and Megowan-Romanowicz 2009; Johnson-Glenberg et al. 2009). A simple form of model augmentation involves the addition of physical sensors to the computer model. A temperature sensor connected to a heat chamber simulation can feed the computer simulation a real-world temperature measurement, which can then be employed as an input by the model. For example, the simulation could use the real-world temperature as a starting point to calculate realistic values of heat transfer. Sensors can also be used to interact with the model. A chemistry simulation could be controlled by gestures—students could mix the solution by physically shaking a sensor-enabled tablet, or add reactants by gesturing in the air and using a gesture capture device such as the Kinect™ sensor. A more complex form of model augmentation involves "actions" on the world as the result of events in a computer model. A gas simulation could make the student feel the collisions between molecules and the container by using a simple haptic paddle—every collision would be translated into a small 'push' in the device. The same simulation could be augmented acoustically: students could, for example, hear the collisions of molecules to get a sense for their frequency. Also, modern materials

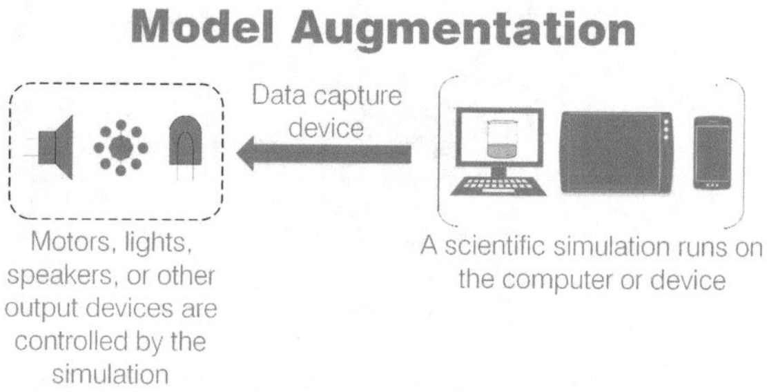

Fig. 2 Computer models and simulations can be augmented using external devices such as motors, lights, sound emitters, and other physical tools

such as Nitinol, a memory alloy that changes shape with variations in electric current flow, could be employed to give the students a dynamic experience of alterations in a spring's constant in response to changes they have made to the computer model.

3.3 Model Mimicking

Another form of enhancing models with tangible technologies is what I call *"model mimicking,"* a relatively common technique in which teachers utilize physical props, such as marbles, ping-pong balls, sticks, and blocks to exemplify otherwise invisible mechanisms (Fig. 3). Through this process teachers or students create a physical version of a virtual or conceptual model and utilize it as a "magnifying glass" to examine a given mechanism or concept. It is very common, for instance, for chemistry teachers to use styrofoam spheres and sticks to demonstrate atomic bonds in molecules. But these models can also be dynamic. For example when studying the gas laws, students or teachers might create a physical, macroscopic instantiation of the gas molecules (for example, a box with moving marbles, or an enclosure containing tiny moving robots which would collide with each other). In studying diffusion, students might create a box with two clusters of ping-pong balls of different colors, which would diffuse into one another as the box is moved, mimicking the mechanism of solid diffusion.

3.4 Bifocal Modeling

Augmentation on both the data and model sides have their own particular educational uses. Bifocal modeling, however, focuses less on providing students a form of data visualization, sensory augmentation, or mixed reality system (Milgram and Kishino 1994); its chief intention is the provision of a hybrid exercise of building analogous systems in dissimilar ontological spaces (the real and the ideal worlds), while *explicitly* engaging students in a comparison of

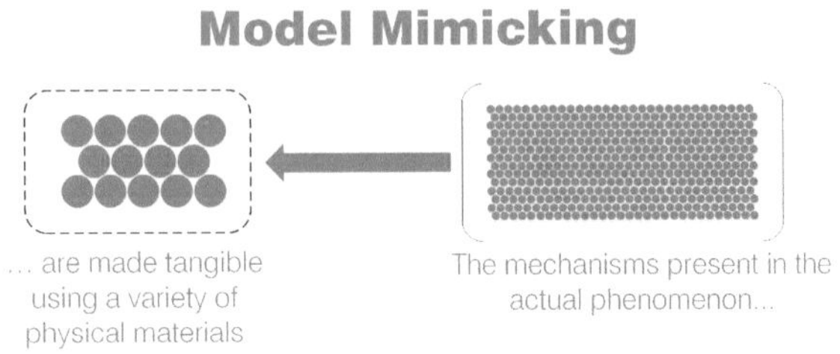

Model Mimicking

... are made tangible using a variety of physical materials

The mechanisms present in the actual phenomenon...

Fig. 3 The micro mechanisms present in a given phenomenon can be "mimicked" macroscopically in the physical world using a variety of materials and techniques. For example, atoms can be represented as metal or styrofoam spheres, and then atomic collision can be demonstrated to students

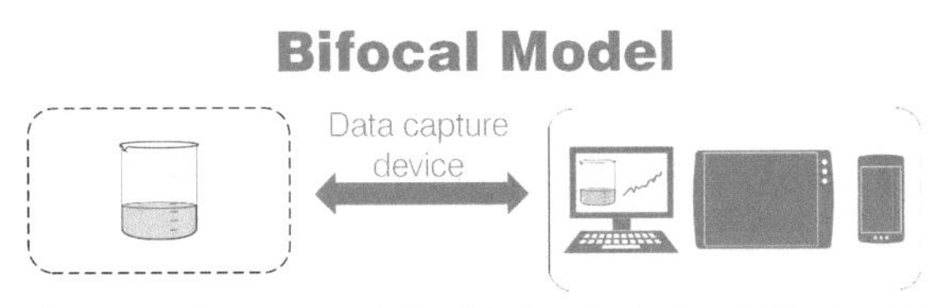

The same phenomenon is "run" in the physical and virtual worlds

Fig. 4 Bifocal modeling: students build or use analogous models in the real and the ideal worlds. In this case, both the virtual and physical models can share data with each other in real-time

models (Fig. 4.) One crucial distinction of bifocal modeling is that it does not compare the learning outcomes of the physical and virtual versions of the same lab. I do not regard lab equipment and models/simulations as two possible equivalents. Rather, the point is that, when properly integrated, these otherwise discrete modalities constitute for the learner a single and fundamentally different epistemic material characterized by its own cognitive and heuristic properties.

By definition, perfectly matching an ideal and a real model is an impossible task—real and ideal systems have fundamentally different internal rules and ontologies. But while this dissimilarity has been considered an obstacle for science learning and dismissed as confusing (Windschitl et al. 2008), or too "cognitively-loaded" (Kirschner et al. 2006), I believe that emphasis on the distinction can provide learners crucial insights into the reality of scientific practice. And while the approach may not adhere strictly to more "classical" formulations of the scientific method, it is, more appropriately, closer to an empirical realization of the actual practices employed in real-world laboratories (Windschitl et al. 2008). My hypothesis is that by struggling with these very differences, students will achieve deeper understandings of the phenomena under scrutiny—for the same reason, Faraday had difficulty separating his papers from his empirical setup for the electromagnetic motor; these apparatuses are also doing cognitive work (Hutchins 1995).

3.5 A Proof of Concept: A Gas Laws Bifocal Model

A prototypical bifocal model is shown in Fig. 5. In the Gas Laws bifocal model, a physical syringe with volume, pressure, and temperature sensors is connected in real-time to a computational model of gas molecules. In the computer model, students may incorporate their own hypotheses about how gas molecules behave and how their interactions change the pressure and temperature within the container as the volume changes. When a student presses the physical syringe, the volume sensor sends the information to the computer and the virtual piston (in red, on the right) moves up and down accordingly. Therefore, as the volume changes simultaneously in the real and virtual chambers, students can examine the changes in pressure and temperature (there are plots comparing the real and the virtual pressures on the model's interface). If the data from the models do not match, the learners can return to their computer model, change it, and repeat the procedure, until a good enough match occurs. Rather than simply representing a perfect match between the two models, this process problematizes the differences between real and ideal systems and leads to reflection and further insight.

4 Examples of Bifocal Models

In this section, I will present several examples of bifocal models built by high-school and college students in a variety of educational scenarios (e.g., an after-school workshop for high-school students and an undergraduate-level class). The goal of this section is to offer concrete instantiations of the framework, and show how it allows students to design interfaces that enable for meaningful explorations into the two *foci*. In the next section, I will present and discuss the research results.

4.1 Heat Transfer

The first student-generated example (Fig. 6) deals with the phenomenon of heat transfer in different materials. The physical model, on the left, has a chamber with a heat source (an incandescent bulb), removable walls (made from metal, wood, and other materials), and a temperature sensor attached to the external part of the wall. The apparatus was connected to a GoGo Board (Sipitakiat et al. 2004), and then to the virtual model programmed in NetLogo (Wilensky 1999, updated 2006). The computational model of heat transfer follows the same "visual language" as the physical model. (Note that the container's wall is represented in yellow, the air inside the container as light blue particles, the air outside as purple particles, and the simulated wall in green). Students were then able to run both models in real-time, experimenting with different types of walls, theorizing about the thermal

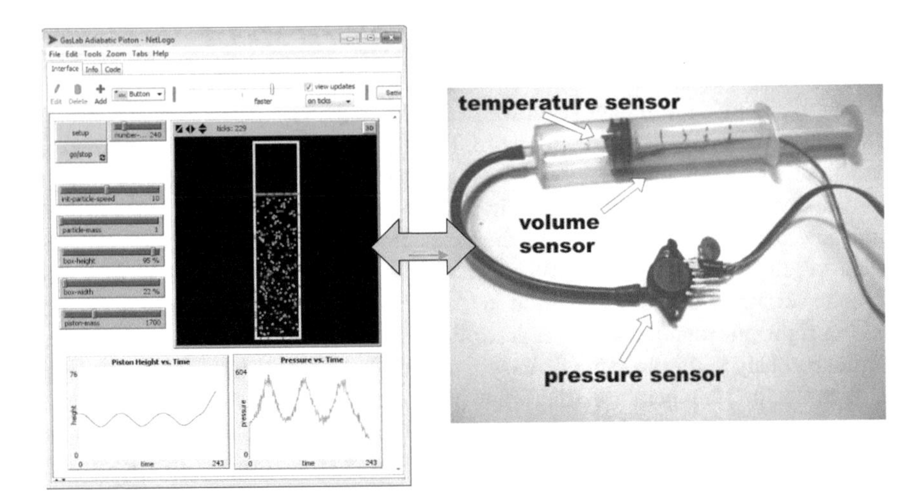

Fig. 5 A model of the gas laws, with the physical sensor-enabled syringe (*right*) and the computer model of the same syringe (*left*), connected in real-time

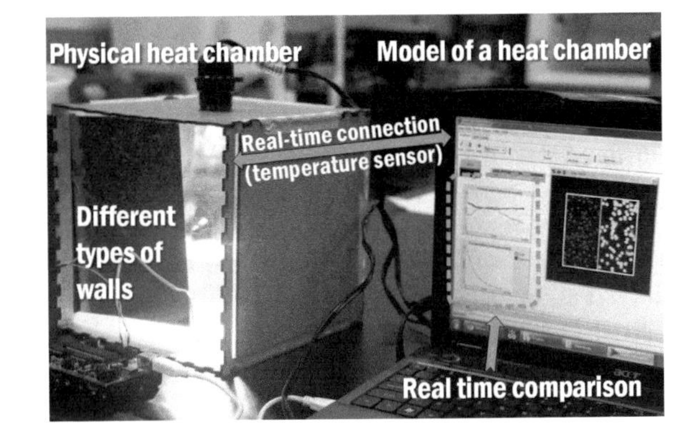

Fig. 6 A bifocal model for studying heat transfer. On the *left*, there is a chamber with a heat source, removable walls with different thermal properties, and a temperature sensor. On the *right* is the analogous model with graphs for side-by-side comparison

conductivity of different materials, and testing their hypotheses about the relationships between the materials, the wall thicknesses, and heat transfer. In this case, differing from a traditional lab experiment in which the goal would be to fit a given curve, the students' goal is to dive deeper into the phenomenon by running multiple comparisons between the real and the ideal systems, making changes in both.

4.2 Refraction

A second example of a bifocal model built by students concerned refraction (Fig. 7). The physical model contained a water chamber, a sensor-enabled laser pointer, and special moveable walls with a metric grid for measurement. The angle

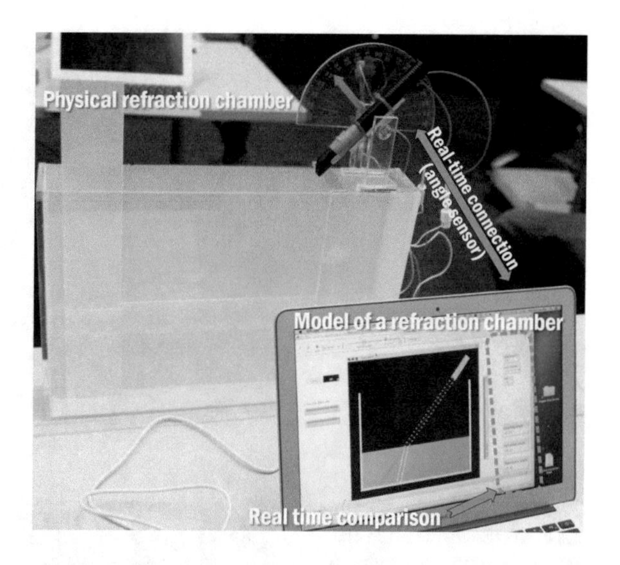

Fig. 7 A bifocal model for refraction. On the *top left* is a water chamber with a computer-controlled laser pointer, and on the *bottom right*, the corresponding computer model. The angle of the physical laser is detected automatically by the computer, which updates the virtual laser in real-time

Fig. 8 Students also built a mimicked model of a photon going through the air/water interface

of the laser pointer was automatically fed to the virtual model, where an analogous system was built, and the virtual and the physical laser moved in tandem. Since the laser pointer controlled the angle of its virtual counterpart, students could see the deflection of the laser in both models at all angles. The light beam of the virtual model was composed of myriad "light" agents (proxies for photons), which would change their velocity upon entering a different medium (water).

Apart from the traditional bifocal setup, students also added a *model mimic*. The group decided to build a physical version of the "light agent" (Fig. 8), comprised of a two-wheeled robot placed in a box with two types of surfaces (smooth, made of wood; and rough, of sand). When the robot crossed the boundary between the two surfaces, the difference in speed of the two wheels would make it turn in a manner analogous to that of the "light agents."

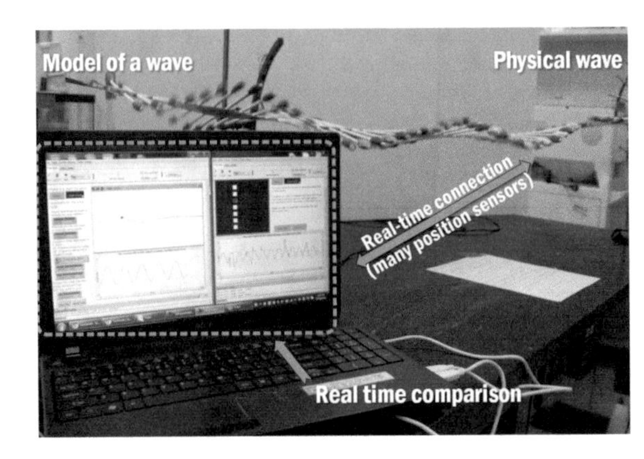

Fig. 9 A bifocal model for wave dynamics. On the *left* is shown a computer with the computational model and the side-by-side comparison. Toward the *right*, the physical wave and the frequency detector

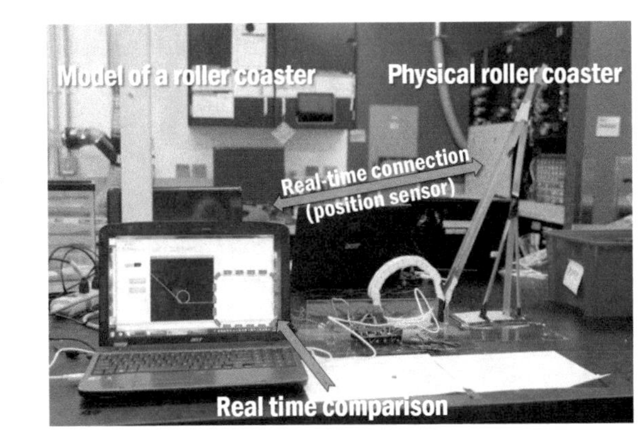

Fig. 10 A bifocal model for classical mechanics. On the *left*, a computer with virtual rollercoaster and the side-by-side comparison is shown. Toward the *right*, the physical rollercoaster equipped with position sensors

4.3 Wave Dynamics

Another group of students built a model of a wave. The physical model consisted of about 100 wooden sticks attached to a flexible "spine," a wave generator (not shown), and an array of infrared sensors to detect the vertical position of one wave element (the black semicircular device partially occluded by the computer screen in Fig. 9). The computational model generated waves based on equations that students programmed, and the comparison between real-world and computational data could be easily seen on screen both as plots and as visualizations of waves. One interesting aspect of this physical model is that at the same time that it generates a "real" wave, it also makes explicit the agent-level behavior within the wave (sticks going up and down). Therefore, it mimics the micro mechanism while *at the same time* showing the aggregate, emergent behavior (Fig. 9).

4.4 Classical Mechanics and Newtonian Physics

The roller coaster model (Fig. 10) is an example of a bifocal model of another class of phenomena: classical mechanics. In this model, a marble roller coaster with several types of interchangeable loops was created, and the final speed of the marble was detected with a pair of infrared sensors. On the computer model, there were similar options for loops and types of marbles, and students could theorize about the laws governing the motion of the marble as well as the influence of friction and air resistance.

5 Case Studies: Framework

5.1 Methods and General Framework

In this section, I present several pilot studies that illustrate the learning outcomes of bifocal modeling. My goal is to give an overview of the different types of activities and learning that take place when students are building or exploring such models, rather than focusing on just one set of empirical evidence. Therefore, the structure of this section will consist of a series of brief case studies with different groups of students in which different formats of implementation are attempted.

In the first study, high-school students built virtual and physical models of bacterial growth in order to learn biology content, computational thinking, and meta-modeling skills. The main research questions concerned (1) students' understandings of the mismatch between idealized virtual and physical models; and (2) how they came to their decisions on which variables and phenomenal factors are necessary to include in their own theoretical models when given real-world data. In the second set of studies, I examined different implementation models for bifocal modeling in classrooms, including more open-ended, generative themes as well as more restrictive tasks. I investigated many different implementation formats and also varied the topic (Biology or Physics) and the degree of model construction required (students building their own models versus students being presented with ready-made models).

Because a full bifocal modeling activity comprises the use of many different tools and techniques, as well as different modes of classroom facilitation, there are a number of formats possible for its implementation. There are five major components of a typical bifocal modeling activity, which may be presented to students or ordered in different ways (see Fig. 11):

A. Content Research. In some studies, students are encouraged to use external learning resources, such as the Internet or books, to gather information about the phenomenon. This option may consist of independent research or be structured as a traditional lecture; the format depends on the context of the study and students' previous knowledge of the phenomenon. The general goal of this phase is to

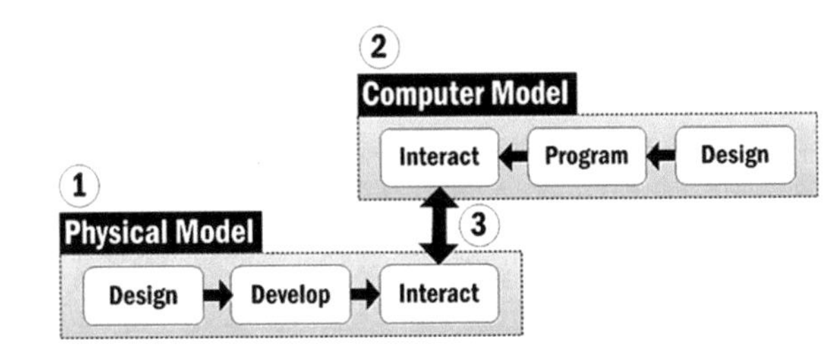

Fig. 11 Components of a bifocal model activity: *1* The physical model, *2* the computer model, and *3* the comparison between the two

provide students with baseline information about the normative representations of the phenomenon under scrutiny. Depending on the learning goals, we may leave this phase to the end or eliminate it altogether.

B. Conceptual Design and Planning. Within a guiding theme/topic (e.g., gas laws), students select variables they wish to explore, make hypotheses about what they are about to observe, and create plans for their physical and virtual models that may potentially address their hypotheses. In designing a virtual model, students typically define the variables and conceptualize micro-rules or equations. In designing a physical model, students create paper or computer drawings and select sensors, building materials, etc.

C. Building/implementation (programming a computer model or developing a physical model). In this stage, students actually construct physical models based on their plans (e.g., a ball and ramp, a Petri dish filled with agar and bacteria, or a system of interconnected beakers containing different substances) and virtual models based on their initial ideas and hypothesis about the phenomenon (e.g., a physics model of gravity and friction, an agent-based model of bacterial growth, or a computer model of diffusion).

D. Interaction and collection of data. Students interact with their physical models via direct observation and collect data using embedded sensors. Similarly, they interact with the virtual model by running the model, changing its parameters, observing the results, and recording data.

E. Comparison. Students compare the physical and virtual data, find discrepancies, reflect on the reasons for the differences, and hypothesize about how to change their models to make them match.

Depending on the format of each of the implementations described in this chapter, we combined these five stages in order to design an activity compatible with the learning goals and the time available. For example, in some implementations, because the time available was short, we provided kits with pre-selected materials that accelerated the process of setting up the experiment. When programming a computer model from scratch was not possible, we provided students with a pre-programmed model or sample code (see more details in the following sections.)

Four pilot studies will be presented. I will describe the first study (Biology) in detail and discuss the three other implementations more briefly (Physics, Chemistry,

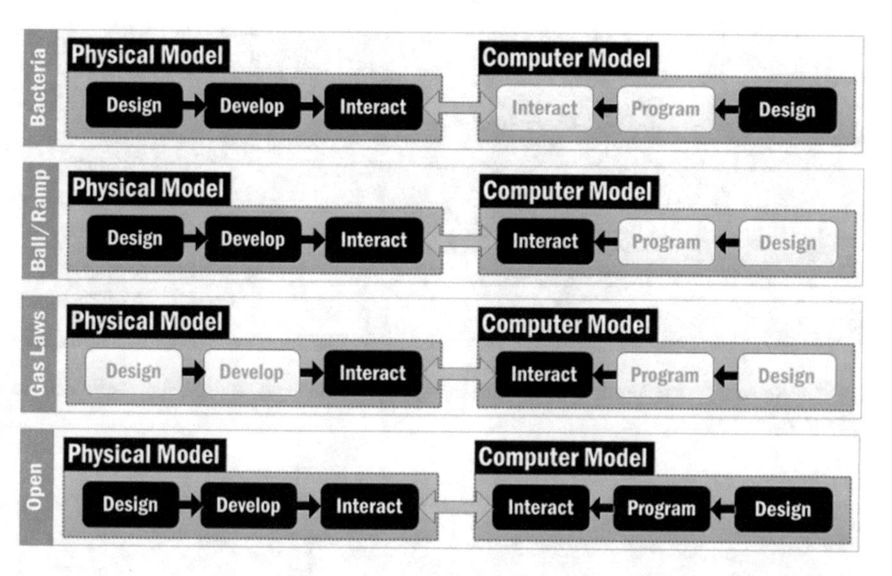

Fig. 12 A visual summary of different bifocal modeling activities showing the blocks used in the four studies (*in black*)

and "open-ended"). I employed a design-based research framework in which implementation, research, data analysis, and redesign are closely connected (Confrey 2005; Edelson 2002). The four modes of implementation are summarized in Fig. 12. The black rectangles indicate the design elements employed. (Note that the "Research" element was not included in the figure; rather, all of the activities began with the students engaging in research).

5.2 Bacterial Growth Study (Biology)

The first study was conducted with four female high-school students. The topic of the workshop was bacterial growth. The study lasted for a total of approximately 5 hours, distributed across three afternoon sessions. The students' first task was to grow real bacteria using Petri dishes and a customizable time-lapse camera setup (Fig. 13). Students first conducted content research on the bacterial growth curve in groups of two. They were also shown a short video about bacterial growth in a Petri dish. In the final task, the authors conducted a variation on a "paper modeling" activity (Blikstein 2009; Blikstein and Wilensky 2009), in which students collectively designed an agent-based model of bacterial growth on a whiteboard (Fig. 14.) In the paper modeling activity, small groups of students created a detailed block diagram of all of the aspects of a computer model and simulated a few runs (without a computer). Students defined all the necessary variables for the model, the required agents, their respective properties and rules, and their possible

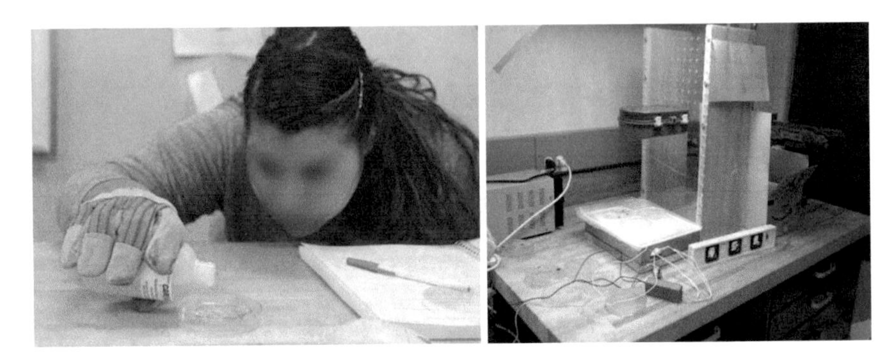

Fig. 13 Students preparing their petri dishes to grow bacteria (*left*), and the time-lapse camera apparatus (*right*)

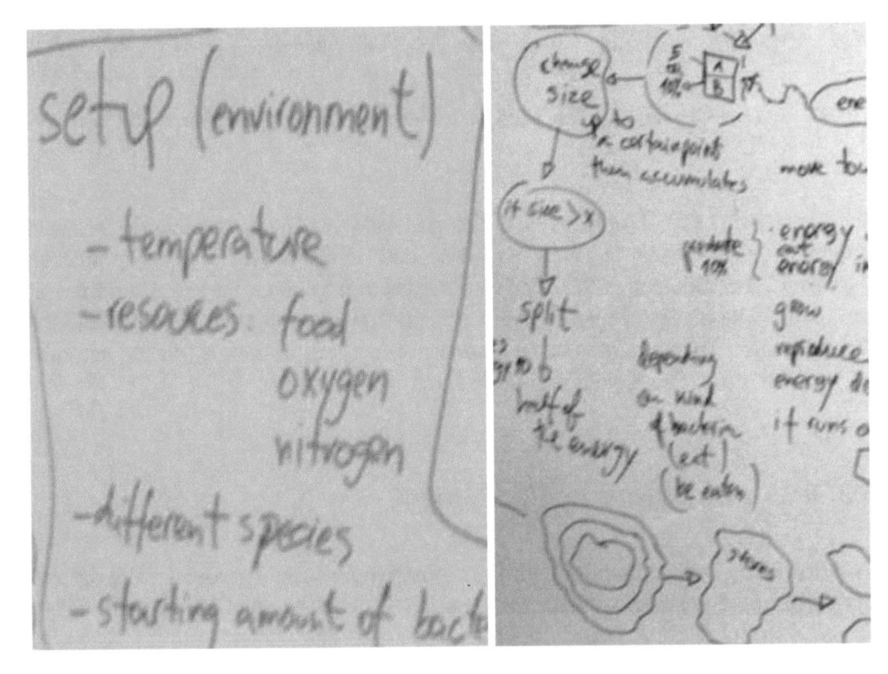

Fig. 14 Some small snapshots of the "whiteboard modeling" activity, in which students collaboratively created a detailed block diagram of the code for an agent-based model of bacterial growth

interactions. The facilitator helped students translate their ideas into the proper "code blocks," but once the students understood the general idea, the facilitator merely documented students' ideas on the whiteboard.

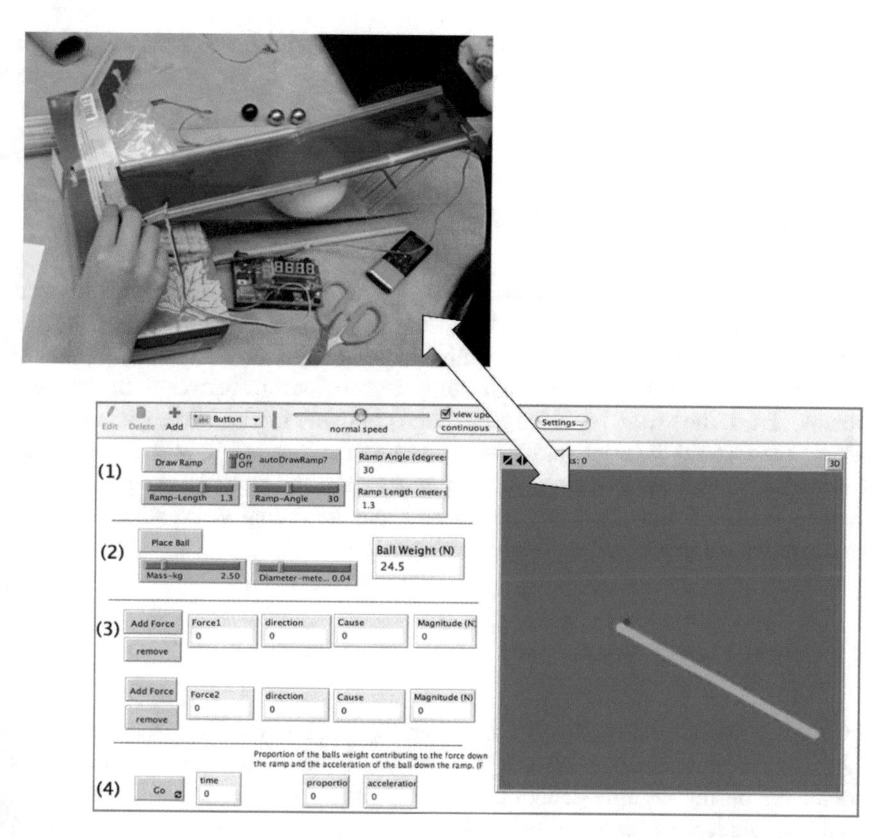

Fig. 15 One of the projects in the "ball and ramp" study, in which students were given kits of parts from which to build their ramps (*above*). The students then compared the final product with a computer model programmed in NetLogo, which allowed the students to change several of its parameters

5.3 Classical Mechanics and Newton's Laws Study (Physics)

In the second study, in a workshop that lasted 6 hours, 11 high-school students (4 females and 7 males) studied classical mechanics and Newton's laws through an investigation of the time required for a ball to travel down a ramp. The students were first asked to build a physical model to investigate the variables affecting the time needed for the ball's descent. In groups of two or three, they constructed their ramps and attached visible light and infrared sensors to detect the position of the moving ball (Fig. 15). In this study, students did not *create* a computer model; they only *interacted with* a model supplied by the facilitators. This model enabled the students to simulate a virtual ball rolling down a ramp and to vary parameters such as ball mass and ramp angle. Finally, the staff led the students in comparisons of the behavior of the virtual and real models—observing, in particular, that the virtual model predicted that a ball's mass would have no effect, whereas the empirical data suggested otherwise.

5.4 Gas Laws Study (Chemistry)

In the third study, over a single 6-hour period, 11 high-school students investigated the relationship among volume and pressure in a closed container. The students were asked to interact with a preassembled syringe system with built-in pressure sensors, and they collected data indicating the relationship between pressure and volume. Next, they were provided a NetLogo computer model of the gas laws with which they could interact and observe representations of how gas molecules behaved in a container with a moving piston. As the volume of the physical syringe changed, the computer model adjusted accordingly, so students could compare values of virtual and real-world pressure.

5.5 Open-Ended Study

In this final study, students freely chose their topic of interest. The group contained 12 freshmen high-school students (11 male, one female), and the experiment was conducted during an after-school program. Over a 3-day workshop (24 hours in total), the students were asked to build a physical model, create a computer model in NetLogo, and write a report comparing the two. Figure 16 shows a typical project, a model investigating liquid diffusion, with the physical model on the left and the computer model on the right.

6 Case Studies: Data and Discussion

This section will begin with a narrative and comments on several episodes centered on the perceived and hypothesized affordances of bifocal modeling. Based on the literature review and previous work (Blikstein 2012; Blikstein and Wilensky 2007), we defined the following recurrent learning outcomes of these activities, which we now illustrate with commented descriptions of episodes and transcriptions of the studies described in the previous section:

a. Students' adoption of more sophisticated strategies to resolve model mismatches.
b. Learners' convergence on a subset of variables relevant to a given phenomenon after evaluating all possible variables.
c. Students' critical evaluation of the assumptions of the models and their validity.
d. Translation between micro and macro perspectives.

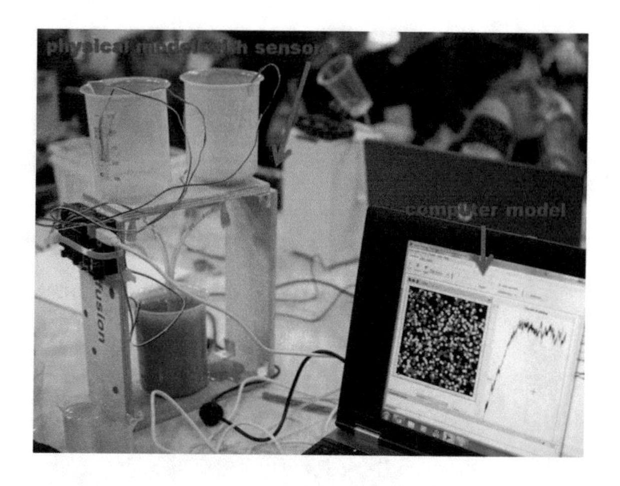

Fig. 16 One of the projects in the "open-ended" workshop. Note that the physical model of diffusion with sensors (on the *left*) and the computer model in NetLogo (on the *right*) are connected in real-time. The colored dots represent molecules of two different liquids mixing, and the graph shows the level of mixing between them over time

6.1 Iteratively Improving the Virtual Model to Resolve Mismatch

The study concerning bacterial growth comprised creating a "white board" model; "running" it to envision how bacteria would multiply according to the model; a comparison of the modeled results to the growth curve that the students obtained from the physical data; and a resolution of the perceived differences between the two processes through the addition to the virtual model of appropriate rules and variables. The group repeated this process a total of four times in the 1.5 hours of the session and developed an increasingly accurate model in the process. In Fig. 17, we present a chronological list of the additions made by the students.

For example, at one point after "running" the virtual model, a student observed that the growth curve was increasing exponentially from the start. She noted that this finding was incorrect because the real growth curve had an initial flat "lag phase" before beginning to grow. After a moment's reflection, she remembered that this outcome had occurred because real bacteria exhibit an initial phase subsequent to inoculation in which they settle into their new environment before multiplying. They had observed this both in their background research, and in time-lapse movies that they generated or watched during the activity. She said, "We need to make a rule that it takes time before the bacteria grow." Another student added that this rule would have to include an exception for the maturation period for subsequent bacteria, because it would apply only to the first bacteria in the dish. After further discussion on how to code the lag phase into their system, the students generated the following rule: "If a bacterium is in the first generation, it has to wait two time steps before reproducing." Running the model a second time, the students could see from the resulting curve that they had successfully generated the lag phase. The students conducted similar processes to add the other relevant variables to their model.

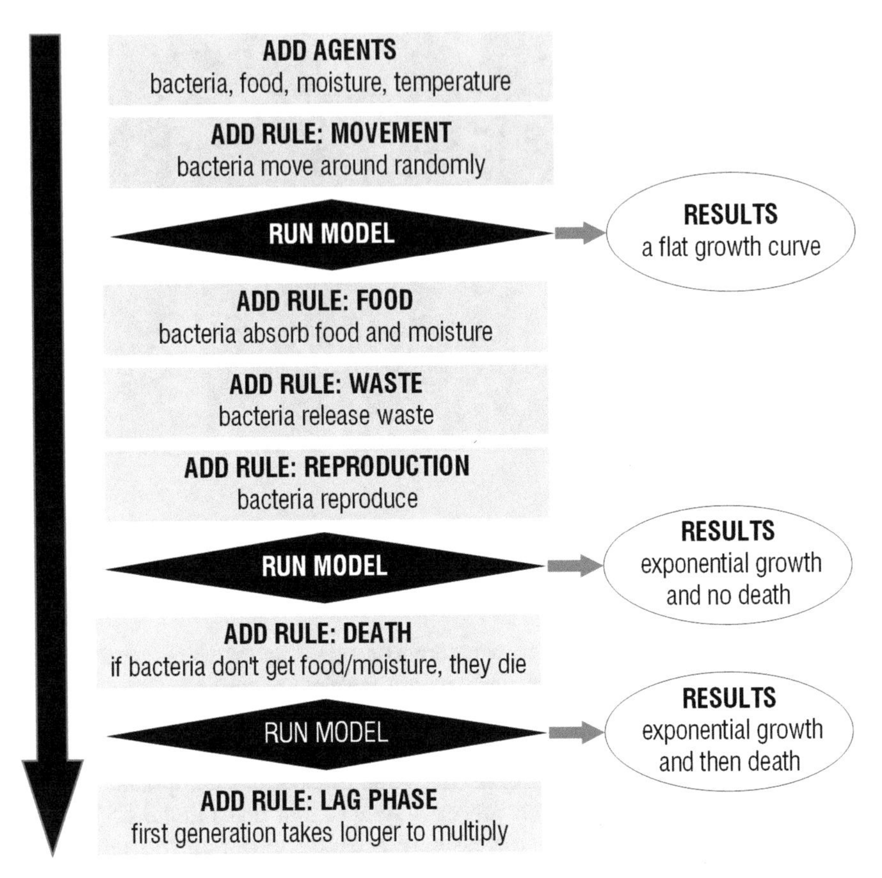

Fig. 17 A chronological list of the additions made by the students to the model and the instances in which they ran it

6.2 Selecting and Converging on Appropriate Variables

When the students searched the Internet for information about bacteria, they collected and recorded a great deal of information that was not necessary for the modeling task that they had been given. For example, some students noted that bacteria are prokaryotes, eat many types of human food, and live in a very broad range of conditions. However, throughout the many iterations of whiteboard modeling, students increasingly privileged variables that were necessary to define the shape of the growth curve of real-world bacteria: food, moisture, waste, and bacterial health. Global variables, such as temperature and oxygen, affect bacterial growth, but the dynamics of the curve assume that these global variables are constant or that the variations are too small to make appreciable differences. That the students excluded these variables without prompting suggests that in this case they implicitly understood the concept of controlled variables. Students also made decisions regarding the level of specificity at which to describe the variables. Beth, for example, took notes about multiple types of bacterial nutrients during her research, but agreed with the group about the need to represent food as a single variable of just one type. When asked for her rationale, she replied that she did not "need to be that specific for this model." The idea of "convergence"—students developing a deeper understanding of the relative importance of different variables and parameters, and focusing on just the most important ones—appears to be one of the core learning outcomes of the study, and of bifocal modeling activities in general. Creating and measuring a physical model forced them to make choices about the data; and by having to select what to measure, what sensors to use, and what to include in their models, students had to problematize and discuss the relative importance of each of the variables and parameters, and understood that certain variables could, for practical purposes, be treated as constants.

6.3 Critically Evaluating Assumptions of Models

In addition to learning about the variables relevant to bacterial growth, students also reflected on the underlying assumptions of the models: in this case, the representations of space and time. Space is represented in NetLogo as a grid of square "patches," units that represent variables such as location or food concentration. This patchwork representation of space was explained to the students at the start of the whiteboard modeling session, but during this introduction, it was only presented as a way to quantify environmental variables such as the availability of food. However, as the session progressed, the students noticed that in their virtual models the bacteria were distributed randomly across the surface and filled the entire surface uniformly as they multiplied. In contrast, the real bacteria formed small circular spots. Seeking to explain the difference, the students engaged in a discussion of how far bacteria can move, which quickly led to questions about the unit of measure and its magnitude as well as the size of the grid system as a whole. As Megan put it:

> This square could be a whole dish, or it could be just a tiny spot in the real Petri dish… if we were looking through a microscope, zooming in, they [the bacteria] will move much more.

At the end of their discussion, the students decided that it was up to them to define the size of the virtual world that they designed, provided that they kept everything in proportion. In previous work (Blikstein and Wilensky 2007), we have observed students reconsidering space in their computer models and engaging in sophisticated discussions about sampling and the size of molecules relative to that of the containers. Students reached similar conclusions, noting that the arbitrary size of the model "seemed" unrealistic, although the model remained useful for their projects, which demonstrated the further understanding that models need not be perfect representations of the physical realities they elucidate.

Time in NetLogo and in the whiteboard model was represented as a series of discrete steps called "ticks." In discussing the proper time delay for the lag phase, one student realized that the group had not agreed on the relationship between ticks and real-time. She asked:

> Do bacteria get food and moisture each minute? Each hour? Each day? Right now we are just doing this with ticks… how can we translate the ticks into real time?

At the end of another discussion about the time scales of bacterial growth in the real world and in the NetLogo model, the students decided that if real-world bacteria multiply every 20 minutes, the same value should apply for the virtual world—so how should they deal with the virtual "ticks," and should the computer model run at the same "speed" as the physical phenomenon? Although the students did not entirely resolve these questions regarding the representation of time and space within their model, their questions indicated that they had achieved a more sophisticated awareness of the assumptions made in the process of creating a theoretical model. This realization is at the heart of critical scientific thinking, and students understood that even seemingly unrealistic assumptions or abstractions can be "good enough" for modeling purposes. At the same time, they engaged in refined discussions about how their own assumptions about time and space could change depending on the focus of the investigation, so a "good enough" model would depend on the goals of the scientist.

6.4 Meta-Modeling

Meta-modeling knowledge entails the understanding of the purpose and benefits of scientific models as well as an acknowledgment of their limitations. My initial hypothesis was that students would change their epistemological stance regarding the credibility of computer models after they developed their own virtual model of a phenomenon and compared its results with the behavior of the physical one. Over the course of the studies, this hypothesis was confirmed. On the mid-test of one of the studies, although 77 % of the 13 participants answered that a virtual model of a phenomenon is not the same as the phenomenon itself, their explanations of the differences between the models were generic and unspecific. (The mid-test was administered after Internet research but prior to model building and comparison.) Asked on the mid-test if they thought the virtual model was similar to what was happening to bacteria in real life, most students responded that the virtual model was just an instrument used to "get closer" to real phenomena. However, by the time of the post-test, the students had developed more specific ideas about the similarities and differences between the virtual model and the real phenomenon. As we see in the following excerpts from the post-test, the students identified more accurately the limitations to the virtual models that emerged in comparisons with physical ones:

Student 1: "[In the physical experiment] we get to see factors that were not used in the simulation."
Student 2: "…there might be factors that we don't know of, that were not included [in the computer model], and are important to bacterial life."
Student 3: "It's more precise [the real experiment] than an actual computer, […] observing things that would actually happen in reality and not in the computer program".
Student 4: "A computer model can make a good estimation, but for real data, one needs to study the actual bacteria."
Student 5: "In the physical model we get to see the actual thing, […] we get to see factors that were not used in the simulation."

We also observed that in their post-test, students had much more specific explanations regarding their models' limitations. For example, some students conjectured that the relatively straight, step-shaped lines they observed in their computational model of bacterial growth resulted from the "lack of randomness" in the model.

Student 6: "We were given a "step" pattern because all bacteria were gathering resources at the same rate, within a species."
Student 7: "The virtual model is not completely the same as the real bacteria, because they all reproduce at once."
Student 8: "…I don't think the virtual model is similar to what is happening in real life, since ours [the virtual model] did not incorporate all of the actions of real bacteria, such as colonies…I believe that one reason for these differences is that the bacteria in our simulation were not very random."

Additionally, students' answers demonstrated that they better understood the added value of investigating both a physical phenomenon and a virtual model. One student, for example, admitted the limitation of virtual models, but emphasized the importance of evaluating such models in order to gain critical perspective:

Student 9: "...However, now that I have experienced the inaccuracies that virtual models produce, I will be able to comprehend in my head that those models are not completely accurate. I think knowing the differences between a virtual model and a real model is important, because it makes people use a skeptical perspective, that will help them understand there is more to a concept than meets the eye; furthermore, it shows them that they should not just accept everything their teacher throws at them, but that they should independently dig deeper into the subject and run experiments, so they can gain a deeper understanding."

6.5 Various Approaches to Resolving Mismatches

One final aspect within this process of developing more complex inquiries into the relationship between virtual and real models is an examination of how students resolve the mismatches. The results of these four pilot studies suggest that changes in the design of a bifocal modeling activity tend to change student approaches to the resolution of such model mismatches (see Fig. 12). In this section, in place of an analysis of particular aspects of the bacterial growth study, I shift the focus to the differences between the implementation models to investigate how the presence of each of the components of bifocal modeling alters student engagement and learning. For example, I was interested in determining how the presence of a longer model-building phase would influence how students conducted the model comparison.

In the first study (bacterial growth), I designed the activities so that the students focused chiefly on creating a conceptual, agent-based model that would match the behaviors to the real-time-lapse video data. Even though they did *not* write the code, the students actively constructed a model on the whiteboard. Overall, the group's method involved "running" a few steps of their whiteboard virtual model to predict how the bacteria grew, followed by a comparison of the results with those obtained when they approximated the growth curve based on the physical model. Then, they worked to resolve the perceived differences between the two by adding to the virtual model rules and variables. In general, as discussed in the following paragraphs, I observed that the students in this study were more engaged in investigating the behavior of the phenomenon due to the intensity of the model-building activity.

In the second study, which focused on the "ball on a ramp" model, I made different design decisions based on the size of the group (11) and the available time (6 hours in total). The students partially designed and developed their own physical model, but I provided the virtual model to them so that the emphasis was simply on their interactions with the premade virtual model. As the students proceeded with the activity, they became more critical about *their own observations*, rather than questioning the *premade model and its assumptions*. For example, in the physical experiment, heavier balls appeared to roll more rapidly than lighter ones, possibly as a result of slipping and the decreased effects of air resistance and friction. When the virtual model appeared to refute the idea that heavier balls would roll faster (which they had observed in their physical experiment), the students were surprised, and they ended up *trusting the provided computer model over their own observations*. For example, when asked which model better represented the scientific phenomenon, one student responded, "…the virtual model! It is computerized and can calculate the time. It is a computer, so we trust it!" In general, the students never questioned whether the computer model could possibly be wrong; they just assumed that because the models were (supposedly) created by experts, and ran on computers, they would be "right." Even in the face of consistently contradictory evidence from their physical experiments, they never distrusted the computer model.

In the "gas laws" study, the emphasis was again on interactions with premade models rather than on creating them. In this case, the students did not build either of the models: they were given both a physical and a virtual model. The students were tasked with collecting real-time data from the pressure sensor and comparing them with the "ideal" data generated by the premade virtual model. The students critically evaluated the scientific experiment with sensors and offered many ideas about how to improve it, attempting to make sense of the discrepancies between the two datasets. For example, when asked about the discrepancies between the pressure/volume graph generated by their own physical measurements and the one generated

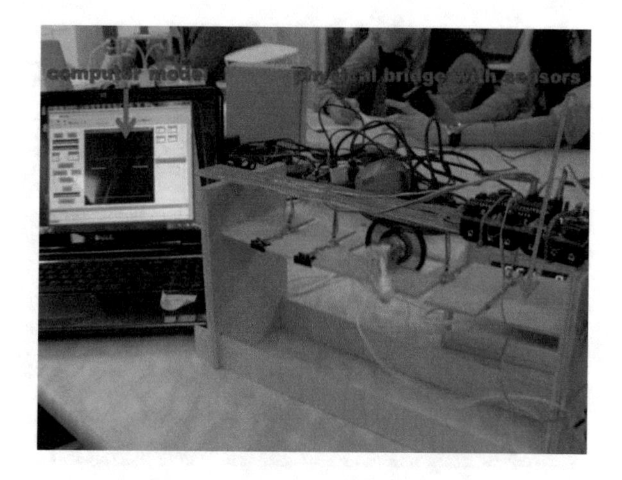

Fig. 18 A bifocal model for bridge harmonics, with the computer model (*left*), and the physical bridge with a wave generator and position sensors (*right*)

by the virtual model, the students suggested potential causes, including the accuracy of the sensor and data-logging software. Again, they never indicated that the virtual models might have been wrong. For example, during an attempt to critique their physical model, the students only looked for technical issues such as sensor limitations or the limited compressibility of the air inside the syringe. Judging by their remarks during the activity, as in the previous study, the students did not even consider that the computer models could have been wrong or even imperfect. They took the accuracy of the models for granted and simply critiqued their own data, even after having repeated the experiment many times with similar results.

The design of the last study was very different. The students not only had to generate their own idea for a project but to design both the physical and the virtual models from scratch. During the construction of the physical model, many dissimilarities became apparent. For example, one group decided to build a bridge to investigate how much vibration it would withstand, together with an accompanying computer model (Fig. 18).

During the construction of the bridge, the facilitators encouraged the students to run systematic experiments with different frequencies of vibration, comparing the results with those achieved using the canonical formula. Even though they were able to construct a plot that approximated the canonical formula, it was not a perfect match due to an intrinsic error in the empirical measures. At first, the students felt that their model was incorrect because it did not fit the theoretical curve perfectly, which was a significant disappointment for them. However, through multiple cycles of measurement and system rebuilding, the students realized that they always obtained "messy" results and that reaching their initial goal of a perfect fit was not just a matter of "getting it right"; it was, in fact, impossible. The students tried to carefully control the voltage source (which, in turn, controlled the vibration frequency) as well as the location of the sensors and the magnetic pieces, but they ultimately realized that they could not make the models exactly match, even after multiple changes to both models.

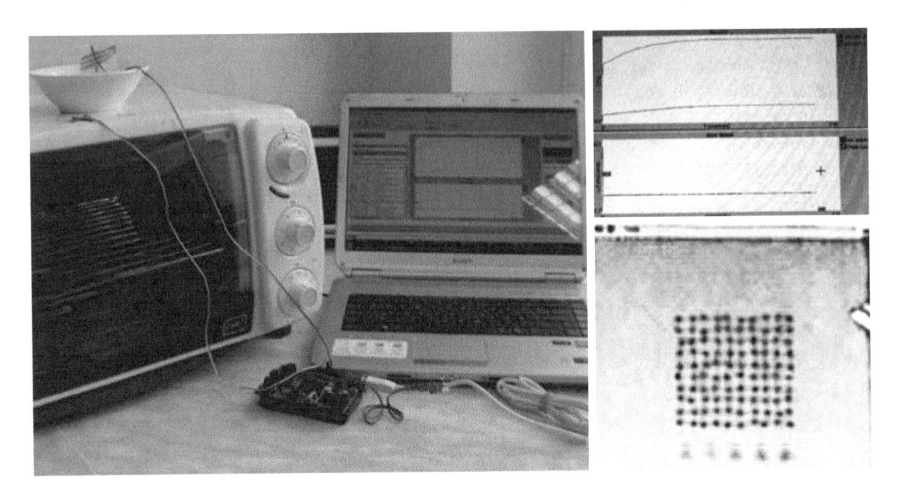

Fig. 19 The melting ice cube model. The toaster oven which was melting the ice and the physical apparatus are on the *left*. Students created two computer models connected to the sensors: a macroscopic (*top right*) and a microscopic one (*bottom right*)

Although this process required significant resources in terms of facilitation and time, I hypothesize that, by constructing both the computer and the physical models and by "glass boxing" the entire process, the students obtained a much more comprehensive set of tools that they were able to use to make sense of the differences and mismatches they observed, as evidenced by the sophistication of their attempts to make the two systems consistent with one another.

One last example from this implementation model sheds some light on the usefulness of full-fledged bifocal modeling activities in which students build both physical and computer models. A group of three students became interested in building a model that would explain how an ice cube melted (Fig. 19). They froze temperature sensors inside an ice cube and placed it on top of a heat source (a toaster oven), which was instrumented with the same type of sensors. The students carefully tracked the temperature of the surface of the heat source and that of the ice cube, and they examined the two plots to generate a tentative equation and curve that related them (see top right plots in Fig. 19). However, after a few hours had passed, they realized that their equation was not a mechanistic model of what was happening with the ice cube—it merely depicted a numerical relationship. Even after several refinements, the students were uneasy about their project because they wanted to go deeper into the melting mechanism to understand *why* the ice was melting rather than simply *how fast* the process was proceeding. They then embarked on a much more ambitious project: the creation of a NetLogo model to describe what was happening *microscopically* inside the ice cube. The students programmed an atomic-level, agent-based model (see the bottom right in Fig. 19) in which the atoms were connected by springs and were allowed to vibrate and eventually break off, and in which "heat waves" (represented by small triangles at the bottom) would collide with the atoms and increase their vibrational energy. In this way, the students arrived at a remarkably complex and accurate mechanistic model of melting by pursuing several iterations of model building and comparison, utilizing physical and computer models. Ultimately, the students realized that the equations that they were given at school (which they had been using as a reference pattern in the initial part of their project) did not reveal the mechanism of the phenomenon, the discovery of which was their main project goal. I would suggest that the bifocal modeling activity gave the technological tools and epistemic framework that led them to the insight into the imperfections of the models and inspired them to attempt increasingly complex endeavors. Their goal, which originally was simply to generate hypotheses about "blackboxed" numerical relationships, became the acquisition of deep insight into the mechanisms behind these relationships.

These last two projects illustrate that despite the significant resource requirements of time, materials, and facilitation, the full-fledged method of bifocal modeling does lead students into uncharted epistemological territory in which they build and compare multiple models using multiple representational media to make sense of micro- and macro-level mechanisms and real-to-virtual reciprocities. These are results that the previous studies could not elicit. However, further research must determine whether there are less resource-intensive ways to achieve similar goals.

7 Conclusion

I started this chapter with the story of Faraday and his "motor kit" to illustrate that real scientific inquiry has been deeply connected with scientific instruments and physical devices. The making of science has always been a complex dialog between the real and the ideal worlds, but this process is absent from science classrooms and labs. Rather, class and lab are compartmentalized and divorced, and the discrepancies between these two worlds are taken as impediments to learning. The bifocal modeling framework hypothesizes that the opposite is true: diving into these discrepancies could actually be quite generative for learners.

In this chapter I offered initial data and insights into the learning outcomes of this new epistemic game (Collins and Ferguson 1993), in which students build and compare physical and virtual models in real-time. In particular, I compared many different implementation models to highlight what was gained with each new activity design element. The first study was a "proof of concept" that established the feasibility of the exercise and indicated some of the learning gains of students engaging in bifocal modeling, which was followed by a comparison of three implementation models. In conducting these four studies, I observed at least one overall pattern across different implementations: resources that are given to students have a different perceived value than do resources that the students build themselves. "Given" resources are nearly always trusted without question because they are perceived as generated by experts and formalized by computers (which "cannot be wrong"). Thus, the only way for a student-constructed resource to have in their eyes the same validity as one provided by an authority would be for it to perfectly match the latter. This asymmetry may be counterproductive for students, especially when the resource in question is an idealized theoretical model, which the physical data will never perfectly match. When manipulating models in this scenario, students tend to distrust their own data, which defeats the purpose of building physical models and collecting sensor data. Since this is how students work with empirical data in most school science labs, the implication of this finding is that "cookbook" labs might be achieving the opposite of their intended outcome: instead of empowering students as scientists, they may make learners misconstrue the relationships between real and idealized systems, and mistrust their own data and empirical results.

A second pattern that can be observed in the findings of this research is that in studies without computational model building (the gas laws and ball/ramp studies), the students' level of sophistication in comparisons of real and virtual models was lower, and their epistemological stance was even more rigid: they would critique their work and their own observations, but they would rarely question the computational models that were provided to them.

I also observed that the benefits of model-building manifested even if the students were not coding but were instead creating "whiteboard models" with paper or a whiteboard as a "computational surface" that allowed them to enact the conjectured agent rules. Within this context, the use of the physical model as a reference pattern in the creation and refinement of the virtual model was generally effective. When the students were instructed to design a virtual model that recreated the bacterial growth curve, they used their previously learned knowledge about the curve and the physical appearance of the bacteria as a reference pattern that indicated what their model should generate. When the model data did not match the observed data, they returned to the model and made changes.

One key conclusion that we can draw from these four implementations is that the full model-building experience (both physical and virtual) was indeed a richer learning experience compared to the other implementation models, especially for the students who built several versions and types of computer models. Further research is needed to determine the degree to which model building can be abbreviated without a significant decrease in learning gains and engagement. It seems that with the proper facilitation and careful choices regarding which modeling phase to abbreviate (e.g., bacteria study), relatively rich learning outcomes could be achieved in a dramatically reduced time frame. Nevertheless, these studies have shown that model building, rather than simple access to sensors and tangible learning tools, was the determining factor in the students' deep engagement with the phenomenon. The data also suggest that students' explorations into real and ideal systems, their affordances, boundaries, and limitations, if properly facilitated, could constitute a generative and rich space for learning *both* the content and the epistemology of science. Rather than dismissing the real-world as too messy and cognitively demanding, or virtual systems as too perfect, we should let students playfully explore, connect, and learn from their incongruities and contradictions.

Acknowledgments Thanks to Shima Salehi, Tamar Furhmann, Bertrand Schneider, and Daniel Greene for their work in the research and earlier versions of this chapter, and Elayne Weissler-Martello for her proofreading work. Special thanks to the students who created the bifocal models shown in this article. This work is funded by the NSF CAREER Award #1055130, NSF DRL 1020101, the Stanford MediaX program, and the Stanford Lemann Center for Educational Entrepreneurship and Innovation in Brazil.

References

Baird D (2006) Thing knowledge: a philosophy of scientific instruments. University of California Press, Berkeley

Birchfield D, Megowan-Romanowicz C (2009) Earth science learning in SMALLab: a design experiment for mixed reality. Int J Comput-Support Collab Learn 4(4):403–421

Blake C, Scanlon E (2007) Reconsidering simulations in science education at a distance: features of effective use. J Comput Assist Learn 23(6):491–502

Blikstein P (2009) An atom is known by the company it keeps: content, representation and pedagogy within the epistemic revolution of the complexity sciences. (PhD. dissertation), Northwestern University, Evanston, IL

Blikstein P (2010) Connecting the science classroom and tangible interfaces: the bifocal modeling framework. In: Proceedings of the 9th International Conference of the Learning Sciences, Chicago, IL, pp 128–130

Blikstein P (2012) Bifocal modeling: a study on the learning outcomes of comparing physical and computational models linked in real time. In: Proceedings of the 14th ACM International Conference on Multimodal Interaction, Los Angeles, CA, pp 257–264

Blikstein P, Wilensky U (2006) 'Hybrid modeling': advanced scientific investigations linking computer models and real-world sensing (an interactive poster). In: Proceedings of the 7th International Conference on Learning Sciences, Bloomington, IN, pp 890–891

Blikstein P, Wilensky U (2007) Bifocal modeling: a framework for combining computer modeling, robotics and real-world sensing. Paper presented at the annual meeting of the American Research Education Association, Chicago, IL

Blikstein P, Wilensky U (2009) An atom is known by the company it keeps: a constructionist learning environment for materials science using agent-based modeling. Int J Comput Math Learn 14(2):81–119

Blikstein P, Wilensky U (2010) MaterialSim: a constructionist agent-based modeling approach to engineering education. In: Jacobson MJ, Reimann P (eds) Designs for learning environments of the future: international perspectives from the learning sciences. Springer, New York, pp 17–60

Blikstein P, Fuhrmann T, Greene D, Salehi S (2012) Bifocal modeling: mixing real and virtual labs for advanced science learning. In: Proceedings of the 11th International Conference on Interaction Design and Children, Bremen, Germany, pp 296–299

Bryson J, Ando Y, Lehmann H (2007) Agent-based modelling as scientific method: a case study analyzing primate social behaviour. Philos Trans R Soc B: Biol Sci 362(1485):1685

Cagnacci F, Boitani L, Powell RA, Boyce MS (2010) Animal ecology meets GPS-based radiotelemetry: a perfect storm of opportunities and challenges. Philos Trans R Soc B: Biol Sci 365(1550):2157–2162

Chen H-T, Tien M-C, Chen Y-W, Tsai W-J, Lee S-Y (2009) Physics-based ball tracking and 3D trajectory reconstruction with applications to shooting location estimation in basketball video. J Vis Commun Image Represent 20(3):204–216

Collins A, Ferguson W (1993) Epistemic forms and epistemic games: structures and strategies to guide inquiry. Educ Psychol 28(1):25–42

Confrey J (2005) The evolution of design studies as methodology. In: Sawyer K (ed) The Cambridge handbook of the learning sciences. Cambridge University Press, Cambridge, pp 135–151

de Jong T, Linn MC, Zacharia ZC (2013) Physical and virtual laboratories in science and engineering education. Science 340(6130):305–308

DeBoer GE (2000) Scientific literacy: another look at its historical and contemporary meanings and its relationship to science education reform. J Res Sci Teach 37(6):582–601

Duschl RA, Grandy RE (2008) Teaching scientific inquiry: recommendations for research and implementation. Sense Publishers, The Netherlands

Edelson DC (2002) Design research: what we learn when we engage in design. J Learn Sci 11(1):105–121

Finkelstein N, Adams W, Keller C, Kohl P, Perkins K, Podolefsky N, Reid S, LeMaster R (2005) When learning about the real world is better done virtually: a study of substituting computer simulations for laboratory equipment. Phys Rev Spec Top-Phys Educ Res 1(1):010103

Gire E, Carmichael A, Chini JJ, Rouinfar A, Rebello S, Smith G, Puntambekar S (2010) The effects of physical and virtual manipulatives on students' conceptual learning about pulleys. In: Proceedings of the 9th International Conference of the Learning Sciences, Chicago, IL, (pp. 937–943)

Grosslight L, Unger C, Jay E, Smith C (1991) Understanding models and their use in science: conceptions of middle and high school students and experts. J Res Sci Teach 28(9):799–822

Hmelo-Silver CE, Marathe S, Liu L (2007) Fish swim, rocks sit, and lungs breathe: expert-novice understanding of complex systems. J Learn Sci 16(3):307–331

Hodson D (1996) Laboratory work as scientific method: three decades of confusion and distortion. J Curriculum Stud 28(2):115–135

Hodson D (1998) Science fiction: the continuing misrepresentation of science in the school curriculum. Curriculum Stud 6(2):191–216

Hofer BK, Pintrich PR (1997) The development of epistemological theories: beliefs about knowledge and knowing and their relation to learning. Rev Educ Res 67(1):88–140

Hutchins E (1995) How a cockpit remembers its speed. Cogn Sci 19(3):265–288

Ingham AM, Gilbert JK (1991) The use of analogue models by students of chemistry at higher education level. Int J Sci Educ 13(2):193–202

Jaakkola T, Nurmi S (2008) Fostering elementary school students' understanding of simple electricity by combining simulation and laboratory activities. J Comput Assist Learn 24(4):271–283

Johnson KS, Needoba JA, Riser SC, Showers WJ (2007) Chemical sensor networks for the aquatic environment. Chem Rev 107(2):623–640

Johnson-Glenberg MC, Birchfield D, Megowan-Romanowicz C, Tolentino L, Martinez C (2009) Embodied games, next gen interfaces, and assessment of high school physics. Int J Learn Media 1(2)

Kirschner PA, Sweller J, Clark RE (2006) Why minimal guidance during instruction does not work: an analysis of the failure of constructivist, discovery, problem-based, experiential, and inquiry-based teaching. Educ Psychol 41(2):75–86

Klahr D, Triona LM, Williams C (2007) Hands on what? The relative effectiveness of physical versus virtual materials in an engineering design project by middle school children. J Res Sci Teach 44(1):183–203

Latour B, Woolgar S (1979) Laboratory life: the social construction of scientific facts. Princeton University Press, Princeton

Lehrer R, Schauble L (2006) Cultivating model-based reasoning in science education. In: Sawyer K (ed) Cambridge handbook of the learning sciences. Cambridge, Cambridge University Press, pp 371–388

Lesh R, Hoover M, Hole B, Kelly A, Post T (2000) Principles for developing thought-revealing activities for students and teachers. In: Kelly A, Lesh R (eds) Handbook of research design in mathematics and science education. Lawrence Erlbaum, Mahwah, pp 591–645

Levy ST, Wilensky U (2008) Inventing a "mid-level" to make ends meet: reasoning through the levels of complexity. Cogn Instr 26(1):1–47

Liu X (2006) Effects of combined hands-on laboratory and computer modeling on student learning of gas laws: a quasi-experimental study. J Sci Educ Technol 15(1):89–100

Milgram P, Kishino F (1994) A taxonomy of mixed reality visual displays. IEICE Trans Inf Syst 77(12):1321–1329

National Research Council (2012) A framework for k-12 science education: practices, crosscutting concepts, and core ideas. The National Academies Press, Washington

Nersessian NJ (2005) Interpreting scientific and engineering practices: integrating the cognitive, social, and cultural dimensions. In: Gorman M, Tweney RD, Gooding D, Kincannon A (eds) Scientific and technological thinking. Lawrence Erlbaum, Hillsdale, pp 17–56

NGSS Lead States (2013) Next generation science standards: for states, by states. The National Academies Press, Washington

Norman D (1991) Cognitive artifacts. In: Carroll JM (ed) Designing interaction: psychology at the human-computer interface. Cambridge University Press, Cambridge, pp 17–38

NRC (1996) National science education standards. National Academy Press, Washington (National Committee on Science Education Standards and Assessment)

Olympiou G, Zacharia ZC (2012) Blending physical and virtual manipulatives: an effort to improve students' conceptual understanding through science laboratory experimentation. Sci Educ 96(1):21–47

Papert S (1980) Mindstorms: children, computers, and powerful ideas. Basic Books, New York

PASCO Scientific. From www.pasco.com

Perkins K, Adams W, Dubson M, Finkelstein N, Reid S, Wieman C, LeMaster R (2006) PhET: Interactive simulations for teaching and learning physics. Phys Teach 44:18

Radder H (2003) The philosophy of scientific experimentation. University of Pittsburgh Press, Pittsburgh

Resnick M, Wilensky U (1998) Diving into complexity: developing probabilistic decentralized thinking through role-playing activities. J Learn Sci 7(2):153–171

Roth K, Garnier H (2006) What science teaching looks like: an international perspective. Educ Leadersh 64(4):16

Rudolph JL (2005) Epistemology for the masses: the origins of "the scientific method" in American schools. Hist Educ Quart 45(3):341–376

Schwarz CV, White BY (2005) Metamodeling knowledge: developing students' understanding of scientific modeling. Cogn Instr 23(2):165–205

Sipitakiat A, Blikstein P (2010) Robotics and environmental sensing for low-income populations: design principles, impact, technology, and results. In: Proceedings of the 9th International Conference of the Learning Sciences, Chicago, IL, pp 447–448

Sipitakiat A, Blikstein P, Cavallo DP (2004) GoGo board: augmenting programmable bricks for economically challenged audiences. In: Proceedings of the 6th International Conference of the Learning Sciences, Santa Monica, CA, pp 481–488

Smith GW, Puntambekar S (2010) Examining the combination of physical and virtual experiments in an inquiry science classroom. Paper presented at the Conference on Computer Based Learning in Science, Warsaw, Poland

Stewart J, Cartier JL, Passmore CM (2005) Developing understanding through model-based inquiry. In: Donovan MS, Bransford JD (eds) How students learn: science in the classroom. National Academy Press, Washington, pp 515–565

Thagard P (2007) Coherence, truth, and the development of scientific knowledge. Philos Sci 74(1):28–47

Tinker R (1991) History of probeware. http://www.concord.org/ccprobeware/probeware_history.pdf

Tinker R (ed) (1996) Microcomputer-based labs: educational research and standards. Springer, Berlin

Treagust DF, Chittleborough GD, Mamiala TL (2002) Students' understanding of the role of scientific models in learning science. Int J Sci Educ 24:357–368

Triona LM, Klahr D (2003) Point and click or grab and heft: comparing the influence of physical and virtual instructional materials on elementary school students' ability to design experiments. Cogn Instr 21(2):149–173

van der Meij J, de Jong T (2006) Supporting students' learning with multiple representations in a dynamic simulation-based learning environment. Learn Instr 16(3):199–212

Vernier Software & Technology. From http://www.vernier.com/

Wark T, Crossman C, Hu W, Guo Y, Valencia P, Sikka P, Corke P, Lee C, Henshall J, Prayaga K, O'Grady J, Reed M, Fisher A (2007) The design and evaluation of a mobile sensor/actuator network for autonomous animal control. In: Proceedings of the 6th International Conference on Information Processing in Sensor Networks, Cambridge, MA

Wilensky U (1999, updated 2006). NetLogo [Computer software]. Center for Connected Learning and Computer-Based Modeling, Evanston, IL. Retrieved from http://ccl.northwestern.edu/netlogo

Wilensky U, Reisman K (2006) Thinking like a wolf, a sheep or a firefly: learning biology through constructing and testing computational theories. Cogn Instr 24(2):171–209

Windschitl M (2004) Folk theories of "inquiry:" How preservice teachers reproduce the discourse and practices of an atheoretical scientific method. J Res Sci Teach 41(5):481–512

Windschitl M, Thompson J, Braaten M (2008) Beyond the scientific method: model-based inquiry as a new paradigm of preference for school science investigations. Sci Educ 92:941–967. doi:10.1002/sce.20259

Zacharia ZC (2007) Comparing and combining real and virtual experimentation: an effort to enhance students' conceptual understanding of electric circuits. J Comput Assist Learn 23(2):120–132

Zacharia ZC, Anderson OR (2003) The effects of an interactive computer-based simulation prior to performing a laboratory inquiry-based experiment on students' conceptual understanding of physics. Am J Phys 71:618

Zacharia ZC, Olympiou G, Papaevripidou M (2008) Effects of experimenting with physical and virtual manipulatives on students' conceptual understanding in heat and temperature. J Res Sci Teach 45(9):1021–1035

Printed in the United States
by Baker & Taylor Publisher Services